Mycorrhizal Symbiosis and Agroecosystem Restoration

Rizwan Ali Ansari • Rose Rizvi •
Irshad Mahmood
Editors

Mycorrhizal Symbiosis and Agroecosystem Restoration

Springer

Editors
Rizwan Ali Ansari
Department of Plant Protection, Faculty of Agricultural Sciences
Aligarh Muslim University
Aligarh, India

Rose Rizvi
Department of Botany
Aligarh Muslim University
Aligarh, India

Irshad Mahmood
Department of Botany
Aligarh Muslim University
Aligarh, India

ISBN 978-981-99-5029-4 ISBN 978-981-99-5030-0 (eBook)
https://doi.org/10.1007/978-981-99-5030-0

This Springer imprint is published by the registered company Springer Nature Singapore Pte Ltd.
The registered company address is: 152 Beach Road, #21-01/04 Gateway East, Singapore 189721, Singapore

Paper in this product is recyclable.

Preface

In recent years, the concept of agroecosystem restoration has gained significant attention among ecologists, environmentalists, and policymakers worldwide. The term refers to the process of restoring the ecological integrity of degraded agricultural landscapes by adopting sustainable and holistic approaches. One of the key challenges in agroecosystem restoration is to rebuild the soil health, which is often compromised due to poor agriculture practices.

Mycorrhiza, a symbiotic association between fungi and plant roots, has emerged as a promising tool for improving soil health and enhancing the sustainability of agroecosystems. The use of mycorrhiza in agroecosystem restoration has gained momentum in recent years due to its potential to improve soil structure, increase nutrient uptake, enhance plant growth, and mitigate the negative impacts of abiotic and biotic stresses. This book, *Mycorrhizal Symbiosis and Agroecosystem Restoration*, is an attempt to provide a comprehensive overview of the current state of knowledge on mycorrhizal symbiosis and its potential applications in agroecosystem restoration. The book is divided into three parts. The chapters presented in this book provide valuable insights into the diversity and function of arbuscular mycorrhizal fungi, the impact of environmental changes on mycorrhizal symbiosis, and the potential for mycorrhizal inoculation in plant health amelioration. The crucial role of mycorrhizal symbiosis in achieving global food security has also been presented. It highlights the benefits of mycorrhizal inoculation in enhancing plant growth, increasing nutrient uptake, and improving soil health. The chapter evaluates different methods of mycorrhizal inoculation and their effectiveness in promoting plant growth, enhancing nutrient uptake, and mitigating the negative impacts of biotic and abiotic stresses. Part II of the book delves into the role of arbuscular mycorrhizal symbiosis in plant nutrient acquisition and disease management. The chapters focus on the importance of root exudates in facilitating nutrient acquisition through arbuscular mycorrhizal symbiosis. The chapters provide insights into the importance of understanding of the interactions between root exudates, mycorrhizal fungi, and soil nutrients in developing sustainable agricultural practices. The authors discuss advancement in mycorrhizal fungi-based sustainable plant disease management. The chapters underscore the potential for mycorrhizal fungi to serve as a sustainable alternative to chemical pesticides for plant disease management. Furthermore, chapter also evaluates the potential of mycorrhizal fungi in phytonematode management in different agro-climatic zones. The literature discusses the role of mycorrhizal fungi in reducing nematode populations through changes in soil microbial communities and plant growth promotion. The book further explores the importance of commercialization of arbuscular mycorrhizal technology in sustainable agriculture, the interaction between mycorrhiza and pathogens during first sight recognition, and the influence of arbuscular mycorrhizal fungi on soil health and plant fitness under hostile environments. These chapters provide insights into the challenges and opportunities in scaling up the use of mycorrhizal technology in agriculture, the mechanisms behind mycorrhiza–pathogen interactions, and the potential for mycorrhizal fungi to improve soil health and plant fitness in stressful environments.

As we bring this book *Mycorrhizal Symbiosis and Agroecosystem Restoration* to its completion, it is incumbent upon us to express our deep sense of gratitude to the numerous individuals and organizations who have played a crucial role in the success of this project. Foremost among them are our families, whose constant and unfaltering support has served as a powerful anchor during the long and often arduous process of researching, writing, and editing this book. Their unwavering presence and emotional backing have been pivotal in sustaining our grit and resilience, and we cannot thank them enough for their unflagging commitment. We would also like to extend our heartfelt appreciation to the many colleagues and collaborators who have generously given their time, expertise, and insights into our research on mycorrhizal ecology and ecosystem restoration. Their contributions have been invaluable in shaping the scientific and practical recommendations presented in this book, and we are indebted to their boundless spirit of inquiry and partnership. We would also like to express our deep appreciation to the reviewers who have rigorously and constructively scrutinized our manuscript, offering insightful comments, critiques, and suggestions that have significantly enhanced the rigor and impact of this book. Their painstaking efforts have been critical in ensuring that the information presented in this book is accurate, up-to-date, and relevant to the scientific and practical communities. Moreover, we would like to express our deep appreciation to the farmers, land managers, and restoration practitioners who have participated in mycorrhiza in agroecosystem restoration driven research, providing us with their rich and diverse practical knowledge, experiences, and challenges. Their invaluable insights and contributions have been critical in developing the various recommendations, and we are immensely grateful for their generosity, commitment, and partnership. Finally, we acknowledge and express our profound gratitude to the readers of this book, who we hope will be inspired and informed by the research and practical skills presented here. We believe this book represents a significant contribution to the field of mycorrhizal ecology and restoration, and we hope it will encourage readers to explore and contribute to this vital and emerging area of research and practice. We are humbled and profoundly grateful for the support and encouragement of the many individuals and organizations who have contributed to the development of this book. Their selfless efforts, unwavering commitment, and visionary support have been pivotal in making this book a reality, and we are deeply grateful for their presence in our lives and in the scientific and practical communities.

We would like to extend our heartfelt gratitude to Prof. Mohammad Gulrez, Vice Chancellor of our esteemed institution, Aligarh Muslim University, Aligarh, for his outstanding support and encouragement to pursue research and development activities. Under his visionary leadership, the university has taken remarkable strides in the field of agricultural sciences besides other sciences and humanities. His unwavering commitment to promoting academic excellence and research culture has been a great source of inspiration for us.

Our heartfelt thanks go to Prof. Akram Ahmad Khan, the Dean of the Faculty of Agricultural Sciences, for his constant support and encouragement throughout this project. We are also indebted to Prof. Mujeebur Rahman Khan, Chairperson, Department of Plant Protection and Prof. M. Badruzzaman Siddiqui, Chairperson, Department of Botany for their valuable suggestions that have been instrumental in shaping the content of this book. Our deepest gratitude goes to Prof. Saghir Ahmad Ansari, who has always motivated us, and made critical suggestions to this work with his vast knowledge and extensive research experience in the field of agricultural sciences. We are also thankful to the esteemed professors, including Prof. Zaki Anwer Siddiqui, Prof. Iqbal Ahmad, Prof. Malik Ahmad, Prof. Saghir Ahmad Khan, Prof. Anwar Shahzad, Prof. Abrar Ahmad Khan, and Prof. Masroor Akhtar Khan, for their invaluable feedback, suggestions, and support throughout this project. Their expertise and knowledge in the field have been instrumental in shaping the content of this book. We would also like to express our gratitude to the talented and dedicated researchers, Dr. Aisha Sumbul, Dr. Safiuddin, Dr. Raees Ullah Khan, Dr. Athar Ali Khan, Dr. Tariq Aftab, Dr. Sana Chaudhary, Dr. Faheem Ahmad, Dr. Syed Kamran Ahmad, Dr. Ziaul Haque, and Dr. Tahir Mohammad Chauhan, for their motivational support during the writing of this work.

We have dreamt of writing this book, but it would not have been possible without a constant source of inspiration from our family members. They deserve special thanks from the den of hearts and our sincere apologies to them if we would have done any unacceptable activity during the writing process.

Aligarh, India Rizwan Ali Ansari
Aligarh, India Rose Rizvi
Aligarh, India Irshad Mahmood

Contents

About the Editors

Rizwan Ali Ansari (Assistant Professor) Dr. Rizwan Ali Ansari, Assistant Professor in the Department of Plant Protection, Faculty of Agricultural Sciences at Aligarh Muslim University in India, is a highly accomplished scholar in the field of Plant Pathology/Nematology. His academic credentials include a Ph.D. from the same institution, where he has actively engaged in cutting-edge research and development strategies that have revolutionized the field.

Dr. Ansari has formulated innovative management modules that leverage a wide array of microorganisms, antagonists, and organic additives to combat a diverse range of soil-borne plant pathogens that infest numerous agricultural crops. He has also attended several prestigious national and international conferences, where his contributions have earned him coveted accolades from renowned scientific societies such as the Society for Plant Protection Sciences (SPSS), Indian Phytopathological Society of India (IPS), and the Nematological Society of India (NSI).

The prolific scholar has authored several highly acclaimed book chapters, research papers, and review articles that delve into the utility of organic additives, mycorrhizal fungi, and plant growth-promoting bacteria in the sustainable management of disease complex forming plant pathogens. His research has been published in several reputable journals and books.

Dr. Ansari's editorial prowess is equally impressive, as he has edited many books which have garnered global appreciation. His current research domains include the judicious application of organic additives and biological agents in the management of plant diseases, improving soil and crop health, and conducting surveys on disease prevalence caused by soil-borne pathogens on various economically significant crops.

Dr. Ansari's multifaceted accomplishments and contributions to the Plant Pathology/Nematology field are exemplary. His pioneering research has the potential to significantly impact the sustainable management of plant pathogens and improve soil and crop health, thereby advancing agricultural productivity and food security.

Rose Rizvi, Ph.D. (Assistant Professor) is currently working as Assistant Professor in the Department of Botany, Aligarh Muslim University, Aligarh, India. She earned her Ph.D. degree from the same department by contributing a lot in Plant Pathology/Nematology. She has thus far published a good number of research, review articles in the journals of great repute. Dr. Rizvi has also participated in various national and international symposia/conferences/workshops and successfully attended the queries of the delegates pertaining to area of her research. She has been the recipient of various awards such as UGC-BSR Research Fellow by the University Grants Commission (UGC); Norman Ernest Borlaug Research Award by Plant Pathology (Photon Foundation); Junior Scientist Award by the National Environmental Science Academy (NESA), New Delhi, India; CSIR-Research Associateship by MHRD, New Delhi; Scientist of the Year Award-2018 by the National Environmental Science Academy (NESA), New Delhi, India; Young Scientist of the Year Award-2018 by the International Foundation for Environment and Ecology (IFEE), Kolkata, India. Her thrust area of research is integrated disease management of phytoparasitic nematodes by the application of various means like biological control fungi, rhizobacteria, organic additives, and nanoparticles.

Irshad Mahmood a former Professor of Plant Pathology and Nematology in the Department of Botany, Aligarh Muslim University, Aligarh. He obtained Ph.D. degree from Aligarh Muslim University in the field of Plant Pathology and Nematology. Promotion of organic farming across the world by utilizing organic additives and potent microorganisms for the sustainable management of phytoparasitic nematodes and plant pathogenic fungi resulting augmented soil and plant health is the domain of his research. He has been engaged with teaching program of undergraduate and post-graduate level students for the last 30 years and has many overseas visits including the USA, France, and the UK. He has attended a significant number of national and international conferences pertaining to wide area of agricultural sciences and published around 150 original research papers, review articles, and book chapters in various refereed national and international publication media, most of them in very high impact factors. He has successfully completed many training courses in various ICAR-sponsored research institute in India and also in North Carolina State University, Raleigh, USA. He is also an active member of national and international scientific organizations, served as experts for selection committee, reviewer of journals, doctoral theses, and funding agencies, and is a recipient of scientist of the year award in the field of Plant Pathology and Nematology. He has guided ten Ph.D., several M. Phil., and a large number of M.Sc. dissertations. He has also been engaged in the establishment of a joint government project with Aligarh Muslim University for the improvement of infrastructural facilities in botanical garden to facilitate ex situ conservation and propagation of rare, endangered, and threatened plants and the plants endemic to the region.

Part I

Introduction to Mycorrhizal Fungi

Diversity of Arbuscular Mycorrhizal Fungi in Mined Land: Distribution and Function in Reclamation of Mined Land Ecosystems

1

Thangavelu Muthukumar and Arumugam Karthikeyan

Abstract

Arbuscular mycorrhizal (AM) fungi are the most widespread and functionally important symbionts of diverse plant species. The AM fungi are likely to be affected by anthropogenic activities like mining. The potential of AM fungi to improve plant growth in stressed environments has led to their inclusion along with plants in the restoration of mined areas. Still, our understanding of the diversity of AM fungi in response to mining activities remains limited. In this chapter, we provide a summary of the diversity of AM fungi in mining and restoration areas, the influence of different mining activities on AM fungal communities and root colonization, the role of AM fungi in improving soil structure and function, and the effect of AM fungi on plant growth in mine substrates under controlled and field conditions. Changes in AM fungal communities have been detected both in roots and soils using conventional and molecular techniques in response to mining activities. However, the majority of studies do report an increased diversity of AM fungi in post-mining sites with time. Generally, moderate to high AM colonization levels are reported in revegetated plant species or those naturally colonizing mined areas. Inoculation of AM fungi individually or along with plant growth-promoting microorganisms in nurseries has been shown to increase growth and nutrient content in plants intended for planting in mining areas. Once transplanted onto mining sites, AM plants exhibit a better survival rate, increased growth, and improved nutrient uptake. This chapter highlights current knowledge on the role of AM fungi in the restoration of

T. Muthukumar (✉)
Root and Soil Biology Laboratory, Department of Botany, Bharathiar University, Coimbatore, Tamilnadu, India

A. Karthikeyan
Forest Protection Division, Institute of Forest Genetics and Tree Breeding, Coimbatore, Tamilnadu, India

R. A. Ansari et al. (eds.), *Mycorrhizal Symbiosis and Agroecosystem Restoration*,
https://doi.org/10.1007/978-981-99-5030-0_1

mine sites and some future research areas that could help in better exploitation of this symbiosis in stressed environments.

Keywords

Heavy metals · Endophytic fungi · Mycorrhization · Arbuscules · Glomeromycota · Nutrient acquisition · Stressed ecosystem

1.1 Introduction

Mining and its related activities remove topsoil, destroy vegetation, and release several harmful inorganic and organic pollutants into the environment. The contaminants originating from the mining areas, in addition to seeping into the soil and contaminating the surface and ground waters, also enter the food chain (He et al. 2005). Mine wastes are dumped into the soil, and abandoned mine sites are one of the most common marginal lands in industrialized countries (Gibbs and Salmon 2015). Environmental constraints that already exist before soil contamination combine synergistically with the properties of mine refuse to aggravate the degradation of the environment, soil and water pollution, loss of biodiversity, and cause health concerns to humans (Confalonieri et al. 2014). Mine spoils are highly unstable with limited cohesion, reduced water holding capacity, high acidity, increased heavy metal toxicity, and low organic matter and nutrient content. These physical, chemical, and mechanical properties render mine spoils unsuitable for biological activities (Yan et al. 2020). The toxicity and the availability of heavy metals in the soils are long-lasting and determined by the total concentrations, chemical form, and soil factors (Rodríguez et al. 2009). Mining activities negatively affect soil microbial communities and drastically affect their population and activities. Moreover, biogeochemical cycles and establishment of plant cover are severely affected in mining areas due to the reduction or removal of microorganisms that are fundamental for these ecosystem processes. Therefore, all these need considerations for effective remediation of a particular site containing mining wastes. Revegetation of mine spoils is a formidable task, and the establishment of natural plant cover or restoration of soil fertility over these degraded areas is a very slow process (Barbosa and Fernando 2018). Still, the use of pioneer plants or native plant species is one of the important steps in establishing plant cover over sites affected by mining activities. The success of vegetation restoration could be increased if the plants can ameliorate the different abiotic and biotic stresses existing like the heavy metal toxicity and extreme soil reactivity. However, the addition to organic matter amendments and inoculating soil microorganisms are essential and useful for the establishment and survival of plants and the formation of plant communities in deteriorated soils (Wang et al. 2017a). Therefore, the aim of this chapter is to integrate information from the literature pertaining to the diversity of AM fungi in mined sites and the role of AM fungi in the restoration of vegetation on sites subjected to mining disturbance.

1.2 Microbial Diversity in Mine Spoils

Soil microorganisms play a pivotal role in many ecological processes. Mining activities drastically alter the populations and activities of soil microorganisms both in the mining and surrounding areas. Studies have shown that the diversity and metabolic activities of bacteria can vary with different levels of mining impact (Fernandes et al. 2018). Mining affects the principal drivers of microbial communities in the soil like the pH, carbon, moisture, and plants. Thus, vegetation development on barren post-mine sites is curtailed by the low abundance or absence of microorganisms that are essential for plant establishment and nutrient cycling (Hart et al. 2019). An analysis of bacterial community structure in response to mercury contamination in a traditional gold mining waste disposal site in North Sulawesi suggested that the high level of soil mercury reduced the diversity and abundance of bacterial phyla (Kepel et al. 2020). In addition, mercury contamination also resulted in a shift in the dominance of bacterial phyla and lower taxa when compared to unpolluted soils (Kepel et al. 2020). Like bacteria, fungal communities in the soil are important as they perform key functions like nutrient turnover, helping plants directly through the formation of mycorrhizas and decomposition of organic matter (Ansari and Mahmood 2017, 2019a, b; Ansari et al. 2017a, b). However, fungal communities are not routinely examined during ecosystem restoration unlike bacteria (Yan et al. 2018). Moreover, information is meager on how soil fungal communities respond to mining activities and post-mining restoration of the disturbed areas despite their direct link with higher trophic levels (Hart et al. 2019). In addition to abiotic conditions, substrate conditions greatly affect the establishment and existence of fungal communities in the soil (Ohsowski et al. 2012). For example, the use of overburden (soil from deep profiles) failed to restore the fungal community in sand mine restoration sites in South Western Australia even after 13 years (Hart et al. 2019). Further, bacterial and fungal communities may respond variedly to a mining disturbance. Chen et al. (2020) studied the composition of fungal and bacterial communities in an opencast coal mine in an Inner Mongolian region. The observations of the study showed that although there was an alteration in the structure of both fungal and bacterial communities by the mining disturbance, there was an increase in the abundance of microorganisms with coal degradation and detoxification of pollutant capabilities. Surprisingly, bacterial communities were less stable than fungal communities (Chen et al. 2020).

Mining activities also affect the population abundance of mycorrhizal fungi that form an association with plants. These fungi are bipartite with part of their thallus within plant roots and the rest in the surrounding soil. The majority of plants (>80%) growing on mining-disturbed sites are mycorrhizal (Wang 2017). Many studies examining the role of mycorrhizal fungi have reported a positive influence of these fungi on habitat restoration. Although there are different types of mycorrhizal symbiosis, the most common types are the ectomycorrhiza and arbuscular mycorrhiza found in more than 75% of the plants in natural and manmade ecosystems (Brundrett and Tedersoo 2018). As ectomycorrhiza is restricted to arborescent forms, arbuscular mycorrhiza occurs in diverse plant forms and plant groups. Studies have shown that mining disturbance could affect both ectomycorrhizal and AM symbiosis (Correia et al. 2021; Suting and Devi 2021), and both these mycorrhizal types can aid in the restoration of disturbed sites (Wang 2017; Policelli et al. 2020).

1.3 AM Fungi

The AM fungi are one of the most widespread soil-borne fungi that form an association with the majority of land plants growing in diverse ecosystems. This symbiosis between plants and AM fungi has evolved some 400 million years ago and played an important role in the colonization of terrestrial habitats by plants (Selosse et al. 2015). All the AM fungi forming a symbiosis with terrestrial plants are placed in the sub-phyla Glomeromycotina and Mucoromycotina of the phylum Mucoromycota (Spatafora et al. 2016). There are approximately 342 validly described species in 43 genera in four orders Archaeosporales, Diversisporales, Glomerales, and Paraglomerales in Glomeromycotina as of May 2021 (http://www.amf-phylogeny.com). The fine root endophyte *Planticonsortium* placed in Mucoromycotina forms an association with diverse plant forms like the glomeromycotean fungi (Orchard et al. 2017). The colonization of plant roots is facilitated by the AM fungal hyphae emerging from a germinating spore or mycorrhizal root. The fungi enter plant roots, and there is a wide array of signal exchanges between the fungi and plant root during the processes of mycorrhization (Gobbato 2015). The entry of the fungus into the plant root system is often characterized by the formation of an appressorium on the root surface (Fig. 1.1). Once inside the roots, the AM fungal hyphae traverse the roots inter- or intra-cellularly. The fungus forms short-living arbuscules in root cortical cells which act as a transit point for nutrient exchange between the host and the fungus (Fig. 1.1). The development of arbuscules is determined by the nutrient demand of the plant host and maybe either be elaborated or rudimentary based on the type of AM colonization (Kobae et al. 2016). Nevertheless, arbuscules are usually absent under conditions where the symbiosis is non-functional and in 8% of the non-mycorrhizal plant species (Brundrett and Tedersoo 2018; Cosme et al. 2018). The fungal hyphae may be either linear or coiled. Based on the hyphal and arbuscular nature, the AM colonization in plant roots is categorized into *Arum*-, *Paris*-, and intermediate-type (Dickson et al. 2007). In addition to these structures, most AM fungi except those in Gigasporaceae of Diversisporales form lipid-filled storage structures in the root called vesicles (Fig. 1.1).

Nevertheless, AM fungi in Gigasporaceae form vesicle-like structures in the soil called auxiliary cells. In addition to these intra-radical structures, the AM fungi also develop spores in the soil and some instances within roots. These soil-borne spores act as the chief perennating organ of the fungus in soils covered by seasonal vegetation. As AM fungi are obligate symbionts, they depend on the host photosynthates that are exchanged for nutrients. The transfer of carbon from the host plant to AM fungus is in the form of hexose sugars and lipids (Andrino et al. 2021). Experimental studies have shown that the host plant transfers around 4–20% of the photosynthetic carbon to the AM fungus (Parihar et al. 2020).

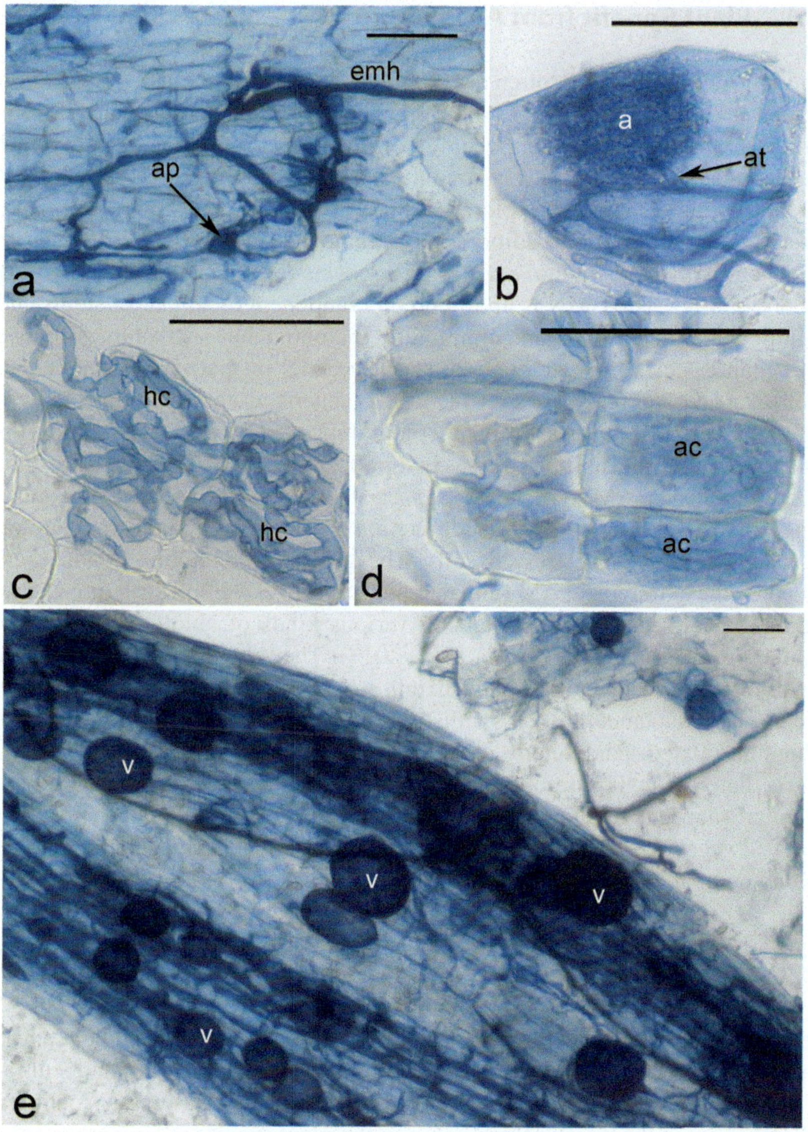

Fig. 1.1 (**a**–**e**) Arbuscular mycorrhizal colonization in roots of forestry species used in reclamation of mine spoils. (**a**) Extramatrical hyphae (emh) and appressorium (ap) in roots of *Dalbergia latifolia*; (**b**) Arbuscule (a) with arbuscular trunk (at) in root cortical cell of *Thespesia populnea*; (**c**) Hyphal coils (hc) in root cortex of *Acacia planifrons*; (**d**) Arbusculate coils (ac) in *A. planifrons*; (**e**) Vesicles (v) in *Casuarina equisetifolia*. Scale bars = 50 μm

1.3.1 Plant Benefit from AM Symbiosis

Although AM symbiosis is often cited as an example of mutualism studies do indicate that the association is more of a mutualism-parasitism continuum (Smith and Smith 2013). The extraradical mycelium of AM fungi originating from mycorrhizal roots extends into the soil well beyond the nutrient depletion zones surrounding the roots and promotes nutrient exploration that is otherwise unavailable to plant roots (Ansari et al. 2019a, b). In addition, the extraradical mycelium also connects roots of coexisting plants in natural and agroecosystems forming a common mycorrhizal network (CMN). The CMN is an important component of soil ecosystems and has a significant influence on plant communities including the establishment of invasive exotic plant species (Pringle et al. 2009). The CMN-mediated interplant exchange of nitrogen (N) and phosphorus (P) among plants plays an important role in maintaining the health and fitness of plant communities (Smith and Read 2008). Plants' tolerance to biotic and abiotic stresses is significantly improved through the nutrient-related benefits of AM symbiosis (Sun et al. 2018). Additionally, AM-mediated changes in soil or plant structure and function greatly benefit plants in their amelioration of anthropogenic stressed soils. The glomalin containing 30–40% carbon exuded by the extraradical AM fungal hyphae improves the soil structure and water-holding capacity in soils experiencing various kinds of abiotic stresses, thereby modulating plant growth (Sharma et al. 2017). The AM symbiosis affects plants' growth-related processes like plant height, root architecture, leaf water potential, stomatal conductance, photosynthetic efficiency, root-to-shoot ratios, and nodulation-related parameters in N-fixing plants (Begum et al. 2019). This AM-mediated growth promotion in AM plants often results from improved macro- and micro-nutrient uptake through various direct or indirect mechanisms. This improved vegetative growth of AM plants often culminates in a greater yield output (Begum et al. 2019).

1.3.2 AM Colonization

The colonization of plant roots is an important step in the establishment and functioning of mycorrhizal symbiosis. A large number of plant, fungal, and soil factors are known to influence the colonization of plant roots by AM fungi. Woźniak et al. (2021) examined the relationships between the plant species assemblages, environmental factors, and AM fungal colonization in *Calamagrostis epigejos* and *Poa compressa*-dominated spontaneous vegetation patches in a post-coal mine heaps chronosequence in Poland. The results of the study indicated a high frequency of mycorrhization in roots of *C. epigejos*- and *P. compressa*-dominated vegetation. However, all the measured mycorrhizal variables were lower in roots of *C. epigejos* than in *P. compressa*. The maximum relative mycorrhizal intensity, and mean arbuscular abundance in the roots of both plant species occurred in *Daucus carota*-dominated vegetation patches (Woźniak et al. 2021). Positive relations that were statistically significant existed between the frequency of mycorrhization, relative mycorrhizal dependency, and arbuscular abundance. However, the mycorrhizal variables were not correlated with mine heaps' age. Nonetheless, statistically significant correlations occurred between the arbuscular abundance and the magnesium (Mg) and potassium (K) content of the substrate (Woźniak et al. 2021).

An examination of AM fungal colonization in roots of plants either planted or growing naturally in a post-coal mining area of Indonesia indicated that roots of all the plant species had medium (10–30%) to high (>30%) levels of colonization (Salim et al. 2020). The percent AM root colonization was more (50–100%) in roots of *Acacia mangium* and *Glirisida sepium* that were growing naturally on the mine site than *Falcataria moluccana* (32–44%) and *Senna siamea* (12–20%) that were widely planted in the revegetated mine area. The AM colonization in roots of plants growing in this post-mine area was positively influenced by organic carbon, total N, cation exchange capacity, calcium (Ca), iron (Fe), and Mg contents of the substrate. Contrarily, total and available P, K, and aluminum (Al) negatively influenced root colonization. The influence of Mg was more on root colonization than other elements (Salim et al. 2020). Similarly, all 84 plant species in 36 plant families growing in two different stabilized Fe ore mines in Goa, India were colonized by AM fungi (Prabhu and Rodrigues 2019). The percentage of AM fungal colonization ranged between 12% (*Adiantum philippense*) and 87% (*Anacardium occidentale*). In addition, AM fungal root colonization was significantly and positively correlated to AM spore numbers in the soil (Prabhu and Rodrigues 2019). Roots of *Vachellia karroo* (=*Acacia karroo*), *Senegalia hereroensis* (=*Acacia hereroense*), and *Vachellia robusta* (=*Acacia robusta*) examined from two gold and uranium mine tailing sites in Welkom, South Africa contained hyphal coils, arbuscules, vesicles, and intraradical spores. Furthermore, root nodules of *Acacia* species examined in these sites were colonized by AM fungi (Buck et al. 2019). An examination of *Acacia nilotica*, *Albizia lebbeck*, and *Pongamia pinnata* roots from plants growing on four various coalmine overburden soils in Telangana of Southern India by Govindu et al. (2020) indicated an AM colonization range of 37–60%. Moreover, AM colonization in these tree species varied between sites with colonization ranging from 40% to 52% in *A. lebbeck*, 37% to 56% in *A. nilotica*, and 43% to 60% in *P. pinnata* (Govindu et al. 2020).

1.3.3 AM Spore Numbers

Spores are important propagules of AM fungi in habitats characterized by seasonal vegetation (Fig. 1.2). Although the soil hyphae and mycorrhizal roots could serve as propagules of AM, fungi spores are still considered to be the major propagule and are routinely enumerated to assess the soil infectivity. Moreover, morphological characterization of AM fungal species is based on the morphology and subcellular characters of spore morphotypes. AM fungal spore numbers of 24–88 spores per 10 g of dry soil were reported in the rhizosphere of plants growing in Seydişehir Al Plant Bauxite mine Deposits in Turkey (Atmaca and Karaca 2019). Spore numbers of 8–304/100 g of soil were reported by Prabhu and Rodrigues (2019) from the two

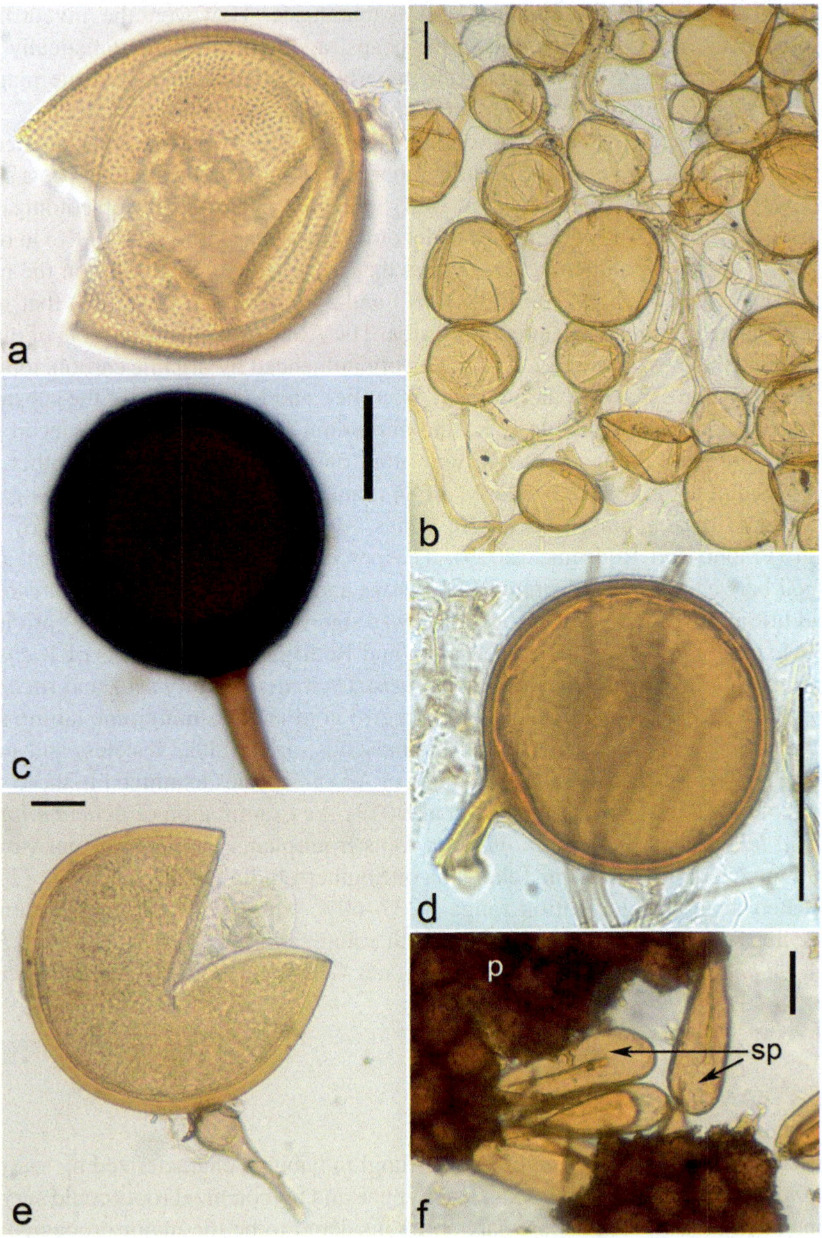

Fig. 1.2 (**a**–**f**) Spores of common arbuscular mycorrhizal encountered in the rhizospheres of plantation tree species. (**a**) Fractured spore of *Acaulospora scrobiculata*; (**b**) Spore cluster of *Rhizophagus aggregatus*; (**c** and **d**) Spore of *Funneliformis geosporus* (**c**) and *Claroideoglomus etunicatum* (**d**); (**e**) Fractured spore of *Scutellospora calospora*; (**f**) Portion of *Sclerocystis sinuosa* sporocarp showing peridium (p) and spores (sp). Scale bars = 50 μm

stabilized Fe-ore dumps in northern Goa, India. The density of AM fungal spores in the topsoil of three opencast coal mines soil stockpiles in South Africa ranged from 40–91/100 g of soil (Ezeokoli et al. 2019). Nonetheless, these spore numbers were numerically lower than those recorded for unmined soils (109 spores/100 g soil). The average AM fungal spores in the rhizosphere of grass (*Melinis minutiflora*) used to revegetate the Brazilian Fe ore tailings piles were >470 spores/100 cm^3 soil (Teixeira et al. 2017).

A low AM spore density was reported from the community's post-gold mining land (10 spores/100 g soil) and artisanal small-scale gold mining tailings (14 spores/100 g soil) in Indonesia (Tuheteru et al. 2020). A later study from Indonesia conducted on a former gold mine site also indicates similar spore numbers (7–13 spores/50 g soil) associated with moderately mycorrhizal plants with colonization levels ranging from 14% to 36% (Pulungan and Nasution 2021). Yang et al. (2015) while assessing the AM status and heavy metal accumulation in tree species in a lead (Pb)-zinc (Zn) mine area of Qinling Mountain, Northwest China found highly varying spore numbers (47–670 spores/100 g dry soil) in the rhizosphere of the tree species. As tree species like *Populus simonii* and *Populus purdomii* harbored large spore numbers (>300–600 spores/100 g soil), other species like *Platycladus orientalis, Eucommia ulmoides,* and *Cotinus coggygria* had spores numbers <100/100 g soil. Moreover, a great variation in AM fungal spore numbers was also found between different areas of the mining site. Generally, the area under former direct mining (mine area and abandoned smelter area) contained less spore numbers compared to areas where mine refuse dumping areas (mine tailing pond) (Yang et al. 2015).

1.3.4 Diversity of AM Fungi on Mined Sites

The influence of mining disturbance on AM fungal communities is often contradictory. Although the majority of the studies do indicate a negative impact of mining on AM fungal communities for several years even after cessation of their activities, other studies have shown either no effect or an increased diversity compared to adjacent unmined areas. The recovery of AM fungal communities in soil or roots of plants growing in mine-disturbed areas is either very slow or rapid depending on the type of mining. Removal and stockpiling of topsoil is a crucial and essential practice in surface mining operations especially in coal mines because topsoil contributes to the successful rehabilitation of the open pit mines after the cessation of mining. In this process, the topsoil from the mining area is stockpiled for future use (Mushia et al. 2016). However, poor management of these topsoil stockpiles reduces the restoration potential of the soils and enhances the restoration costs. An assessment of AM fungal spore density and viability at different depths of three opencast coal mines soil stockpiles varying in age from 1 to 18 years in South Africa (Ezeokoli et al. 2019) suggested that the spore numbers and the diversity of AM fungal morphotypes were more in the topsoil (<20 cm) than in the sub-soil (>20 cm) of the stockpiles. Moreover, there was a reduction in AM spore numbers in the topsoil of stockpiles when compared to unmined soils (Ezeokoli et al. 2019). This clearly suggests disturbance and storing soil for extended period of time inevitably reduces the AM inoculum potential of the soils.

Evidence does exist that certain AM fungi could tolerate the high concentrations of heavy metals in the mining sites. For example, Sánchez-Castro et al. (2017) examined the ability of AM fungi to tolerate and survive in an extreme metal-polluted (total Zn 97,333 ppm and total Pb 31,333 ppm) tailing basin and a less contaminated waste heap in an abandoned mining site in southern France. The Glomeromycota diversity in mycorrhizal roots determined through sequencing of the ribosomal large subunit (LSU) indicated the presence of many AM fungal ribotypes from Glomerales and a few from Diversisporales and Paraglomerales. The diversity of AM fungi and mycorrhization of roots was lower in the tailing basin than in waste heaps. A species of *Paraglomus* that was common in mining sites in Poland occurred in the roots of various plant species (Sánchez-Castro et al. 2017). This indicates that certain AM fungal isolates could tolerate the extreme heavy metal concentrations in mine sites and help in the tolerance and fitness of plants selected for the efficient rhizostabilization of extreme environments.

Many studies have reported high diversity of AM fungi in rehabilitated mine sites. An examination of AM fungi associated with 84 plant species (in 36 families) growing on two stabilized Fe ore mine dumps in northern Goa, India indicated the presence of spore morphotypes of 19 AM fungal species belonging to eight genera (Prabhu and Rodrigues 2019). The AM fungal community in the mine dumps was dominated by *Acaulospora* (six species) followed by *Gigaspora* (four species), *Sclerocystis* (three species), and *Scutellospora*, *Funneliformis*, *Glomus*, *Racocetra,* and *Rhizophagus* with one species each. A total of 31 AM spore morphotypes in 12 genera were reported from rehabilitated Fe ore tailings piles in Brazil. More than 40% of the spore morphotypes belonged to the AM fungal genera *Acaulospora*, and AM species richness was higher in the tailing piles when compared to unmined habitats (Teixeira et al. 2017). This study also showed that spore morphotypes of certain AM fungal species like *Acaulospora alpina*, *Acaulospora nivalis*, *Acaulospora scrobiculata*, *Acaulospora tuberculata*, and *Septoglomus viscosum* were exclusively present in mine tailings (Teixeira et al. 2017).

Certain studies have shown that the abundance and diversity of AM fungi tend to increase with time after disturbance. In a recent trap culture study involving soils of different ages (<2 years to >14 years) after disturbance from limestone mine sites of Meghalaya, India indicated the occurrence of 77 AM fungal species in 17 genera (Suting and Devi 2021). The diversity of AM fungi was maximum in soils that were >14 years old after disturbance and was even higher when compared to soils from an unmined forest area. Contrarily, minimum AM fungal species diversity occurred in the recent disturbed site (<2 years). Moreover, AM fungal species belonging to *Acaulospora, Funneliformis, Gigaspora, Glomus, Rhizophagus,* and *Septoglomus* were most frequent in limestone mines of different ages (Suting and Devi 2021). The high frequency of occurrence of certain AM fungi in limestone spoils of all ages suggests that these resilient AM fungal species could be useful in the revegetation of the mine spoils and aid plant establishment in the disturbed ecosystems.

An assessment of the composition and diversity of soil AM fungal communities using a high-throughput sequence analysis in open-cast coal mining dumps that were revegetated and were under reclamation for different periods (<1–20 years) indicated the presence of 156 operational taxonomic units (OTUs) belonging to 12 genera and nine families (Bi et al. 2021). Although the nonmetric multidimensional scaling analyses and network analysis indicated a significant variation in the AM fungal diversity between the initial (<1 year) and end of the reclamation period (20 years), the diversity remained similar during the intermediate stages of reclamation (5–15 years). Moreover, the fungal diversity during different stages of reclamation was related to total N, available soil P, and N/P and carbon/P ratios (Bi et al. 2021). This clearly shows that revegetation of mining areas could aid in the quick reestablishment of AM fungal communities in disturbed ecosystems.

A comparison between the revegetated Fe-mining site and some natural ecosystems in Brazil suggests that the magnitude of AM fungal diversity in the mining site was almost similar to natural vegetation. A total of 35 species of AM fungi were recorded in the mining site, whereas the diversity of AM fungal species in the natural vegetation ranged from 25 to 32 (Vieira et al. 2018). Furthermore, species belonging to *Acaulospora*, *Glomus*, and *Racocetra* constituted 54% of the AM fungal diversity in the mine-disturbed areas. Sequences generated from total soil DNA using Illumina revealed AM fungal OTUs belonging to Ambisporaceae, Claroideoglomeraceae, Gigasporaceae Glomeraceae, and Paraglomeraceae. In mine soils, OTUs of AM fungal community were dominated by uncultured Glomerales during dry and rainy seasons. Both molecular analysis and trap culture technique yielded less AM fungal diversity than direct observation of spores from the field soil (Vieira et al. 2018). A similar comparative study using morphological and molecular techniques on AM fungal communities in a mine-tail and natural forest in Korea suggested the presence of eight AM fungal species belonging to seven genera in the post-mining area and six AM fungal species belonging to five genera from the natural forest (Park et al. 2016). An assessment of AM fungal diversity in roots also indicated the presence of seven OTUs in the post-mine roots and five OTUs in roots of forest plants (Park et al. 2016).

1.4 Mining Subsidence and AM Fungal Communities

Subsurface mining or excavation results in the sudden settlement of the overlying soil called subsidence. This result in cracks on the ground surface, damages soil structure, destruct plant roots, reduces soil moisture, available nutrients, and microbial communities, thereby decreasing soil fertility and deferring the establishment of natural vegetation (Wang et al. 2017b). Therefore, restoration of vegetation and critical management is important for ensuring ecological sustainability in the mining areas threatened by subsidence. Transplanting shrubs and trees in the mine subsidence area is often adopted to restore green cover (Sun et al. 2019; Xiao et al. 2019). The establishment and survival of transplanted plants in the mine subsidence area are often increased by inoculating plants with various microbial inoculants

(Guo et al. 2020). Nonetheless, studies on the influence of subsidence on AM fungal communities are very limited.

A study conducted in the coal mining subsidence area of the Chinese northwestern city of Lingwu indicated that the AM fungal communities in the subsidence area were significantly affected by disturbance, seasons, and soil depths (Huang et al. 2020). The number of AM fungal OTUs recorded during spring was higher than in the winter. Similarly, the surface soils (0–15 cm) contained fewer AM fungal OTUs compared to the subsurface region (15–30 cm). The AM fungi in this coal mining subsidence area belonged to six genera in six families: *Glomus* (Glomeraceae), *Paraglomus* (Paraglomeraceae), *Diversispora* (Diversisporaceae), *Claroideoglomus* (Claroideoglomeraceae), *Ambispora* (Ambisporaceae), and *Scutellospora* (Gigasporaceae). In addition, the abundance of each taxon differed with soil depth as the abundance of *Glomus* was more in the surface soil and reduced with soil depth (15–30 cm). Contrarily, *Claroideoglomus* was more abundant in the 15–30 cm soil and this soil region also contained the most unclassified sequences at the genus level. The abundance of *Diversispora* and *Paraglomus* was considerably higher than the *Glomus* in the subsidence than the undisturbed area. *Ambispora* was absent in the subsidence soils, and *Scutellospora* was least abundant among all the AM fungal genera (Huang et al. 2020).

Subsidence in mining areas also affects the diversity of AM fungi colonizing the roots. Examination of roots of *Artemisia sphaerocephala*, *Caragana korshinskii*, and *Salix psammophila* growing in subsidence area of the Chinese Shendong Bulianta coal mining region by pyrosequencing analysis indicated the symbiosis of AM fungi belonging to Diversisporaceae and Glomeraceae (Bi et al. 2019c). Of the 57 OTUs present in roots of the examined plant species 81% belonged to *Glomus*, 4% to *Diversispora*, 12% to *Rhizophagus*, and 2% each to *Septoglomus* and *Otospora*. Among the five AM fungal genera recorded, species belonging to *Rhizophagus* and *Glomus* were dominant in roots but significantly varied in proportion in all three plant species growing in different mining areas (Bi et al. 2019c).

Guo et al. (2020) conducted a field study on the influence of subsidence and tree transplantation on AM fungal communities in the Shendong coal mining area of the Chinese Daliuta Town. The AM fungal diversity based on spore morphology indicated that 60–93% of the spores belonged to *Glomus* followed by *Diversispora* (1–33%), *Scutellospora* (1–7%), and others (0–0.1%). The molecular analysis of the rhizosphere and nonrhizosphere subsidence soils also showed a higher AM fungal diversity associated with subsidence areas under transplantation compared to natural vegetation regeneration areas (Guo et al. 2020).

Inoculation of AM fungi and the nature of revegetation significantly affect not only the AM fungal diversity in the subsidence soils but also influence nutrient availability and microbial activities in the soil. For example, revegetation type and inoculation with the AM fungus *Funneliformis mosseae* significantly influenced the soil nutrients, microbial biomass, and soil enzyme activities in the rhizospheres of *Hippophae rhamnoides*, *Amorpha fruticosa*, *Cerasus humilis*, and *Xanthoceras sorbifolium* transplanted in the subsidence area in Dongshan coal mine site in northwestern China (Xiao et al. 2019). Moreover, *F. mosseae* inoculation also was related to N acquisition and N cycling in the coal mine subsidence soils. In *F. mosseae* inoculated subsidence soils, microbial activities were affected by nitrate-N, total N, and dissolved organic N (Xiao et al. 2019). Similarly, inoculation of *F. mosseae* in the coal mine subsidence area in the west Chinese County Shenmu enhanced the survival rate of *H. rhamnoides* over 4 years. The AM fungus inoculation also improved the plant height, crown, and ground diameter of sea buckthorn as well as soil organic carbon after 4 years of growth (Zhang et al. 2020). These studies clearly show that inoculation of AM fungi can improve the soil quality and promote plant growth for the long term which is essential for the restoration of the damaged subsidence mine areas.

1.4.1 Aggregate Mining and AM Fungi

Aggregates refer to a broad spectrum of materials (stone crush, gravel, sand, etc.) used in constructions in urban and semi-urban areas and are one of the widely mined materials worldwide. It is estimated that 50 billion tons of aggregates are mined annually from natural regions causing destruction to landscape, shrinking of agricultural and grazing lands, breakdown of river banks, polluting waterways with heavy metals, reducing ground water table, and deforestation (Steinberger et al. 2010; Ako et al. 2014). Restoration of these excavated mining sites is a crucial environmental issue. Many of the excavated mine pits especially sand pits mostly consist of subsoil and rock material and lack organic matter which affects the growth and activity of microorganisms. In two experiments spanning over two growing seasons, Ohsowski et al. (2018) examined the effects of biochar, compost, and AM fungus *Rhizophagus irregularis* inoculation/amendment individually or in combination on plant growth and establishment in an active sand mine pit using two different approaches. In the first approach, the authors transplanted variously treated plugs prepared in the greenhouse, and in the second approach, seeds were directly sown in the field sites in various treatments. The experimental plants included C4 grasses (*Andropogon gerardii, Panicum virgatum*), C3 grasses (*Elymus canadensis, Bromus kalmii*), N-fixing forbs (*Desmodium canadense, Lespedeza capitata*), and composite forbs (*Liatris cylindracea, Symphyotrichum laeve*). The results of the study indicated that although individual inoculation of *R. irregularis* or biochar application failed to significantly influence plant growth, simultaneous application of biochar and compost amendment along with AM fungus inoculation resulted in the largest plant response in the seed trial study (Ohsowski et al. 2018). The significant response to AM fungus inoculation and organic amendments in the seed trial study was attributed to the positive influence of the AM fungus on seed germination and seedling establishment in the nutrient-stressed sandpits (Ohsowski et al. 2018).

1.4.2 AM Fungi Improves Soil Structure

Erosion of mining substrate and their deposition in the surrounding areas and waterways is one of the major causes of higher heavy metal loading of soil and surface water systems worldwide. For example, a study conducted in the north Mongolian gold mining region has shown that the sediment input from the mining site, into Tuul River was substantially more than from erosion from natural causes (Jarsjö et al. 2017). The contaminating metals originated directly from the erosion of waste heaps or exposed mine tailings as mining and processing activities frequently produce considerable quantities of liquid and solid wastes. Moreover, barren surface mines cover significant areas in different parts of the world, and disturbances in these mining sites resulting from road construction, changes in topography, and infrastructure development can increase the rate of erosion (Jaramillo et al. 2015). Previous studies suggest that AM fungi can deter soil erosion by improving the soil structure (Chen et al. 2018). The extraradical mycelial network of AM fungi forms a three-dimensional matrix entangling the soil particles. In addition to physical binding, aggregation and stabilization of soil particles also happen in response to the production of a glycoprotein called glomalin by AM fungal hyphae in the soil (Singh et al., 2013). This soil protein is not a well-established gene product and is chemically a heterogeneous molecular species and therefore is also termed as glomalin-related soil proteins (GRSPs). Glomalin is a soil fraction defined by its immune-reactive properties and extractability with unclear origin (Rillig 2004). Although the ecological role of glomalin is far from clear, it is known to improve the physical properties of soil, enhance soil nutrient content, carbon sequestration, stabilize pollutants, positively influence microbial activities, and thereby help in ecological restoration (Singh et al. 2020). Glomalin-related soil proteins can account for 2–5% of the total soil organic carbon with an estimated half-life of several decades (Wilson et al. 2009). The benefits of GRSPs are particularly important in dry sandy soils of arid regions where soils are low in fertility and are highly prone to water and wind erosion. In such circumstances, transplanting mycorrhizal plants is a sustainable way to prevent erosion and enhance soil fertility. Kumar et al. (2018) studied the GRSP fractions in six reclaimed coal mine fields' chronosequence ranging from one to 26 years and an unmined forest site in West Bengal. The results of the study demonstrated an increase in GRSPs content with the increasing age of the mine sites. Further, the strong association between GRSPs, soil organic carbon, and AM spore density as revealed by multivariate analysis suggests that factors favoring the accumulation of soil organic carbon improve GRSPs content and proliferation of AM fungal propagules during the rehabilitation process (Kumar et al. 2018).

In a mining soil, restoration system constructed in a coal mine involving microorganisms-complex substrate-plant in Xuzhou, China, sludge was amended to enhance the reclamation process and the system was monitored for 10 years (Li et al. 2021). The results of the study showed that sludge amendment increased soil aggregate stability. Soil GRSPs increased between 1 and 5 years and did not differ much thereafter until the tenth year. Moreover, GRSPs were significantly and positively correlated to soil organic carbon and soil structure-related parameters (Li et al. 2021). The GRSPs content in polluted soils tends to vary with pollutants and soil depth. Yang et al. (2017) investigated the influence of heavy metals (Pb and Zn) in polluted soils on soil aggregate distribution and stability at various soil depths (0–40 cm). The results of the study showed that the GRSP content in the soil was inhibited more by Pb than Zn in the 0–20 cm of soil. This shows the greater toxicity of Pb than Zn in the topsoil profile layers. Moreover, GRSP and soil organic matter were positively related to mean weight diameter and soil large macro-aggregates (>2000 μm) suggesting their role in the binding of soil particles (Yang et al. 2017). Govindu et al. (2020) showed that the easily extractable glomalin fraction and total glomalin content in the rhizosphere soil of three tree species (*A. nilotica, A. lebbeck, P. pinnata*) in a North Telangana coal mine region were effective in instigating seed germination and the glomalin percentage was directly correlated to populations of AM fungi.

1.4.3 AM Fungal Inoculation and Habitat Restoration

Inoculation of AM fungi not only improves the growth of nursery-raised forestry seedlings used in plantings at mining sites but also increases the outplant survival and performance of the inoculated plants. Under nursery conditions, AM fungal inoculation significantly improves seedling growth (Muthukumar and Udaiyan 2018), thereby decreasing the time for seedlings to achieve the plantable size and significantly reducing the use of chemical fertilizers in seedling production (Fig. 1.3).

Nursery and field studies conducted at the Institute of Forest Genetics and Tree Breeding, Coimbatore, India have shown that inoculation of forestry species with AM fungi and other plant growth-promoting microbes can improve seedling growth in nurseries as well as the establishment and survival of these seedlings in mine sites (Figs. 1.4, 1.5, and 1.6).

Inoculation of *Eucalyptus camaldulensis* with *Rhizophagus aggregatus* individually or along with *Azospirillum* and phosphate solubilizing bacteria (PSB) increased seedling growth in the forest nursery (Karthikeyan and Prakash 2008). Moreover, transplantation of microbial inoculated *E. camaldulensis* onto bauxite mine spoils significantly increased the survival rate of multimicrobial inoculated seedlings (AM fungus + *Azospirillum* + PSB) by twofolds and growth by fivefolds than uninoculated plants 2 years after transplantation (Karthikeyan and Prakash 2008). Similarly, inoculation of *Eucalyptus tereticornis* with a combined AM fungal inocula consisting of *R. aggergatus*, *R. fasciculatus*, and *Funneliformis geosporus* increased growth and nutrient content of the seedlings when compared to uninoculated seedlings under nursery conditions. The AM-inoculated *E. tereticornis* seedlings exhibited a 95% survival rate compared to 40% of uninoculated seedlings 2 years after transplantation onto a bauxite mine spoil. The growth performance of AM-inoculated *E. tereticornis* seedlings was three folds higher than uninoculated plants during the same period (Karthikeyan and Krishnakumar 2012). Simultaneous inoculation of *Casuarina equisetifolia* seedlings with *R. aggregatus*, PSB, and *Frankia*

Fig. 1.3 Nursery raised *Acacia auriculiformis* seedlings in the presence (+M) and absence (−M) of arbuscular mycorrhizal fungi for transplantation onto mine sites

significantly increased the growth of seedlings when compared to seedlings inoculated with these microbes alone or uninoculated seedlings raised in bauxite mine spoils under nursery conditions (Karthikeyan et al. 2009). When transplanted onto bauxite mine spoils, multimicrobial inoculated *C. equisetifolia* seedlings exhibited a survival rate of 90–100%, increased growth, and nutrient uptake at the end of 2 years (Karthikeyan et al. 2009). Wang (2017) summarized a large number of laboratory experiments and field studies reporting the influence of AM fungi on the restoration of mines. This showed that 95% of the studies reported a positive influence of AM fungi like increased plant growth, enhanced survival, improved plant and soil nutrients, better soil quality and structure, and plant establishment. Several studies published after 2017 also indicates a positive response of AM fungal inoculation on plant growth and habitat restoration (Table 1.1).

Fig. 1.4 Restoration of vegetation through transplantation of mycorrhized forestry species on a bauxite mine spoils over a 4-year period. (**a**) Barren mine spoil; (**b–e**) mine spoil 1–4 years after restoration

1.5 Conclusions and Future Perspectives

Rehabilitation of degraded sites is one important area where AM fungi and other plant growth-promoting microorganisms could play an important role as these substrates either lack or contain extremely low populations of these microorganisms. The absence or reduced populations of microorganisms in mining areas defer the natural reestablishment of vegetation in these sites. Moreover, the harsh environmental conditions in the mining sites further complicate the establishment of plant communities for several years even after the cessation of the mining activities. There is already ample evidence to show that AM fungi could quickly recolonize mine sites if inoculated or other environmental conditions are favorable. As AM fungi are obligate symbionts, they fail to survive in the absence of host plants. Therefore, habitats detrimental to plant growth are also deleterious to AM fungal presence and functioning. Thus to revegetate mine sites, it is important to select plant-AM fungal combinations that are resilient and could tolerate harsh environmental conditions.

Recent evidence clearly indicates that certain levels of AM fungal diversity with special characteristics could be found hidden even in extreme ecosystems. The use of novel cutting-edge technologies like the new sequence methods along with

Fig. 1.5 Transplanted *Casuarina equisetifolia* inoculated with arbuscular mycorrhizal fungi growing at magnesite-mined outlands

conventional approaches would help in better understanding of the changes in AM fungal communities in the soil and plant roots during the restoration of mine sites. Most information on the influence of mining activities on AM fungal diversity stems from a few mining types, and some widespread mining types like aggregate mining are largely ignored. Moreover, a critical examination of AM fungal inoculation studies generally reveals the use of a few AM fungal species and host plants. Moreover, in field transplantation studies, monitoring the populations of inoculated AM fungi in mine soils is generally missing. An observation of AM symbiosis in plant species that reestablishes naturally in revegetated mining areas is important to acknowledge the existence of succession in AM symbiosis. However, soil properties like the presence of heavy metals and extraordinary high or low pH, electrical conductivity, organic matter, and nutrient levels could greatly influence the mycorrhization of roots and functioning of AM symbiosis in mining sites. Identification of plant-fungal species combination through rigorous screening procedures is important as AM fungal communities could modify the composition of plant communities. In addition, co-inoculation of other plant growth-promoting microorganisms along with AM fungi could be beneficial as many of these microorganisms could help in the establishment of symbiosis. Increasing the diversity of host plant species and their densities may favor AM-mediated mine restoration. Further understanding of AM fungal-mediated tolerance mechanisms in plants and the various signal

Fig. 1.6 *Casuarina equisetifolia* inoculated with arbuscular mycorrhizal fungi and plant growth promoting microorganisms planted growing on limestone mine areas

exchanges between the symbionts that regulate plant growth and development may help to improve plant's performance in mining areas. Even so disengaging several intricate processes that are involved in the development of symbiosis, as well as the interaction of AM fungi with coexisting microorganisms and soil factors would reveal a potent strategy that could aid in rapid restoration of degraded mine sites.

Table 1.1 Response of plants growing on mine substrates to AM fungal inoculation individually or along with other bioinoculants and soil amendments

Site/substrate origin	Study type	Plant species	AM fungi	Coinoculated organisms/ amendments[a]	Response[b]	Reference
Brown coal post-mining site	Microcosm experiment	*Poa compressa, Festuca rubra, Centaurea jacea,* and *Lotus corniculatus*	*Rhizophagus intraradices*	Bacteria, saprophytic fungi, protists	Lowered above- and below-ground plant biomass	Ardestani et al. (2019)
Coal mine	Pot experiment	*Zea mays*	*R. intraradices, Funneliformis mosseae, Claroideoglomus etunicatum*		Increased arbuscular richness, colonization rate, colonization density, extraradical hyphae, sporulation, soil P fractions (Olsen-P, H_2O-P, $NaHCO_3$-P)	Luo et al. (2019)
Coal mine waste	Microcosm	*Medicago sativa*	*F. mosseae*	PSB strain (*Pantoesstewarti*)	Increased phytate mineralization, enhanced plant biomass	Bi et al. (2019a)
Coalmine waste	Pot experiment	*Chrysopogon zizanioides*	*Acaulospora colombiana, Acaulospora morrowiae, Acaulospora scrobiculata, Dentiscutata heterogama, Gigaspora margarita, Rhizophagus clarus*		Increased growth, P, Cu, and Zn	Meyer et al. (2017)
Coal mining subsidence area	Field study	*Prunus sibirica*	*F. mosseae*		Increased biomass, stomatal conductance, transpiration rate, N, P, K contents in roots and shoots	Bi et al. (2021)

Site/substrate origin	Study type	Plant species	AM fungi	Coinoculated organisms/ amendments[a]	Response[b]	Reference
Feldspar	Pot experiment	*Cyamopsis tetragonoloba*	*F. mosseae*	*Azotobacter*, *Azospirillum*, organic and inorganic amendments	Increased rhizosphere enzymes, growth, nutrient uptake	Junia et al. (2021)
Gold mine tailing soil	Greenhouse	*Nauclea orientalis*	*Rhizophagus aggregatus*, *Glomus* sp., *Acaulospora delicata*		Increased plant height, stem girth, leaf length, and width, P levels in roots and shoots, reduced Pb in the root and shoots	Tuheteru et al. (2020)
Iron mining tailing	Greenhouse	*Urochloa ruziziensis*, *Crotalaria spectabilis*, and *Guizotia abyssinica*	*A. morrowiae*		Increased root colonization, reduced root dry weights, microbial carbon biomass, and basal respiration	Zanchi et al. (2021)
Iron ore mined overburden dump	Nursery	*Holoptelea integrifolia*	*Unidentified native AM fungal consortium*	Fluorescent *Pseudomonas* and *Azospirillum* sp.	Increased seed germination, plant height, root length, leaf N, P, and Mg content	Verma and Verma (2019)
Lead-zinc mine wasteland	Pot experiment	*Cynodon dactylon*	*F. mosseae, Diversispora spurcum*		Increased soil pH, uptake of P, S, heavy metals (Pb, Cd, Zn), decreased available Pb and Zn in soils, and shoot Pb	Zhan et al. (2019)

(continued)

Table 1.1 (continued)

Site/substrate origin	Study type	Plant species	AM fungi	Coinoculated organisms/ amendments[a]	Response[b]	Reference
Opencast coal mine field	Nursery/ mine conditions	*Albizia saman* and *Paraserianthes falcataria*	*R. clarus*, *Gigaspora decipiens*, *Scutellospora* sp.		Nursery: Increased shoot P and dry weight Mine site: Higher stem diameter, shoot N content, shoot P content, shoot dry weight, and survival rate	Wulandari et al. (2016)
Open gold pit mine dumpsite	Screen house experiment	*Pterocarpus indicus*	Commercial AM inoculum MYKORICH® and MYCOVAM® (containing 12 species belonging to *Acaulospora, Entrophospora, Glomus, Gigaspora, Scutellospora)* or indigenous AM fungi (*G. margarita, C. etunicatum* and *Glomus macrocarpum*)		Increased seedling height, stem diameter, dry weight, nodulation, decreased Cu content in roots	Aggangan and Cortes (2018)
Open-pit coal mine	Greenhouse	*Pongamia pinnata*	*R. clarus*		Improved seedling growth, and Fe absorption	Agus et al. (2019)
Reclaimed coal mining subsidence	Field study	*Amorpha fruticose*	*F. mosseae*		Higher AM colonization, glomalin-related soil protein, soil organic carbon, soil nutrients, and enzyme activities	Qiu et al. (2019)

Site/substrate origin	Study type	Plant species	AM fungi	Coinoculated organisms/ amendments[a]	Response[b]	Reference
Rehabilitated feldspar mine spoil	Pot experiment	*Cassia angustifolia* and *Cyamopsis tetragonoloba*	*F. mosseae*	*Azotobacter chroococcum, Azospirillum brasilense*, organic matter, NPK fertilization	Increased stem girth, plant biomass, leaf chlorophyll content, shoot nutrients (N, P, K, Ca, Mg, Fe, Cu, Mn, Zn), soil enzyme activities (dehydrogenase, phosphatase)	Junia et al. (2020)
Stimulated coal mining subsidence	Microcosms	*Z. mays*	*F. mosseae*		Increased plant growth, AM colonization, extramatrical hyphal density, higher N, P, K, Ca, and Mg contents in shoots and the roots, higher indole-3-acetic acid, gibberellin, and cytokinin contents in roots	Bi et al. (2019b)
Stimulated mine subsidence	Ground fissure simulation device	*Z. mays*	*F. mosseae*		Increased growth, indole-3-acetic acid, cytokinin, but decreased abscisic acid levels in roots and leaves	Zhang et al. (2021)
Silica post mining	Poly bags	*Albizia chinensis* and *P. pinnata*	*F. moseae*, *Acaulospora* sp., *G. margarita*	Lime and compost	Increased total chlorophyll content, seedling height, stem diameter, biomass, root colonization and P accumulation	Budi et al. (2020)

[a]*PSB* Phosphate solubilizing bacteria
[b]*Pb* lead, *Zn* zinc, *Cd* cadmium, *Fe* iron, *N* nitrogen, *P* phosphorus, *K* potassium, *S* sulfur, *Cu* copper, *Ca* calcium, *Mg* magnesium, *Mn* manganese

Acknowledgements This work was partly supported by the Department of Science and Technology (DST) Fund for Improvement of Science and technology Infrastructure (FIST) [Grant No: SR/FST/LS-II/2017/106 (C)]. We also thank the Indian council of Forestry Research and Education, Dehra Dun for the necessary support of this work.

References

Aggangan NS, Cortes AD (2018) Screening mined-out indigenous mycorrhizal fungi for the rehabilitation of mine tailing areas in the Philippines: rehabilitation of mine tailings. Reforesta 28:71–85

Agus C, Primananda E, Faridah E, Wulandari D, Lestari T (2019) Role of arbuscular mycorrhizal fungi and *Pongamia pinnata* for revegetation of tropical open-pit coal mining soils. Int J Environ Sci Technol 16:3365–3374

Ako TA, Onoduku US, Oke SA, Essien BI, Idris FN, Umar AN, Ahmed AA (2014) Environmental effects of sand and gravel mining on land and soil in Luku, Minna, Niger State, North Central Nigeria. J Geosci Geomat 2:42–49

Andrino A, Guggenberger G, Sauheitl L, Burkart S, Boy J (2021) Carbon investment into mobilization of mineral and organic phosphorus by arbuscular mycorrhiza. Biol Fertil Soils 57:47–64

Ansari RA, Mahmood I (2017) Optimization of organic and bio-organic fertilizers on soil properties and growth of pigeonpea. Sci Hortic 226:1–9

Ansari RA, Mahmood I (2019a) Plant health under biotic stress. In: Ansari RA, Mahmood I (eds) Microbial interactions, vol II. Springer Nature, Singapore. https://doi.org/10.1007/978-981-13-6040-4

Ansari RA, Mahmood I (2019b) Plant health under biotic stress. In: Ansari RA, Mahmood I (eds) Organic strategies, vol I. Springer Nature, Singapore. https://doi.org/10.1007/978-981-13-6043-5

Ansari RA, Mahmood I, Rizvi R, Sumbul A (2017a) PGPR: current vogue in sustainable crop production. In: Kumar V (ed) Probiotics and plant health. Springer Nature, Singapore, pp 455–472

Ansari RA, Mahmood I, Rizvi R, Sumbul A, Safiuddin (2017b) Siderophores: augmentation of soil health and crop productivity. In: Kumar V (ed) Probiotics in agroecosystem. Springer Nature, Singapore, pp 291–312

Ansari RA, Sumbul A, Rizvi R, Mahmood I (2019a) Potential role of plant growth promoting Rhizobacteria in alleviation of biotic stress. In: Ansari RA, Mahmood I (eds) Plant health under biotic stress, Microbial interactions, vol II. Springer Nature, Singapore, pp 177–188

Ansari RA, Sumbul A, Rizvi R, Mahmood I (2019b) Organic soil amendments: potential tool for soil and plant health management. In: Ansari RA, Mahmood I (eds) Plant health under biotic stress, Organic strategies, vol I. Springer Nature, Singapore, pp 1–35

Ardestani MM, Jílková V, Bonkowski M, Frouz J (2019) The effect of arbuscular mycorrhizal fungi *Rhizophagus intraradices* and soil microbial community on a model plant community in a post-mining soil. Plant Ecol 220:789–800

Atmaca E, Karaca UÇ (2019) The relationship between certain microbiological and some arbuscular mycorrhizal parameters of plants prevalent around an aluminum bauxite mine deposit. Appl Ecol Environ Res 17:10941–10961

Barbosa B, Fernando AL (2018) Aided phytostabilization of mine waste. In: Vara Prasad MN, Campos Favas PJ, Maiti SK (eds) Bio-geotechnologies for mine site rehabilitation. Elsevier, pp 147–157

Begum N, Qin C, Ahanger MA, Raza S, Khan MI, Ashraf M, Ahmed N, Zhang L (2019) Role of arbuscular mycorrhizal fungi in plant growth regulation: implications in abiotic stress tolerance. Front Plant Sci 10:1068. https://doi.org/10.3389/fpls.2019.01068

Bi Y, Xiao L, Liu R (2019a) Response of arbuscular mycorrhizal fungi and phosphorus solubilizing bacteria to remediation abandoned solid waste of coal mine. Int J Coal Sci Technol 6:603–610

Bi Y, Xiao L, Sun J (2019b) An arbuscular mycorrhizal fungus ameliorates plant growth and hormones after moderate root damage due to simulated coal mining subsidence: a microcosm study. Environ Sci Pollut Res 26:11053–11061

Bi Y, Xie L, Wang J, Zhang Y, Wang K (2019c) Impact of host plants, slope position and subsidence on arbuscular mycorrhizal fungal communities in the coal mining area of north-central China. J Arid Environ 163:68–76

Bi Y, Xie L, Wang Z, Wang K, Liu W, Xie W (2021) Arbuscular mycorrhizal symbiosis facilitates apricot seedling (*Prunus sibirica* L.) growth and photosynthesis in Northwest China. Int J Coal Sci Technol 2:1–10

Brundrett MC, Tedersoo L (2018) Evolutionary history of mycorrhizal symbioses and global host plant diversity. New Phytol 220:1108–1115

Buck MT, Straker CJ, Mavri-Damelin D, Weiersbye IM (2019) Diversity of arbuscular mycorrhizal (AM) fungi colonising roots of indigenous *Vachellia* and *Senegalia* trees on gold and uranium mine tailings in South Africa. S Afr J Bot 121:34–44

Budi SW, Arty B, Wasis B, Wibowo C, Sukendro A (2020) Influence of arbuscular mycorrhizal fungi and soil ameliorants on the mycorrhizal colonization, chlorophyll content, and performance growth of two tropical tree seedlings grown in soil media with high aluminum content. Malays Appl Biol 49:41–53

Chen M, Arato M, Borghi L, Nouri E, Reinhardt D (2018) Beneficial services of arbuscular mycorrhizal fungi–from ecology to application. Front Plant Sci 9:1270. https://doi.org/10.3389/fpls.2018.01270

Chen J, Mo L, Zhang Z, Nan J, Xu D, Chao L, Zhang X, Bao Y (2020) Evaluation of the ecological restoration of a coal mine dump by exploring the characteristics of microbial communities. Appl Soil Ecol 147:103430. https://doi.org/10.1016/j.apsoil.2019.103430

Confalonieri UE, Margonari C, Quintão AF (2014) Environmental change and the dynamics of parasitic diseases in the Amazon. Acta Trop 129:33–41

Correia M, Espelta JM, Morillo JA, Pino J, Rodríguez-Echeverría S (2021) Land-use history alters the diversity, community composition and interaction networks of ectomycorrhizal fungi in beech forests. J Ecol 109:2856. https://doi.org/10.1111/1365-2745.13674

Cosme M, Fernández I, Van der Heijden MG, Pieterse CM (2018) Non-mycorrhizal plants: the exceptions that prove the rule. Trends Plant Sci 23:577–587

Dickson S, Smith FA, Smith SE (2007) Structural differences in arbuscular mycorrhizal symbioses: more than 100 years after Gallaud, where next? Mycorrhiza 17:375–393

Ezeokoli OT, Nwangburuka CC, Adeleke RA, Roopnarain A, Paterson DG, Maboeta MS, Bezuidenhout CC (2019) Assessment of arbuscular mycorrhizal fungal spore density and viability in soil stockpiles of South African opencast coal mines. S Afr J Plant Soil 36:91–99

Fernandes CC, Kishi LT, Lopes EM, Omori WP, Souza JA, Alves LM, Lemos EG (2018) Bacterial communities in mining soils and surrounding areas under regeneration process in a former ore mine. Braz J Microbiol 49:489–502

Gibbs HK, Salmon JM (2015) Mapping the world's degraded lands. Appl Geogr 57:12–21

Gobbato E (2015) Recent developments in arbuscular mycorrhizal signaling. Curr Opin Plant Biol 26:1–7

Govindu D, Duvva A, Podeti S (2020) The impact of arbuscular mycorrhizal fungi on glomalin-related soil protein distribution and their relationship with soil properties of agroforestry plants in coal mine region of North Telangana. Indian Phytopathol 73:737–740

Guo Y, Liu X, Tsolmon B, Chen J, Wei W, Lei S, Yang J, Bao Y (2020) The influence of transplanted trees on soil microbial diversity in coal mine subsidence areas in the Loess Plateau of China. Glob Ecol Conserv 21:e00877

Hart M, Gorzelak MA, McAmmond BM, Van Hamme JD, Stevens J, Abbott L, Whiteley AS, Nevill P (2019) Fungal communities resist recovery in sand mine restoration. Front For Glob Change 2:78. https://doi.org/10.3389/ffgc.2019.00078

He ZL, Yang XE, Stoffella PJ (2005) Trace elements in agroecosystems and impacts on the environment. J Trace Elem Med Biol 19:125–140

Huang H, Guo J, Zhang Y (2020) The response of arbuscular mycorrhizal fungal communities to the soil environment of underground mining subsidence area in Northwest China. Int J Environ Res Public Health 17:9157. https://doi.org/10.3390/ijerph17249157

Jaramillo F, Baccard M, Narinesingh P, Gaskin S, Cooper V (2015) Assessing the role of a limestone quarry as sediment source in a developing tropical catchment. Land Degrad Dev 27:1064–1074

Jarsjö J, Chalov SR, Pietroń J, Alekseenko AV, Thorslund J (2017) Patterns of soil contamination, erosion and river loading of metals in a gold mining region of northern Mongolia. Reg Environ Chang 17:1991–2005

Junia R, Jain N, Sohal J, Sharma D, Khare N, Aseri GK (2020) Comparative impact of bioinoculants on nutrient uptake, enzyme activities and growth of *Cassia angustifolia* Vhal (Senna) and *Cyamopsis tetragonoloba* (L.) (Guar) in Feldspar mine spoil. S Afr J Bot 140:434. https://doi.org/10.1016/j.sajb.2020.12.006

Junia R, Kasana RC, Jain N, Aseri GK (2021) Guar (*Cyamopsis tetragonoloba* L.): a potential candidate for the rehabilitation of feldspar mine spoil amended with bioinoculants. Indian J Agric Res 55:129–136

Karthikeyan A, Krishnakumar N (2012) Reforestation of bauxite mine spoils with *Eucalyptus tereticornis* Sm. seedlings inoculated with arbuscular mycorrhizal fungi. Ann For Res 55:207–216

Karthikeyan A, Prakash SM (2008) Effects of arbuscular mycorrhizal fungi, *Phosphobacterium* and *Azospirillum* sp. on the successful establishment of *Eucalyptus camaldulensis* Dehn. in bauxite mine spoils. For Trees Livelihoods 18:183–191

Karthikeyan A, Deeparaj B, Nepolean P (2009) Reforestation in bauxite mine spoils with *Casuarina equisetifolia* Frost. and beneficial microbes. For Trees Livelihoods 19:153–165

Kepel BJ, Gani MA, Tallei TE (2020) Comparison of bacterial community structure and diversity in traditional gold mining waste disposal site and rice field by using a metabarcoding approach. Int J Microbiol 2020:1858732. https://doi.org/10.1155/2020/1858732

Kobae Y, Ohmori Y, Saito C, Yano K, Ohtomo R, Fujiwara T (2016) Phosphate treatment strongly inhibits new arbuscule development but not the maintenance of arbuscule in mycorrhizal rice roots. Plant Physiol 171:566–579

Kumar S, Singh AK, Ghosh P (2018) Distribution of soil organic carbon and glomalin related soil protein in reclaimed coal mine-land chronosequence under tropical condition. Sci Total Environ 625:1341–1350

Li D, Yin N, Xu R, Wang L, Zhang Z, Li K (2021) Sludge amendment accelerating reclamation process of reconstructed mining substrates. Sci Rep 11:1–9

Luo Y, Hao X, Zhang K (2019) Effect of inoculation of AM fungi on different phosphorus morphology of discarded soil in coal mining area. Southwest China J Agric Sci 32:381–388

Meyer E, Londoño DM, de Armas RD, Giachini AJ, Rossi MJ, Stoffel SC, Soares CR (2017) Arbuscular mycorrhizal fungi in the growth and extraction of trace elements by *Chrysopogon zizanioides* (vetiver) in a substrate containing coal mine wastes. Int J Phytoremediation 19:113–120

Mushia NM, Ramoelo A, Ayisi KK (2016) The impact of the quality of coal mine stockpile soils on sustainable vegetation growth and productivity. Sustainability 8:546

Muthukumar T, Udaiyan K (2018) Coinoculation of bioinoculants improve *Acacia auriculiformis* seedling growth and quality in a tropical Alfisol soil. J For Res 29:663–673

Ohsowski BM, Klironomos JN, Dunfield KE, Hart MM (2012) The potential of soil amendments for restoring severely disturbed grasslands. Appl Soil Ecol 60:77–83

Ohsowski BM, Dunfield K, Klironomos JN, Hart MM (2018) Plant response to biochar, compost, and mycorrhizal fungal amendments in post-mine sandpits. Restor Ecol 26:63–72

Orchard S, Standish RJ, Nicol D, Dickie IA, Ryan MH (2017) Fine root endophytes under scrutiny: a review of the literature on arbuscule-producing fungi recently suggested to belong to the Mucoromycotina. Mycorrhiza 27:619–638

Parihar M, Rakshit A, Meena VS, Gupta VK, Rana K, Choudhary M, Tiwari G, Mishra PK, Pattanayak A, Bisht JK, Jatav SS (2020) The potential of arbuscular mycorrhizal fungi in C cycling: a review. Arch Microbiol 202:1581–1596

Park H, Lee EH, Ka KH, Eom AH (2016) Community structures of arbuscular mycorrhizal fungi in soils and plant roots inhabiting abandoned mines of Korea. Mycobiology 44:277–282

Policelli N, Horton TR, Hudon AT, Patterson T, Bhatnagar JM (2020) Back to roots: the role of ectomycorrhizal fungi in boreal and temperate forest restoration. Front For Glob Change 3:97. https://doi.org/10.3389/ffgc.2020.00097

Prabhu TN, Rodrigues BF (2019) Arbuscular mycorrhizal (AM) fungal diversity on stabilized iron ore mine dumps in Goa, India. Kavaka 53:34–41

Pringle A, Bever JD, Gardes M, Parrent JL, Rillig MC, Klironomos JN (2009) Mycorrhizal symbioses and plant invasions. Annu Rev Ecol Evol Syst 40:699–715

Pulungan AS, Nasution MY (2021) Biodiversity arbuscular mycorrhizal fungi in the former gold mine area in North Sumatra. J Phys Con Ser 1819:012045. https://doi.org/10.1088/1742-6596/1819/1/012045

Qiu L, Bi Y, Jiang B, Wang Z, Zhang Y, Zhakypbek Y (2019) Arbuscular mycorrhizal fungi ameliorate the chemical properties and enzyme activities of rhizosphere soil in reclaimed mining subsidence in northwestern China. J Arid Land 11:135–147

Rillig MC (2004) Arbuscular mycorrhizae, glomalin, and soil aggregation. Can J Soil Sci 84:355–363

Rodríguez L, Ruiz E, Aloso-Azcárate J, Rincón J (2009) Heavy metal distribution and chemical speciation in tailing and soils around a Pb–Zn mine in Spain. J Environ Manag 90:1106–1116

Salim MA, Budi RSW, Setyaningsih L, Iskandar, Wahyudi I, Kirmi H (2020) Root colonization by arbuscular mycorrhizal fungi (AMF) in various age classes of revegetation post-coal mine. Biodiversitas 21:5013–5022

Sánchez-Castro I, Gianinazzi-Pearson V, Cleyet-Marel JC, Baudoin E, van Tuinen D (2017) Glomeromycota communities survive extreme levels of metal toxicity in an orphan mining site. Sci Total Environ 598:121–128

Selosse MA, Strullu-Derrien C, Martin FM, Kamoun S, Kenrick P (2015) Plants, fungi and oomycetes: a 400-million year affair that shapes the biosphere. New Phytol 206:501–206

Sharma S, Prasad R, Varma A, Sharma AK (2017) Glycoprotein associated with *Funneliformis coronatum*, *Gigaspora margarita* and *Acaulospora scrobiculata* suppress the plant pathogens *in vitro*. Asian J Plant Pathol 11:192–202

Singh PK, Singh M, Tripathi BN (2013) Glomalin: an arbuscular mycorrhizal fungal soil protein. Protoplasma 250:663–669

Singh AK, Zhu X, Chen C, Wu J, Yang B, Zakari S, Jiang XJ, Singh N, Liu W (2020) The role of glomalin in mitigation of multiple soil degradation problems. Crit Rev Environ Sci Technol 13:1–35

Smith S, Read D (2008) Mycorrhizal symbiosis, 3rd edn. Academic Press, New York

Smith FA, Smith SE (2013) How useful is the mutualism-parasitism continuum of arbuscular mycorrhizal functioning? Plant Soil 363:7–18

Spatafora JW, Chang Y, Benny GL, Lazarus K, Smith ME, Berbee ML, Bonito G, Corradi N, Grigoriev I, Gryganskyi A, James TY (2016) A phylum-level phylogenetic classification of zygomycete fungi based on genome-scale data. Mycologia 108:1028–1046

Steinberger JK, Krausmann F, Eisenmenger N (2010) Global patterns of materials use: a socioeconomic and geophysical analysis. Ecol Econ 69:1148–1158

Sun Z, Song J, Xin XA, Xie X, Zhao B (2018) Arbuscular mycorrhizal fungal 14-3-3 proteins are involved in arbuscule formation and responses to abiotic stresses during AM symbiosis. Front Microbiol 9:91

Sun S, Sun H, Zhang D, Zhang J, Cai Z, Qin G, Song Y (2019) Response of soil microbes to vegetation restoration in coal mining subsidence areas at Huaibei coal mine, China. Int J Environ Res Public Health 16:1757

Suting EG, Devi NO (2021) Occurrence and diversity of arbuscular mycorrhizal fungi in trap cultures from limestone mining sites and un-mined forest soil of Mawsmai, Meghalaya. Trop Ecol 31:1–3

Teixeira AF, Kemmelmeier K, Marascalchi MN, Stürmer SL, Carneiro MA, Moreira FM (2017) Arbuscular mycorrhizal fungal communities in an iron mining area and its surroundings: inoculum potential, density, and diversity of spores related to soil properties. Ciênc e Agrotecnologia 41:511–525

Tuheteru FD, Arif A, Husna H, Mansur I, Tuheteru EJ, Jusniar J, Basrudin B, Albasri A, Hadijah MH, Karepesina S (2020) Arbuscular mycorrhizal fungal inoculation improves *Nauclea orientalis* L. growth and phosphorus uptake in gold mine tailings soil media. J Degraded Min Lands Manage 7:2193–2200

Verma P, Verma RK (2019) Role of growth promoting microbes in growth improvement of *Holoptelia integrifolia* seedling for plantation on iron ore mine overburden. Bull Env Pharmacol Life Sci 8:68–77

Vieira CK, Marascalchi MN, Rodrigues AV, de Armas RD, Stürmer SL (2018) Morphological and molecular diversity of arbuscular mycorrhizal fungi in revegetated iron-mining site has the same magnitude of adjacent pristine ecosystems. J Environ Sci 67:330–343

Wang F (2017) Occurrence of arbuscular mycorrhizal fungi in mining-impacted sites and their contribution to ecological restoration: mechanisms and applications. Crit Rev Envion Sci Technol 47:1901–1957

Wang J, Wang P, Qin Q, Wang H (2017a) The effects of land subsidence and rehabilitation on soil hydraulic properties in a mining area in the loess plateau of China. Catena 159:51–59

Wang J, Zhao F, Yang J, Li X (2017b) Mining site reclamation planning based on land suitability analysis and ecosystem services evaluation: a case study in Liaoning province, China. Sustainability 9:890. https://doi.org/10.3390/su9060890

Wilson GWT, Rice CW, Rillig MC, Springer A, Hartnett DC (2009) Soil aggregation and carbon sequestration are tightly correlated with the abundance of arbuscular mycorrhizal fungi: results from long-term field experiments. Ecol Lett 12:452–461

Woźniak G, Chmura D, Małkowski E, Zieleźnik-Rusinowska P, Sitko K, Ziemer B, Błońska A (2021) Is the age of novel ecosystem the factor driving arbuscular mycorrhizal colonization in *Poa compressa* and *Calamagrostis epigejos*? Plan Theory 10:949

Wulandari D, Cheng W, Tawaraya K (2016) Arbuscular mycorrhizal fungal inoculation improves *Albizia saman* and *Paraserianthes falcataria* growth in post-opencast coal mine field in East Kalimantan, Indonesia. For Ecol Manag 376:67–73

Xiao L, Bi Y, Du S, Wang Y, Guo C (2019) Effects of re-vegetation type and arbuscular mycorrhizal fungal inoculation on soil enzyme activities and microbial biomass in coal mining subsidence areas of Northern China. Catena 177:202–209

Yan D, Mills JG, Gellie NJ, Bissett A, Lowe AJ, Breed MF (2018) High-throughput eDNA monitoring of fungi to track functional recovery in ecological restoration. Biol Conserv 217:113–120

Yan A, Wang Y, Tan SN, Mohd Yusof ML, Ghosh S, Chen Z (2020) Phytoremediation: a promising approach for revegetation of heavy metal-polluted land. Front Plant Sci 11:359

Yang Y, Liang Y, Ghosh A, Song Y, Chen H, Tang M (2015) Assessment of arbuscular mycorrhizal fungi status and heavy metal accumulation characteristics of tree species in a lead–zinc mine area: potential applications for phytoremediation. Environ Sci Pollut Res 22:13179–13193

Yang Y, He C, Huang L, Ban Y, Tang M (2017) The effects of arbuscular mycorrhizal fungi on glomalin-related soil protein distribution, aggregate stability and their relationships with soil properties at different soil depths in lead-zinc contaminated area. PLoS One 12:e0182264. https://doi.org/10.1371/journal.pone.0182264

Zanchi CS, Batista ER, Silva AO, Barbosa MV, Pinto FA, dos Santos JV, Carneiro MA (2021) Recovering soils affected by iron mining tailing using herbaceous species with mycorrhizal inoculation. Water Air Soil Pollut 232:1–3

Zhan F, Li B, Jiang M, Li T, He Y, Li Y, Wang Y (2019) Effects of arbuscular mycorrhizal fungi on the growth and heavy metal accumulation of bermudagrass [*Cynodon dactylon* (L.) Pers.] grown in a lead–zinc mine wasteland. Int J Phytoremediation 21:849–856

Zhang Y, Bi Y, Shen H, Zhang L (2020) Arbuscular mycorrhizal fungi enhance sea buckthorn growth in coal mining subsidence areas in Northwest China. J Microbiol Biotechnol 30:848–855

Zhang J, Bi Y, Song Z, Xiao L, Christie P (2021) Arbuscular mycorrhizal fungi alter root and foliar responses to fissure-induced root damage stress. Ecol Indic 127:107800. https://doi.org/10.1016/j.ecolind.2021.107800

Arbuscular Mycorrhizal Fungi and Attainment of Food Security

2

Tesleem Taye Bello and Oluwatoyin Adenike Fabiyi

Abstract

The reciprocal give-and-take relationship of a fungus that colonises the plant root is termed mycorrhizae. The prime partner in the association is the fungus, whose obligation is to provide food and growth hormones to the plant. The fungus is also saddled with the responsibility of shielding the plant from pathogens. Reciprocal alliances are formed by several plants and crops with arbuscular mycorrhizal fungi. This interchangeable relationship has propelled the occurrence of new techniques in crop breeding and agricultural methods to buttress and advance arbuscular Mycorrhizal fungi (AMF) in agroecosystems. The beneficial effect of AMF could be highly favourable to crops and the ecosystem at large in several ways, such as furnishing great resilience to plant diseases and enabling amelioration of the soil structure. The colonisation does not automatically bring about amplified plant burgeoning and high crop yield because land management customs, which motivate mycorrhiza–crop relationships, do not favour higher yield from crops. Land management traditions that could stimulate AMF, such as low tillage and a reduction in the use of chemicals. The functions of mycorrhiza such as its positive effect on nutrition and growth of the host plant, are shifted through intensive agricultural exercise. In general, the role of mycorrhizas is brought down by the high availability of nutrients from synthetic fertil-

T. T. Bello
Department of Plant Soil and Microbial Sciences, Michigan State University, East Lansing, MI, USA

Department of Agricultural Science Education, Federal College of Education, Abeokuta, Ogun, Nigeria

O. A. Fabiyi (✉)
Department of Crop Protection, Faculty of Agriculture, University of Ilorin, Ilorin, Kwara, Nigeria

R. A. Ansari et al. (eds.), *Mycorrhizal Symbiosis and Agroecosystem Restoration*, https://doi.org/10.1007/978-981-99-5030-0_2

isers to common symbionts, and in some cases. The tropical ecosystems of the world are blessed with a high abundance and diversity of AMF, which have great potential for improving agroecosystems. This chapter aspires to examine crop development and growth in conjunction with AMF in tropical agroecosystems, particularly in the area of attaining food security. The key issues surrounding the non-nutritional and nutritional tasks of AMF bonding with the crops as influenced by the environment, with a focus on optimising and employing the symbiotic consortium in viable management policies for excellent crop production have also been taken into account.

Keywords

Arbuscular mycorrhizal fungi · Agroecosystems · Management · Food security

2.1 Introduction

The tropical regions serve as the major production sites for most agriculture-based raw materials in the world, including food, fibre and fuel (Laurance et al. 2014). Yet, commercial agriculture within this region has faced with a lot of challenges, especially a relatively high incidence of several pests and pathogens that are largely responsible for the prevalence of high yield losses in the region (Aghale et al. 2017; Atolani and Fabiyi 2020; Bello et al. 2021; Fabiyi and Olatunji 2021a). It is pertinent to save crops from attack of pests and pathogens; the salvaged fraction will be used to support the undernourished in society and the world at large (Fabiyi 2021a). The potential of pests and pathogens to impede successful agricultural production and a bountiful yield is very high if proper control or protection is not instituted in place, and this may prompt to food insecurity (Fabiyi et al. 2020a,b, 2021a,b; Fabiyi 2021b). The significance of pests, pathogens and diseases in crop development and grain storage is very high, and the damage of their menace has shot up monumentally, with great retardation to the provision of sufficient quantity and the right standard of food to the population throughout the world (Fabiyi et al. 2018a,b, 2019; Fabiyi 2020; Fabiyi and Olatunji 2021b). Crops and their products provide around 57% of the complete quota of production in the agricultural units of most underdeveloped countries; however, the production is thwarted by several pests (Atolani et al. 2014a,b; Fabiyi and Olatunji 2018). It has become increasingly difficult to sustain an intensive agricultural system based on crops because of the threat occasioned by the build-up of several soil-borne microorganisms (Fabiyi 2019, 2021c,d, 2022a,b,c,d; Fabiyi et al. 2020c, 2022a,b). There is a necessity to utilise uncultivated lands in a dependable manner to facilitate a tremendous increase in food production for future use and demands (Godfray et al. 2010). One of the potentially feasible approaches is the no-till agriculture. Results in the literature have established that field trials have presented dense soil fungal hyphae (Paul et al. 2013), high macroaggregates and soil carbon with the no-till approach in comparison to the traditional and customary tilled system (Paul et al. 2013). Many dependable farming methods, such as conservation and intercropping, could be employed to maintain soil biodiversity (Palm et al. 2014; Brooker et al. 2015), with accruing benefits like nutrient retention, good capacity for water storage, sequestration of carbon and detoxification (Bardgett and van der Putten 2014; Puschel et al. 2016; Bender and van der Heijden 2015; Cavagnaro et al. 2015; Stirling et al. 2016). The potential of these occurrences has been recognised to be of great importance to yield increase and reliability to crop production improvement (Godfray et al. 2010; Plassard and Dell 2010).

Almost every agricultural activity depends directly or indirectly on the soil. Since time immemorial, soil has been associated with the production of arable crops, cash crops and raw materials for satisfying a variety of human needs. The soil microbial biosphere is the largest store of biodiversity on earth (Edwards et al. 2012). They remain the most viable indicators of soil health. Hence, the tropical soil is seen as the most important economic industry for the vast majority of people, especially within the tropics (Sachs et al. 2001). Conventional agriculture, which formed the basis of tropical crop production, prioritises production and profit maximisation with little consideration for long-term consequences on biodiversity and the dynamics of agroecological systems (Erenstein et al. 2008). Furthermore, the increasing demand for food and raw materials due to the rapid increase in the human population has mounted significant pressure on the soil. To improve production, farmers use chemical fertilisers and chemical pesticides, which have negative effects on soil biodiversity and soil health. The soil rhizosphere is known to contain a high abundance of microorganisms, mostly bacteria, fungi, actinomycetes and protoza (Gottel et al. 2011). The most significant association is the symbiotic relationship between fungi and plant roots. This association influences the competition for photosynthetates, nutrients and colonisation sites, which in turn stimulates host plant defence mechanisms against pests and soil-borne pathogens (Jung et al. 2012). Soil organisms that are involved in a symbiotic relationship with plants, like mycorrhizal fungi, could be a source of these advantageous effects and thus end up being significant constituents of maintainable intensified agriculture with tolerable capacity for abiotic and biotic factors (Godfray et al. 2010). Plant health, nutrition and growth are key factors in crop development that are functionally affected by AM fungi. The majority of plant taxa on the land have customary symbiotic associations with AM fungi which are obligate organisms (Philippot et al. 2013). Several plant species of economic relevance have an entrenched association with AMF. Medicinal plants, cereals, fruit trees, pulses and vegetables also develop symbiotic relationships (Philippot et al. 2013). The AMFs are endowed with extra-radical mycelium (ERM), which spreads from the colonised roots of host plants to the soil and is used for the uptake of nutrients and water for the host (Smith and Smith 2011a). The interconnectedness, extent and structure of the ERM affect the flow of nutrients to the host. It is a critical tool of the fungal structure and a significant factor in the symbiotic union (Smith and Smith 2011a). The ERM contains genes responsible for nutrient translocation from the soil to the host roots (Smith and Smith 2011a). This assists the ERM to take nutrients farther away from the host's rhizosphere; the absorbing process is highly structured with a good surface-to-volume ratio (Smith and Smith 2011a; Puschel et al. 2016). Apart from providing improved nutrition to the host

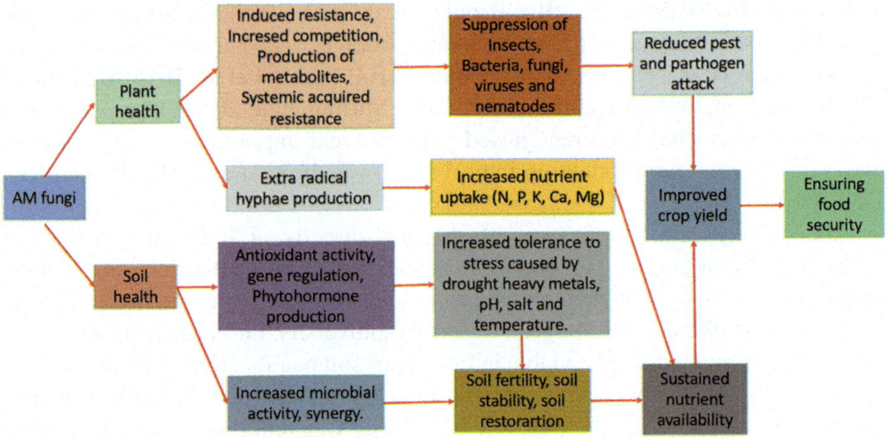

Fig. 2.1 A schematic representation of various processes and the impact of AM fungi on achieving food security. These essential processes contribute significantly to improved plant health

plant, the biogeochemical cycles are completed with the aid of AMF (Puschel et al. 2016), the forbearance of host plants to abiotic and biotic stress is improved with an increase in plant phytochemical content (Fig. 2.1), there is an increase in soil aggregation, and carbon sequestration is also facilitated (Puschel et al. 2016) with great support during any form of environmental stress (Table 2.1). Studies have shown and established that the multifarious aid furnished by AMF is connected with the synergistic life of the multiple bacterial communities residing in the mycorrhizosphere (Rouphael et al. 2015). They are associated sternly with extra-radical mycelium and spores, which has enabled them to play numerous plant growth-promoting responsibility that range from the production of antibiotics (Rouphael et al. 2015), siderophores and indole acetic acid to mineralisation, phosphorus solubilisation and fixation of nitrogen without intervening factors or oblique processes as depicted in Fig. 2.1 (Rouphael et al. 2015). The bacterial strains linked with AMF could be employed as bio-stimulants and bio-fertilisers for tenable food production systems (Ortas 2012; Turrini et al. 2018). This chapter aspires to examine crop development/growth in conjunction with AMF in tropical agroecosystems, particularly in the area of attaining food security. The key issues surrounding the non-nutritional and nutritional tasks of AMF bonding with crops as influenced by the environment, with a focus on optimising and employing the symbiotic consortium in viable management policies for excellent crop production have also been taken into account.

2.2 Biology of AM Fungi

AMF are members of the phylum Glomeromycota, which consists of several families including *Gigasporaceae, Acaulosporaceae, Glomeraceae, Ambisporaceae Pacisporaceae, Archaesporacea, Paraglomeraceae, Claroidoglomeraceae,*

Table 2.1 Effect of AM fungi on plant health amelioration under different environmental conditions

S. No.	Environmental stress	Plant species	AM fungi	Plant response	References
1.	Heat	Triticum aestivum	*Funneliformis Geosporum*	Increased grain number and nutrient allocation	Cabral et al. (2016)
2.	High temperature	*Zea mays*	*Rhizophagus intraradices*	Application of the mycorrhizal fungi enhanced the photosynthetic capacity of the plants and increased overall biomass and yield	Mathur et al. (2016)
3.	Metals	*Sesbania rostrata*	*Glomus mosseae*	Stimulated formation of root nodules and increase in nitrogen and phosphorus content	Lin et al. (2007)
4.	Salinity	*Oryza zativa*	*Claroideoglomus Etunicatum*	Improved quantum yield of PSII	Porcel et al. (2015)
5.	Drought	*Fragaria ananessa*	*F. Geosporus*	Increased root and shoot fresh weights	Boyer et al. (2014)
6.	Drought	*Glycine max*	*Glomus aggregatum*	Improved water content with phosphorus and nitrogen levels	Grümberg et al. (2015)
7.	Drought	*Hordeum vulgare*	*Glomus intraradices*	Increased root volume, phosphorus content and activity of the phosphatase enzyme	Bayani et al. (2015)

Sacculosporaceae and Diversisporaceae (http://www.amf-phylogeny.com/, accessed on 7 August 18 2021). The AMF life cycle can only be completed within a host plant because biotrophs are obligate. Many hyphae of limited life span are produced through spore germination in accordance with the presence of physical factors like pH, temperature and moisture; this is termed the asymbiotic phase of their development (Kiers et al. 2011). Another phase of development known as pre-symbiotic occurs while the root exudate of the host plant is present (Kiers et al. 2011). Here, the morphogenesis of differential hyphae is initiated, a reorientation by the germling hyphae is seen, forming the elongation direction, and thus many branching patterns are formed (Jiang et al. 2017). After this stage, a physical contact is established in the middle of the hyphae and roots of the host, appressoria differentiation occurs, intracellular growth of hyphae is established in the root cortex, the root cells are penetrated, and tree structures in the form of hyphae are instituted in place, with several branches appearing like haustoria and generally called arbuscules (Jiang et al. 2017). The arbuscules are the main organs in the host plant and mycorrhizal association where nutrient exchange is achieved. Carbon and lipids are obtained from the host by AMF; in return, the absorbed mineral nutrients are released and then translocated by ERM (Luginbuehl et al. 2017). There are two types of mycorrhizal formations, namely the *Paris type* and the *Arunm type* (Kiers et al. 2011). The colonisation pattern of the *Paris type* is characterised by the direct growth of the fungus from cell to cell in the root cortex. Thus, hyphal coils are formed intracellularly, while arbuscules are seen as intercalary formations through the coils (Luginbuehl et al. 2017). On the other hand, the *Arum-type arbuscles* are formed terminally on hyphal intercellular branches, and the symbionts equally spread in the root cortical cells in an intercellular manner (Jiang et al. 2017). In general, available data in the literature principally originate from the *Arunm type*; they occur widely in natural habitats and are distributed widely in agricultural farming systems (Jiang et al. 2017). Structures that act as storage facilities filled with lipids made of spores are common to AMF species; these could be termed intraradical vesicles apart from the arbuscules. Once the AMF collects carbon from the host, the ability of the symbiont to develop extra radically begins (Luginbuehl et al. 2017). The surroundings are completely colonised, the host receives nutrients regularly, the AMF interacts with microorganisms around the soil rhizosphere (Jiang et al. 2017) and the roots of nearby plants are equally colonised, and nutrients are exchanged between different hosts, whether they belong to the same species, families, genera or not (Luginbuehl et al. 2017). Later, asexual spores are formed by the ERM, depending on how the soil can maintain the growth of mycorrhiza and also the fertility of the soil, this stage then closes the life cycle of the AMF (Luginbuehl et al. 2017).

2.3 Ecology of AM Fungi in Tropical Environments

Mycorrhizal fungi are known to have a widespread presence in almost all known environments. Although, on a global scale, AMF communities within the temperate regions have been extensively studied (Smith and Read 2008; Raviv 2010; Ansari et al. 2020a; Solanki et al. 2020; Ansari and Mahmood 2019a), marked variations are known to exist in terms of their composition and diversity among different ecosystem types worldwide. In this regard, variations in AMF communities due to vegetation types have been reported (Velázquez et al. 2013), between grasslands and forest ecosystems (Öpik et al. 2010; Kivlin et al. 2011; Davison et al. 2012; Goldmann et al. 2020), as well as due to the different agro-ecological scales, especially in terms of land-use intensity (Oehl et al. 2010; Moora et al. 2014), soil types (Oehl et al. 2010; Zhu et al. 2020) and altitudinal differences (Coutinho et al. 2015; Vieira et al. 2019). A large studies on AMF community structures are available from the semi-natural and anthropogenic systems of temperate North America and Europe (Öpik et al. 2010; Kivlin et al. 2011; Davison et al. 2012), data from most natural ecosystems of the tropical and subtropical regions remains scanty (Moreira et al. 2007). However, the abundance and diversity of AMF in India are well documented (Gupta et al. 2014, 2016; Yadav and Pandey 2016; Jha and Songachan 2020). Reports from tropical and subtropical ecosystems in India indicate a high abundance of AMF species existing in different agricultural systems (Hindumathi and Reddy 2011; Dessai and Rodrigues 2012; Parihar et al. 2019) and natural ecosystems (Suchitra et al. 2012; Gupta et al. 2017; Jha and Songachan 2020). Other reports from the tropics include the studies of Husband et al. (2002), who recorded a high diversity of AMF from the tropical forests of Panama in Central America. Reports from the natural tropical ecosystems of Africa that revealed a high number of novel taxa largely addressed agro-ecosystems and some pastoral farmlands in West Africa (Tchabi et al. 2008; Alori et al. 2012; Adeyemi et al. 2019; Olubode et al. 2020), Central Africa (Mathurin et al. 2022) and East Africa (Mathimaran et al. 2007; De Beenhouwer et al. 2015; Utaile et al. 2021). Jemo et al. (2018) further reported a variation in the AMF community structure under fallow and cropping conditions in the tropical humid soil of Cameroon. Furthermore, research on tropical, natural ecosystems focused on specific host plants from shrublands and forests (Öpik et al. 2013; Gazol et al. 2016), savanna (Tchabi et al. 2008), forests (Wubet et al. 2006), and arid regions (Yamato et al. 2009). Available studies from the tropical and subtropical ecosystems of South America revealed a high diversity of AMF, especially in the rupestrian fields of Brazil (De Carvalho et al. 2012; Covacevich et al. 2012; Pagano and Cabello 2012). In addition to these, several indigenous AMF species were identified from agricultural fields (Miranda et al. 2011) as well as from different vegetation gradients in Brazil (da Silva et al. 2015). Despite the high abundance of AMF species in tropical regions of the world, their inherent potential is yet to be fully harnessed to improve agricultural productivity and ultimately ensure food security within the region.

2.4 AMF Function, Environmental Modulation and Potential Applications in Agriculture

The AMF-plant relationship is ubiquitous, spanning many phylogenies of plants growing on the land and developing gradually from 475MYA (Field et al. 2015). Micronutrients such as zinc, nitrogen, copper, and other major elements like phosphorus from soil are greatly accessed via the multiplication of the mycelial in an extra-radical fashion (Liu et al. 2000). DNA and proteins, which are organic sources of phosphorus, could be acquired by the extra radical hyphae when acid phosphates are excluded in the hyphae (Sato et al. 2015). The Mucoromycota and Glomeromycota are very regular, and they establish symbiosis intracellularly with nearly all food crops (Smith and Smith 2011a; Spatafora et al. 2016). The apparent advantage of AMF to crop development through the supply of nutrition has led lately to substantial attraction in their inherent ability to enable a cutback in the application of synthetic fertiliser in the absence of any accompanying loss of yield (Navarro et al. 2014; Alqarawi et al. 2014a,b; Berruti et al. 2016). Environmental factors such as the concentration of atmospheric CO_2 and nutrient availability coupled with the species of fungi and host plants, are known to impact the degree of nutrient uptake by AMF (Field et al. 2012). However, the exceedingly fluctuating end results of the factors involved present a significant obstacle against the application of AMF in sustainable agriculture. Johnson (2010) opined that symbiotic relationship between plant and AM fungi is advantageous based on the expanse of nutrient exploitable and obtainable for growth and development of plants given the level of soil nitrogen and phosphorus comparable to accessibility (Johnson 2010). Plants in a bond with AMF are known to have displayed amplified growth in phosphorus-exhausted soils in parallel to plants not connected to AMF (Bender et al. 2016). The transmission of nitrogen to plant associates by AMF through the extra radical hyphae is extensively studied (Hodge and Fitter 2010). It has been also observed that increased biomass of the plants are observed while the phosphorus and nitrogen contents of such plants are equally high (Thirkell et al. 2016). These discoveries reveal that AMF has the capacity to diversify the origin of nutrients for acquisition by the host plants and also boost uptake (Thirkell et al. 2016; Ansari and Mahmood 2017a,b, 2019b; Ansari et al. 2020b). However, the level to which plants utilise the advantages of the AMF alliance will depend on fertiliser usage and other management practises on the land (Johnson 2010). The attitude of cereal crops may be particularly different. Cynically, the growth of cereals is not affected by AMFs in soils with a phosphorus concentration that is equal to the level obtained with fertiliser application (Li et al. 2016). Thus, suggesting that a foremost equilibrium exists between the usage of fertiliser and the satisfaction obtained from AMF interrelation by plants (Li et al. 2016). Some plants have exhibited reduced growth in tie-ups with AMF in spite of the supported phosphorus uptake. High phosphorus conveyance by fungi exudes plant carbon to the fungi; the outcome of this process is that the plant is deficient in carbon, which then translates into slow and reduced growth (Kiers et al. 2011). Reports by Walder et al. (2012) indicated that the collection of nutrients and minerals from a common mycelium is determined by the host specification and has nothing to do with carbon availability in the mycelium network (Walder et al. 2012; Ansari et al. 2019a,b; Sumbul et al. 2017). The normal straight pathway for phosphorus uptake by plant roots through the naturally designated transporters of phosphates in the roots is usually impaired and reduced with the AMF alliance (Smith and Smith 2011b; Walder et al. 2012). Phosphorus collection is reduced in AMF-allied plants if the AMF phosphorus delivery course does not compensate for the low phosphorus use by plants (Smith and Smith 2011b; Walder et al. 2012). The dissimilarity and comparative part played by the phosphorus uptake routes could be accountable for discrepancies in phosphorus usage among several plant cultivars and species (Li et al. 2005; Walder et al. 2012). Although, AMF associations have been regarded lately as having a huge positive impact on the growth of plants in the families Solanaceae, Alliaceae and Fabaceae (Li et al. 2005). The reaction of other crop families, predominantly cereals, is uncertain (Li et al. 2005). Some cultivars of the same species also exhibit contrast in reaction to growth under the AMF alliance (Tawaraya 2003). The example was revealed in wheat, where some cultivars responded positively to the alliance with AMF (Hu et al. 2014) and some behaved negatively (Li et al. 2016). Lately, some findings have established that new cultivars of crops may be more sensitive to AMF association than the earlier varieties (Lehmann et al. 2012). The exact cause of this divergence is not known, thus portraying a noteworthy barrier to the use of AMF on a broad scale in production of cereal. The physiology, morphology and traits of plant roots are perhaps responsible for the degree and level of nutrient collection advantages in AMF associations (Navarro-Fernandez et al. 2016). For instance, an immensely branched cereal root allows a systematised uptake of nutrients, so there is no need for an AMF alliance because the roots are naturally designed to perform functions close to those of AMF (Smith and Smith 2011a). Several factors must be considered for the optimum maximisation of nutrients for crop development and growth (Smith and Smith 2011a; Sumbul et al. 2017; Rizvi et al. 2015; Ansari et al. 2017a,b). In order of importance, the genetics of the plant, consistency of nutrient accessibility, knowledge of root characteristics and quality decisive factors of root architecture, tendency for symbiont alliance and finally the effect of pathogens and pests should always be considered (Li et al. 2016).

2.5 AMF as Moderators of Soil Functions for Sustainable Agriculture

The AM fungi communities in the subsoil are unique and abundant (Higo et al. 2013; Sosa-Hernández et al. 2018). The characteristics of the subsoil, like compaction, large bulk density, limited pores and minimal concentration of oxygen in totality, provide a semi-optimal habitat for plant roots (Moll et al. 2016). The supposition is that the AMF are well suited to the subsoil, even though there is no verifiable proof of specific traits of subsoil of AMF (Chagnon et al. 2015). It is hypothesised that the AMF inhabiting the subsoil will have the ability to colonise soil spores, however small they may be, their tolerance to anaerobic situations and the production of long-lasting mycelium (Chagnon et al. 2015). It is anticipated that the AMFs inhabiting the subsoil will be tolerant to stress, going by the CSR framework (Chagnon et al. 2015). The phylotypes are envisaged to display high efficiency in the use of resources with long-lasting biomass, amounting in cost/advantage savings for the plant (Chagnon et al. 2015). Although at the initial stage, mycelium growth might be slow, without instantaneous benefit, and most likely a carbon sink for the host (Säle et al. 2015). However, once the network of mycelium is entrenched, the services provided to the plant will be long-lasting without any cost (Säle et al. 2015). The dispersal of AMF spores could be by arthropods, earthworms, small mammals or wind (Egan et al. 2014), but these vectors are not likely to be germane in the subsoil, with the earthworm being a deviation. Once microbial activity is low and there is little disruption, a long-lasting mycelium is produced, which could serve as a dispersal agent on its own (Egan et al. 2014). The formation of the mycelium is a probable gain for the host plant because the AMF spores are a storage for carbohydrates and lipids (Walder et al. 2012). In this relationship, while the host provides energy and carbon, the mycelium probes the soil and extracts nutrients for the plant's use (Walder et al. 2012). Succinctly put, the benefits that accrue to the plant for each unit of carbon are very high in the subsoil compared to the top soil (Walder et al. 2012). Some schools of thought have identified the significance of biological elements in pedogenesis, and the AMFs are known to have played important functions in the formation of soil (Liu et al. 2016). The AMFs are involved in rock weathering, and the bacteria group equally has very high geochemical competence (Berruti et al. 2016). In reality, it is very arduous to comprehend pedogenesis in the history of the earth without giving consideration to AMF and the evolution of plant roots (Leake and Read 2017). The ectomycorrhizal fungi (EM) have the ability to exude organic chelators of low molecular weight, which supports the weathering of minerals; this characteristic is absent in AMF (Leake and Read 2017). However, the AMF contributes to the weathering of minerals through indirect routes like improved soil stabilisation and respiration, plus intensified exudation and evapotranspiration (Lehmann et al. 2012). However, the differences in the weathering patterns of AM and EM are not as obvious as expected (Koele et al. 2014). The deeper layers of the soil do not support biological activity, in spite of the clay minerals that dominate the upper layers. They are made up of primary minerals, which have great potential for weathering and nutrient delivery (Koele et al. 2014). The mycorrhizosphere is largely developed with soil volume, which comes under the sway of the symbionts, and thus microbial activity is fostered in the subsoil with huge soil volume (Koele et al. 2014). This integrated activity of AMF, roots and the affiliated community of soil microorganisms favours soil enlargement (Higo et al. 2013).

2.6 AM Fungi for Improving Agricultural Sustainability: Future Direction

Mycorrhizal functions could shift with intensive agricultural activities, despite the fact that AMFs have been reported to have a progressive effect on the growth and nutrition of the host plant (Chagnon et al. 2015; Zou et al. 2016). The chemical fertiliser releases high nutrients, which might affect the role of AMFs and turn them into ordinary symbionts or, at times, pathogens or pests (Chagnon et al. 2015; Johnson et al. 2015). The exemplification by Rillig et al. (2015) emphasises that it is pertinent to appraise the land history and land practices employed over the years before the introduction of AMF as an inoculum. For instance, the addition of AMF may be positive for wheat, while it may turn out to be species-specific for barley (Navarro-Fernandez et al. 2016). This may be a constraint to the use of AMFs in mixed cultivations (Navarro-Fernandez et al. 2016). The majority are short-run experiments conducted in controlled situations without specific links or associations with abiotic and biotic factors (Li et al. 2016). There is usually no experimental field data, which is a great setback in determining the possibility of scaling up AMF applications on a large scale (Li et al. 2016). Similarly, the nutrient exchange for carbon among the symbionts has not been fully quantified, nor has it been evaluated on arable crops for commercial purposes (Li et al. 2016). This also opens up a knowledge gap in research on the use of AMFs for the cultivation of significant food crops. It is important to address these deficiencies before the utilisation of AMF-centred approach, that is, employing AMF totally as part of a crop development process (Li et al. 2016). It is pertinent to focus on the particular functions of AMF in the systems of crops, especially in response to 'mycorrhizal application' (Li et al. 2016). Application rate, method and doses of AM fungi is also an important factors that influence the quantitative and qualitative characters of the plants (Field et al. 2012). The effect of the AMF alliance on crops cannot be overlooked, though we may say that it is not a significant factor in nutrient exchange (Field et al. 2012). The level of importance of AMF to the host could be measured via the carbon received. Also, carbon, phosphorus and nitrogen exchange in the AMF plant relationship measurement is hardly not substantially recorded (Field et al. 2012; Ansari and Khan 2012a,b; Ansari et al. 2017b). Some schools of thought have opined that over-saturation of the atmosphere with CO_2 reduces the carbon-nutrient exchange between symbionts (Field et al. 2012), but the state of this in relation to agricultural crops is not certain. A look at the projections of the IPCC regarding high concentrations of atmospheric CO_2 and its effect on crop productivity in AM fungi amended environment is crucially important (Challinor et al. 2016). Comprehending the response of these symbionts to climate change, especially with regards to increased atmospheric CO_2, is necessary (Challinor et al. 2016). The effect of AMFs on crop yield globally and their prospective ability to assuage environmental issues in the future should be of prime concern (Challinor et al. 2016). AMFs may be used for some environmental issues that are easily projected, and they can also be used specifically in a region for problems that arise (Field et al. 2012; Challinor et al. 2016). It is crucial to recognise the attributes of plants below and above the ground that offer malleability functionally to AMF and crop alliances in response to interchanging environmental situations, which paves the way for them to have flexibility to keep up yield in changing climates (Challinor et al. 2016). Among dissimilar crop species, carbon exchange for nutrients may be evaluated similarly to soil types incorporated into varying techniques of land management. These are unidentified attributes that could function optimally within a fluctuating environment and an be taken into account (Challinor et al. 2016). The temperate regions of the world account for basically most of the information on AMF relationships with crops with regards to food production (Field et al. 2012). This opens up a wide gap of missing information on plant interactions with AMF in tropical climates, especially in underdeveloped nations that are facing serious challenges of food insecurity (Field et al. 2012; Challinor et al. 2016). The mechanism of functioning of the relationship between AMF and a plant of a single species at the physiological and molecular level should be discerned in simple model experiments (Sharma and Bhutani 2000; Sharma et al. 2005). Information is little on mycorrhiza effects at ecosystem, landscape or field levels. At these levels, the ecosystem is dynamic and fungal, and plants are connected in multiple ways (Begum et al. 2019). New mechanisms in ecology could be applied to food production systems, by evaluating and understanding the customs and complexity of such systems (Begum et al. 2019). Constant field experiments are needed to measure the exchange of nutrients, discover fungal connectivity and pick out their behaviour relative to instability in weather and climate so as to ensure the application of AMF (Birhane et al. 2012). The measurement of carbon, phosphorus and nitrogen outflow among symbionts under normal and projected high levels of CO_2 is a vital inquest that has to be addressed (Birhane et al.

2012). Presently, the benefits of the AMF association are not limited to nutrient supply. The influence of the symbiotic relationship on plant defence, soil function and structure appears to be the significant gain from the association (Salam et al. 2017). Future research is expected to concentrate on optimisation of nutrient uptake with an effect on yield actualisation and dependable output. By optimally recognising, the traits relevant to AMF flexibility in plant species, crucial progress can be achieved towards food security in supportable systems of agriculture (Abdel Latef 2011; Abdel Latef and Chaoxing 2011, 2014).

2.7 Conclusions and Future Prospects

The task ahead in the use of mycorrhizal fungi for reliable crop production is to adjust and make effective use of several factors, such as means of inoculation, soil properties, the combination of AM fungi inoculum and the plant. Inoculation of mycorrhiza is crucial for the efficiency, health and good development of seedlings. The impact of mycorrhizae on plant physiology will be the mechanism path to wholesome food for humans. Innumerable ecological factors are accountable the effectiveness of AMFs; these include but are not limited to the inoculums, species of the plant, soil type and inoculation mode. Indigenous soil mycorrhiza can infect plants successfully and improve the plant's growth and yield characters. In the field, phosphorus supply is known to augment mycorrhiza inoculation, but with some differences on a yearly basis, with high concentrations of phosphorus in the soil, the reliance on mycorrhiza declines. To manage the traditional mycorrhizae in the soil on a long-term basis, it will be necessary to evaluate the effects of soil management practices and crop rotation activities very scientifically and wisely. For the control of soil-borne diseases, compost and solarisation could be used to augment the inoculation of mycorrhizae to achieve disease-free plant growth and desirable nutrition for the mycorrhizal-dependent plants.

References

Abdel Latef AA (2011) Influence of arbuscular mycorrhizal fungi and copper on growth, accumulation of osmolyte, mineral nutrition and antioxidant enzyme activity of pepper (*Capsicum annuum* L.). Mycorrhiza 21:495–503. https://doi.org/10.1007/s00572-010-0360-0

Abdel Latef AA, Chaoxing H (2011) Arbuscular mycorrhizal influence on growth, photosynthetic pigments, osmotic adjustment and oxidative stress in tomato plants subjected to low temperature stress. Acta Physiol Plant 33:1217–1225. https://doi.org/10.1007/s11738-010-0650-3

Abdel Latef AA, Chaoxing HJ (2014) Does the inoculation with *Glomus mosseae* improve salt tolerance in pepper plants? J Plant Growth Regul 33:644–653. https://doi.org/10.1007/s00344-014-9414-4

Adeyemi NO, Atayese MO, Olubode AA (2019) Identification and relative abundance of native arbuscular mycorrhizal fungi associated with oil-seed crops and maize (Zea mays L.) in derived savannah of Nigeria. Acta fytotechn zootechn 22(3):84–89

Aghale DN, Egbucha KC, Umeh OJ (2017) Plant parasitic nematode threat to Nigerian Agricultural Food Security. Int J Adv Res Bot 3(3):15–21. https://doi.org/10.20431/2455-4316.0303003. www.arcjournals.org

Alori ET, Fawole O, Afolayan A (2012) Characterization of arbuscular mycorrhizal spores isolated from Southern Guinea Savanna of Nigeria. J Agric Sci 4(7):13–19

Alqarawi AA, Abd-Allah EF, Hashem A (2014a) Alleviation of saltinduced adverse impact *via* mycorrhizal fungi in *Ephedra aphylla* Forssk. J Plant Interact 9(1):802–810. https://doi.org/10.1080/17429145.2014.949886

Alqarawi AA, Hashem A, Abd Allah EF, Alshahrani TS, Huqail AA (2014b) Effect of salinity on moisture content, pigment system, and lipid composition in *Ephedra alata Decne*. Acta Biol Hung 65(1):61–71. https://doi.org/10.1556/ABiol.65.2014.1.6

Ansari RA, Khan TA (2012a) Parasitic association of root-knot nematode, *Meloidogyne incognita* on guava. e J Sci Technol 5(12):65–67

Ansari RA, Khan TA (2012b) Diversity and community structure of Phytonematodes associated with guava in and around Aligarh, Uttar Pradesh. Trends Biosci 5(3):202–204

Ansari RA, Mahmood I (2017a) Optimization of organic and bio-organic fertilizers on soil properties and growth of pigeonpea. Sci Hortic 226:1–9

Ansari RA, Mahmood I (2017b) Determination of disease incidence caused by *Meloidogyne* spp. and or *fusarium udum* on pigeonpea in Aligarh district: a survey. Trends Biosci 10(24):5239–5243

Ansari RA, Mahmood I (2019a) Plant health under biotic stress. In: Ansari RA, Mahmood I (eds) Microbial interactions, vol II. Springer Nature, Singapore. https://doi.org/10.1007/978-981-13-6040-4

Ansari RA, Mahmood I (2019b) Plant health under biotic stress. In: Ansari RA, Mahmood I (eds) Organic strategies, vol I. Springer Nature, Singapore. https://doi.org/10.1007/978-981-13-6043-5

Ansari RA, Mahmood I, Rizvi R, Sumbul A (2017a) PGPR: current vogue in sustainable crop production. In: Kumar V (ed) Probiotics and plant health. Springer Nature, Singapore, pp 455–472

Ansari RA, Mahmood I, Rizvi R, Sumbul A, Safiuddin (2017b) Siderophores: augmentation of soil health and crop productivity. In: Kumar V (ed) Probiotics in agroecosystem. Springer Nature, Singapore, pp 291–312

Ansari RA, Sumbul A, Rizvi R, Mahmood I (2019a) Potential role of plant growth promoting Rhizobacteria in alleviation of biotic stress. In: Ansari RA, Mahmood I (eds) Plant health under biotic stress, Microbial interactions, vol II. Springer Nature, Singapore, pp 177–188

Ansari RA, Sumbul A, Rizvi R, Mahmood I (2019b) Organic soil amendments: potential tool for soil and plant health management. In: Ansari RA, Mahmood I (eds) Plant health under biotic stress, Organic strategies, vol I. Springer Nature, Singapore, pp 1–35

Ansari RA, Rizvi R, Mahmood I (2020a) Management of phytonematodes: management of phytonematodes: recent advances and future challenges. Springer Nature, Singapore. https://doi.org/10.1007/978-981-15-4087-5

Ansari RA, Rizvi R, Sumbul A, Mahmood I (2020b) Plant-growth-promoting Rhizobacteria (PGPR)-based sustainable management of phytoparasitic nematodes: current understandings and future challenges. In: Ansari RA et al (eds) Management of phytonematodes: recent advances and future challenges. Springer Nature, Singapore. https://doi.org/10.1007/978-981-15-4087-5_3

Atolani O, Fabiyi OA (2020) Plant parasitic nematodes management through natural products: current progress and challenges. In: Ansari RA, Rizvi R, Mahmood I (eds) Management of Phytonematodes: recent advances and future challenges. Springer, Singapore, pp 297–315

Atolani O, Fabiyi OA, Olatunji GA (2014a) Isovitexin from *Kigelia pinnata*, A potential eco-friendly Nematicidal agent. Trop Agric 91(2):67–74

Atolani O, Fabiyi OA, Olatunji GA (2014b) Nematicidal Isochromane glycoside from *Kigelia pinnata* leaves. Acta Agric Slovenica 104(1):25–31

Bardgett RD, van der Putten WH (2014) Belowground biodiversity and ecosystem functioning. Nature 515:505–511

Bayani R, Saateyi A, Faghani E (2015) Influence of arbuscular mycorrhiza in phosphorus acquisition efficiency and drought-tolerance mechanisms in barley *Hordeum vulgare* L. Int J Biosci 7:86–94. https://doi.org/10.12692/ijb/7.1.86-94

Begum N, Qin C, Ahanger MA, Raza S, Khan MI, Ashraf M, Ahmed N, Zhang L (2019) Role of arbuscular mycorrhizal fungi in plant growth regulation: implications in abiotic stress tolerance. Front Plant Sci 10:1068. https://doi.org/10.3389/fpls.2019.01068

Bello TT, Coyne DL, Fourie H (2021) Reproduction potential of Nigerian Meloidogyne spp. and the response of six commercial watermelon cultivars to predominant species. J Plant Dis Protect 128(3):831–842

Bender SF, van der Heijden MGA (2015) Soil biota enhance agricultural sustainability by improving crop yield, nutrient uptake and reducing nitrogen leaching losses. J Appl Ecol 52:228–239

Bender SF, Wagg C, van der Heijden MGA (2016) An underground revolution: biodiversity and soil ecological engineering for agricultural sustainability. Trends Ecol Evol 31:440–452

Berruti A, Lumini E, Balestrini R, Bianciotto V (2016) Arbuscular mycorrhizal fungi as natural biofertilizers: let's benefit from past successes. Front Microbiol 6:1559

Birhane E, Sterck F, Fetene M, Bongers F, Kuyper T (2012) Arbuscular mycorrhizal fungi enhance photosynthesis, water use efficiency, and growth of frankincense seedlings under pulsed water availability conditions. Oecologia 169:895–904. https://doi.org/10.1007/s00442-012-2258-3

Boyer LR, Brain P, Xu XM, Jeffries P (2014) Inoculation of drought stressed strawberry with a mixed inoculum of two arbuscular mycorrhizal fungi: effects on population dynamics of fungal species in roots and consequential plant tolerance to water. Mycorrhiza 25(3):215–227. https://doi.org/10.1007/s00572-014-0603-6

Brooker RW, Bennett AE, Cong WF et al (2015) Improving intercropping: a synthesis of research in agronomy, plant physiology and ecology. New Phytol 206:107–117

Cabral C, Sabine R, Ivanka T, Bernd W (2016) Arbuscular mycorrhizal fungi modify nutrient allocation and composition in wheat (Triticum aestivum L.) subjected to heat-stress. Plant Soil 408(1–2):385–399. https://doi.org/10.1007/s11104-016-2942-x

Cavagnaro TR, Bender SF, Asghari HR, van der Heijden MGA (2015) The role of arbuscular mycorrhizas in reducing soil nutrient loss. Trends Plant Sci 20:283–290

Chagnon PL, Bradley RL, Klironomos JN (2015) Trait-based partner selection drives mycorrhizal network assembly. Oikos 124:1609–1616. https://doi.org/10.1111/oik.01987

Challinor AJ, Koehler AK, Ramirez-Villegas J, Whitfield S, Das B (2016) Current warming will reduce yields unless maize breeding and seed systems adapt immediately. Nat Clim Chang 6:954–958

Coutinho ES, Fernandes GW, Berbara RLL, Valério HM, Goto BT (2015) Variation of arbuscular mycorrhizal fungal communities along an altitudinal gradient in rupestrian grasslands in Brazil. Mycorrhiza 25(8):627–638

Covacevich F, Echeverria HE, Pagano MC (2012) Arbuscular mycorrhizal fungi: essential belowground organisms for earth life but sensitive to a changing environment. Afr J Microbiol Res 6(27):5523–5535

da Silva DKA, de Souza RG, de Alencar Velez BA, da Silva GA, Oehl F, Maia LC (2015) Communities of arbuscular mycorrhizal fungi on a vegetation gradient in tropical coastal dunes. Appl Soil Ecol 96:7–17

Davison J, Öpik M, Zobel M, Vasar M, Metsis M, Moora M (2012) Communities of arbuscular mycorrhizal fungi detected in forest soil are spatially heterogeneous but do not vary throughout the growing season. PLoS One 7:e41938

De Beenhouwer M, Van Geel M, Ceulernans T, Muleta D, Lievens B, Honnay O (2015) Changing soil characteristics alter the arbuscular mycorrhizal fungi communities of Arabica coffee (*Coffea arabica*) in Ethiopia across a management intensity gradient. Soil Biol Biochem 91:133–139. https://doi.org/10.1016/j.soilbio.2015.08.037

De Carvalho F, De Souza FA, Carrenho R, de Souza Moreira FM, da Conçeição Jesus E, Fernandes GW (2012) The mosaic of habitats in the high-altitude Brazilian rupestrian fields is a hotspot for arbuscular mycorrhizal fungi. Appl Soil Ecol 52:9–19

Dessai SA, Rodrigues BF (2012) Diversity studies on arbuscular mycorrhizal fungi in vegetable crop plants of Goa, India. Plant Pathol Quar 2:87–101

Edwards KJ, Becker K, Colwell F (2012) The deep, dark energy biosphere: intraterrestrial life on earth. Annu Rev Earth Planet Sci 40:551–568

Egan C, Li DW, Klironomos J (2014) Detection of arbuscular mycorrhizal fungal spores in the air across different biomes and ecoregions. Fungal Ecol 12:26–31. https://doi.org/10.1016/j.funeco.2014.06.004

Erenstein O, Sayre K, Wall P, Dixon J, Hellin J (2008) Adapting no-tillage agriculture to the conditions of smallholder maize and wheat farmers in the tropics and sub-tropics. In: Goddard T, Zoebisch M, Gan Y, Ellis W, Watson A, Sombatpanit S (eds) No-till farming systems. World Association of Soil and Water Conservation (WASWC), pp 253–278

Fabiyi OA (2019) Management of groundnut (*Arachis hypogea*) root-knot nematode (*Meloidogyne incognita*): effect of *Prosopis africana* pods. Indian J Nematol 49(2):214–216

Fabiyi OA (2020) Growth and yield response of groundnut *Arachis hypogaea* (Linn.) under *Meloidogyne incognita* infection to furfural synthesized from agro-cellulosic materials. J Trop Agric 58(2):241–245

Fabiyi OA (2021a) Application of furfural in sugarcane nematode pest management. Pak J Nematol 39(2):151–155. https://doi.org/10.17582/journal.pjn/2021.39.2.151.155

Fabiyi OA (2021b) Sustainable Management of *Meloidogyne incognita* infecting carrot: green synthesis of silver nanoparticles with *Cnidoscolus aconitifolius*: (*Daucus carota*). Vegetos 34(2):277–285

Fabiyi OA (2021c) Evaluation of Nematicidal activity of *Terminalia glaucescens fractions* against *Meloidogyne incognita* on *Capsicum chinense*. J Hort Res 29(1):67–74

Fabiyi OA (2021d) Evaluation of plant materials as root-knot nematode (*Meloidogyne incognita*) suppressant in okro (*Abelmuscous esculentus*). Agric Conspec Sci 86(1):51–56

Fabiyi OA (2022a) Fractions from *Mangifera indica* as an alternative in *Meloidogyne incognita* management. Pak J Nematol 40(1):65–74

Fabiyi OA (2022b) Cytotoxicity and Nematicidal potential of leaf extracts of *Adansonia digitata* and *Khaya senegalensis* on root-knot nematode (*Meloidogyne incognita*) associated with cabbage *(Brassica oleracea)*. J Agric Sci Sri Lanka 17(3):425–436

Fabiyi OA (2022c) Evaluation of weeds against root-knot nematode (*Meloidogyne incognita*) in vegetables. Sarhad J Agric 38(4):1289–1299

Fabiyi OA (2022d) Application of composited municipal refuse dump site soil, orange, potato and pineapple peels in the control of root knot nematode (*Meloidogyne incognita*) infecting carrots (*Daucus carota* L.). J Solid Waste Manag 48(3):474–485

Fabiyi OA, Olatunji GA (2018) Application of green synthesis in Nano particles preparation: *Ficus mucoso* extracts in the Management of *Meloidogyne incognita* parasitizing groundnut *Arachis hypogea*. Indian J Nematol 48(1):13–17

Fabiyi OA, Olatunji GA (2021a) Environmental sustainability: bioactivity of *Leucaena leucocephala* leaves and pesticide residue analysis in tomato fruits. Acta Univ Agric Silvic Mendel Brun 69(4):473–480

Fabiyi OA, Olatunji GA (2021b) Toxicity of derivatized citrulline and extracts of water melon rind (*Citrullus lanatus*) on root-knot nematode (*Meloidogyne incognita*). Trop Agric 98(4):347–355

Fabiyi OA, Olatunji GA, Osunlola OS, Umar KA (2018a) Efficacy of agricultural wastes in the control of rice cyst nematode (*Heterodera sacchari*). Agric Conspec Sci 83(4):329–334

Fabiyi OA, Olatunji GA, Saadu AO (2018b) Suppression of *Heterodera sacchari* in rice with agricultural waste-silver nano particles. J Solid Waste Manage 44(2):87–91

Fabiyi OA, Olatunji GA, Daodu IO (2019) Nematicidal effect of organic extract metal complex on *Meloidogyne incognita* infecting groundnuts (*Arachis hypogea*). Sci Agric Bohem 50(3):191–196

Fabiyi OA, Atolani O, Olatunji GA (2020a) Toxicity effect of *Eucalyptus globulus* to *Pratylenchus* spp of *Zea mays*. Sarhad J Agric 36(4):1244–1253

Fabiyi OA, Olatunji GA, Atolani O, Olawuyi RO (2020b) Preparation of bio-nematicidal nanoparticles of *Eucalyptus officinalis* for the control of cyst nematode (*Heterodera sacchari*). J Anim Plant Sci 30(5):1172–1177

Fabiyi OA, Saliu OD, Claudius-Cole AO, Olaniyi IO, Oguntebi OV, Olatunji GA (2020c) Porous starch citrate biopolymer for controlled release of Carbofuran in the management of root knot nematode *Meloidogyne incognita*. Biotechnol Rep 25(e00428):1–9

Fabiyi OA, Claudius-Cole AO, Olatunji GA, Abubakar DO, Adejumo OA (2021a) Response of Meloidogyne javanica to silver nanoparticle liquid from agricultural wastes. Agrivita J Agric Sci 43(3):524–534. https://doi.org/10.17503/agrivita.v43i3.1936

Fabiyi OA, Claudius-Cole AO, Olatunji GA (2021b) In vitro assessment of n-phenyl imides in the management of *Meloidogyne incognita*. Sci Agric Bohem 52(3):60–65

Fabiyi OA, Baker MA, Olatunji GA (2022a) Application of fatty acid esters on *Meloidogyne incognita* infected Jew's mallow. Pak J Nematol 40(2):127–137

Fabiyi OA, Baker MT, Claudius-Cole AO, Alabi RO, Olatunji GA (2022b) Application of cyclic imides in the management of Root-knot nematode (*Meloidogyne incognita*) on cabbage. Indian Phytopathol 75:457. https://doi.org/10.1007/s42360-022-00480-1

Field KJ, Cameron DD, Leake JR, Tille S, Bidartondo MI, Beerling DJ (2012) Contrasting arbuscular mycorrhizal responses of vascular and non-vascular plants to a simulated Palaeozoic CO2 decline. Nat Commun 3:835

Field KJ, Rimington WR, Bidartondo MI, Allinson KE, Beerling DJ, Cameron DD, Duckett JG, Leake JR, Pressel S (2015) First evidence of mutualism between ancient plant lineages (Haplomitriopsida liverworts) and Mucoromycotina fungi and its response to simulated Palaeozoic changes in atmospheric CO2. New Phytol 205:743–756

Gazol A, Zobel M, Cantero JJ, Davison J, Esler KJ, Jairus T, Moora M (2016) Impact of alien pines on local arbuscular mycorrhizal fungal communities—evidence from two continents. FEMS microbiology ecology 92(6):fiw073

Godfray HCJ, Beddington JR, Crute IR et al (2010) Food security: the challenge of feeding 9 billion people. Science 327:812–818

Goldmann K, Boeddinghaus RS, Klemmer S, Regan KM, Heintz-Buschart A, Fischer M et al (2020) Unraveling spatiotemporal variability of arbuscular mycorrhizal fungi in a temperate grassland plot. Environ Microbiol 22(3):873–888

Gottel NR, Castro HF, Kerley M, Yang Z, Pelletier DA, Podar M et al (2011) Distinct microbial communities within the endosphere and rhizosphere of Populus deltoides roots across contrasting soil types. Appl Environ Microbiol 77(17):5934–5944

Grümberg BC, María UC, Shroeder A, Vargas-Gil S, Luna CM (2015) The role of inoculum identity in drought stress mitigation by arbuscular mycorrhizal fungi in soybean. Biol Fertil Soils 51:1–10. https://doi.org/10.1007/s00374-014-0942-7

Gupta MM, Naqvi NS, Singh VK (2014) The state of arbuscular mycorrhizal fungal diversity in India: an analysis. Sydowia 66(2):265–288

Gupta N, Bihari KM, Sengupta I (2016) Diversity of arbuscular mycorrhizal fungi in different salinity of mangrove ecosystem of Odisha, India. Adv Plants Agric Res 3:85–90

Gupta MM, Naqvi N, Kumar P (2017) i AMF—centralized database of arbuscular mycorrhizal distribution, phylogeny and taxonomy. J Recent Adv Appl Sci 30:1–5

Higo M, Isobe K, Yamaguchi M, Drijber RA, Jeske ES, Ishii R (2013) Diversity and vertical distribution of indigenous arbuscular mycorrhizal fungi under two soybean rotational systems. Biol Fertil Soils 49:1085–1096. https://doi.org/10.1007/s00374-013-0807-5

Hindumathi A, Reddy BN (2011) Occurrence and distribution of arbuscular mycorrhizal fungi and microbial flora in the rhizosphere soils of mungbean [Vigna radiata (L.) Wilczek] and soybean [Glycine max (L.) Merr.] from Adilabad, Nizamabad and Karimnagar districts of Andhra Pradesh state, India. Adv Biosci Biotechnol 2:275–286

Hodge A, Fitter AH (2010) Substantial nitrogen acquisition by arbuscular mycorrhizal fungi from organic material has implications for N cycling. Proc Natl Acad Sci U S A 107:13754–13759

Hu J, Wang H, Wu F, Wu S, Cao Z, Lin X, Wong MH (2014) Arbuscular mycorrhizal fungi influence the accumulation and partitioning of Cd and P in bashfulgrass (Mimosa pudica L.) grown on a moderately Cdcontaminated soil. Applied Soil Ecology 73:51–57

Husband R, Herre EA, Turner SL, Gallery R, Young JPW (2002) Molecular diversity of arbuscular mycorrhizal fungi and patterns of host association over time and space in a tropical forest. Molecular Ecology 11(12):2669–2678

Jemo M, Dhiba D, Hashem A et al (2018) Mycorrhizal fungal community structure in tropical humid soils under fallow and cropping conditions. Sci Rep 8:17061. https://doi.org/10.1038/s41598-018-34736-6

Jha SS, Songachan LS (2020) Research on diversity and community composition of arbuscular mycorrhizal fungi species in India: a review

Jiang Y, Wang W, Xie Q, Liu N, Liu L, Wang D, Zhang X, Yang C, Chen X, Tang D (2017) Plants transfer lipids to sustain colonization by mutualistic mycorrhizal and parasitic fungi. Science 356:1172–1175

Johnson NC (2010) Resource stoichiometry elucidates the structure and function of arbuscular mycorrhizas across scales. New Phytol 185:631–647

Johnson NC, Wilson GWT, Wilson JA, Miller RM, Bowker MA (2015) Mycorrhizal phenotypes and the law of the minimum. New Phytol 205:1473–1484

Jung SC, Martinez-Medina A, Lopez-Raez JA, Pozo MJ (2012) Mycorrhiza-induced resistance and priming of plant defenses. J Chem Ecol 38(6):651–664

Kiers ET, Duhamel M, Beesetty Y, Mensah JA, Franken O, Verbruggen E, Fellbaum CR, Kowalchuk GA, Hart MM, Bago A et al (2011) Reciprocal rewards stabilize cooperation in the mycorrhizal symbiosis. Science 333:880–882

Kivlin SN, Hawkes CV, Treseder KK (2011) Global diversity and distribution of arbuscular mycorrhizal fungi. Soil Biol Biochem 43:2294–2303. https://doi.org/10.1016/j.soilbio.2011.07.012

Koele N, Dickie IA, Blum JD, Gleason JD, de Graaf L (2014) Ecological significance of mineral weathering in ectomycorrhizal and arbuscular mycorrhizal ecosystems from a field-based comparison. Soil Biol Biochem 69:63–70. https://doi.org/10.1016/j.soilbio.2013.10.041

Laurance WF, Sayer J, Cassman KG (2014) Agricultural expansion and its impacts on tropical nature. Trends Ecol Evol 29(2):107–116

Leake JR, Read DJ (2017) Mycorrhizal symbioses and pedogenesis throughout earth's history. In: Mycorrhizal mediation of soil. Elsevier, Amsterdam, pp 9–33. https://doi.org/10.1016/B978-0-12-804312-7.00002-4

Lehmann A, Barto EK, Powell JR, Rillig MC (2012) Mycorrhizal responsiveness trends in annual crop plants and their wild relatives—a meta-analysis on studies from 1981 to 2010. Plant and Soil 355:231–250

Li HY, Zhu YG, Marschner P, Smith FA, Smith SE (2005) Wheat responses to arbuscular mycorrhizal fungi in a highly calcareous soil differ from those of clover, and change with plant development and P supply. Plant and Soil 277:221–232

Li XX, Zeng RS, Liao H (2016) Improving crop nutrient efficiency through root architecture modifications. J Integr Plant Biol 58:193–202

Lin AJ, Zhang XH, Wong MH, Ye ZH, Lou LQ, Wang YS (2007) Increase of multi-metal tolerance of three leguminous plants by arbuscular mycorrhizal fungi colonization. Environ Geochem Health 29:473–481. https://doi.org/10.1007/s10653-007-9116-y

Liu A, Hamel C, Hamilton RI, Ma BL, Smith DL (2000) Acquisition of Cu, Zn, Mn and Fe by mycorrhizal maize (Zea mays L.) grown in soil at different P and micronutrient levels. Mycorrhiza 9:331–336

Liu S, Guo X, Feng G, Maimaitiaili B, Fan J, He X (2016) Indigenous arbuscular mycorrhizal fungi can alleviate salt stress and promote growth of cotton and maize in saline fields. Plant and Soil 398:195–206

Luginbuehl LH, Menard GN, Kurup S, Van Erp H, Radhakrishnan GV, Breakspear A, Oldroyd GED, Eastmond PJ (2017) Fatty acids in arbuscular mycorrhizal fungi are synthesized by the host plant. Science 356:1175–1178

Mathimaran N, Ruh R, Jama B, Verchot L, Frossard E, Jansa J (2007) Impact of agricultural management on arbuscular mycorrhizal fungal communities in Kenyan ferralsol. Agric Ecosyst Environ 119(1–2):22–32

Mathur S, Sharma MP, Jajoo A (2016) Improved photosynthetic efficacy of maize *Zea mays* plants with arbuscular mycorrhizal fungi (AMF) under high temperature stress. J Photochem Photobiol B 180:149–154. https://doi.org/10.1016/j.jphotobiol.2018.02.002

Mathurin DD, Hubert DBE, Tahir AD, Denis ON (2022) Diversity of arbuscular mycorrhizal fungi in the three agroecological zones of the Central African Republic. Afr J Biotechnol 21(1):26–34

Miranda D, Fischer G, Ulrichs C (2011) The influence of arbuscular mycorrhizal colonization on the growth parameters of cape gooseberry (Physalisperuviana L.) plants grown in a saline soil. J Soil Sci Plant Nutr 11(2):18–30

Moll J, Hoppe B, König S, Wubet T, Buscot F, Krüger D (2016) Spatial distribution of fungal communities in an arable soil. PLoS One 11:1–17. https://doi.org/10.1371/journal.pone.0148130

Moora M, Davison J, Öpik M, Metsis M, Saks Ü, Jairus T et al (2014) Anthropogenic land use shapes the composition and phylogenetic structure of soil arbuscular mycorrhizal fungal communities. FEMS Microbiol Ecol 90(3):609–621

Moreira M, Baretta D, Tsai SM, Gomes-da-Costa SM, Nogueira Cardoso EJB (2007) Biodiversity and distribution of arbuscular mycorrhizal fungi in Araucaria angustifolia forest (Piracicaba, Braz.). Sci Agric 64:393–399

Navarro JM, Perez-Tornero O, Morte A (2014) Alleviation of salt stress in citrus seedlings inoculated with arbuscular mycorrhizal fungi depends on the root stock salt tolerance. J Plant Physiol 171(1):76–85. https://doi.org/10.1016/j.jplph.2013.06.006

Navarro-Fernandez CM, Perez-Ramos IM, de la Riva EG, Vera JR, Roumet C, Villar R, Maranon T (2016) Functional responses of Mediterranean plant communities to soil resource heterogeneity: a mycorrhizal trait-based approach. J Veg Sci 27:1243–1253

Oehl F, Laczko E, Bogenrieder A, Stahr K, Bösch R, van der Heijden M, Sieverding E (2010) Soil type and land use intensity determine the composition of arbuscular mycorrhizal fungal communities. Soil Biol Biochem 42(5):724–738

Olubode AA, Babalola OA, Dare OM, Adeyemi NO, Aderibigbe SG, Okonji CJ, Sakariyawo OS (2020) Diversity of indigenous arbuscular mycorrhizal fungi in rhizosphere of upland rice (Oryza sativa L.) varieties in Southwest Nigeria. Acta Fytotechnica et Zootechnica 23(2):42–48

Öpik M, Vanatoa A, Vanatoa E, Moora M, Davison J, Kalwij JM, Reier Ü, Zobel M (2010) The online database MaarjAM reveals global and ecosystemic distribution patterns in arbuscular mycorrhizal fungi (Glomeromycota). New Phytol 188:223–241

Öpik M, Zobel M, Cantero JJ, Davison J, Facelli JM, Hiiesalu I, Jairus T, Kalwij JM, Koorem K, Leal ME, Liira J, Metsis M, Neshataeva V, Paal J, Phosri C, Põlme S, Reier Ü, Saks Ü, Schimann H, Thiéry O, Vasar M, Moora M (2013) Global sampling of plant roots expands the described molecular diversity of arbuscular mycorrhizal fungi. Mycorrhiza 23:411–430

Ortas I (2012) The effect of mycorrhizal fungal inoculation on plant yield, nutrient uptake and inoculation effectiveness under long-term field conditions. Field Crops Res 125:35–48. https://doi.org/10.1016/j.fcr.2011.08.00

Pagano MC, Cabello MN (2012) Occurrence of mycorrhizas in highlands fields. In: Pagano MC (ed) Mycorrhiza: occurrence in natural and restored environments, USA. Nova Science Publishers, Hauppauge, NY, pp 87–98

Palm C, Blanco-Canqui H, DeClerck F, Gatere L, Grace P (2014) Conservation agriculture and ecosystem services: an overview. Agriculture, Ecosystems & Environment 187:87–105

Parihar M, Rakshit A, Singh HB, Rana K (2019) Diversity of arbuscular mycorrhizal fungi in alkaline soils of hot sub humid eco-region of Middle Gangetic Plains of India. Acta Agric Scand B Soil Plant Sci 69(5):386–397. https://doi.org/10.1080/09064710.2019.1582692

Paul BK, Vanlauwe B, Ayuke F et al (2013) Medium-term impact of tillage and residue management on soil aggregate stability, soil carbon and crop productivity. Agric Ecosyst Environ 164:14–22

Philippot L, Raaijmakers JM, Lemanceau P, Van Der Putten WH (2013) Going back to the roots: the microbial ecology of the rhizosphere. Nat Rev Microbiol 11:789–799

Plassard C, Dell B (2010) Phosphorus nutrition of mycorrhizal trees. Tree Physiol 30:1129–1139. https://doi.org/10.1093/treephys/tpq063

Porcel R, Redondogómez S, Mateosnaranjo E, Aroca R, Garcia R, Ruizlozano JM (2015) Arbuscular mycorrhizal symbiosis ameliorates the optimum quantum yield of photosystem II and reduces non-photochemical quenching in rice plants subjected to salt stress. J Plant Physiol 185:75–83. https://doi.org/10.1016/j.jplph.2015.07.006

Puschel D, Janouskova M, Hujslova M, Slavikova R, Gryndlerova H, Jansa J (2016) Plant-fungus competition for nitrogen erases mycorrhizal growth benefits of Andropogon gerardii under limited nitrogen supply. Ecol Evol 6:4332–4346

Raviv M (2010) The use of mycorrhiza in organically-grown crops under semi arid conditions: a review of benefits, constraints and future challenges. Symbiosis 52(2):65–74

Rillig MC, Aguilar-Trigueros CA, Bergmann J, Verbruggen E, Veresoglou SD, Lehmann A (2015) Plant root and mycorrhizal fungal traits for understanding soil aggregation. New Phytol 205:1385–1388

Rizvi R, Singh G, Safiuddin ARA, Tiyagi SA, Mahmood I (2015) Sustainable management of root-knot disease of tomato by neem cake and *Glomus fasciculatum*. Cogent Food Agric 1:1008859

Rouphael Y, Franken P, Schneider C, Schwarz D, Giovannetti M, Agnolucci M, De Pascale S, Bonini P, Colla G (2015) Arbuscular mycorrhizal fungi act as biostimulants in horticultural crops. Sci Hortic 196:91–108

Sachs JD, Mellinger AD, Gallup JL (2001) The geography of poverty and wealth. Sci Am 284(3):70–75

Salam EA, Alatar A, El-Sheikh MA (2017) Inoculation with arbuscular mycorrhizal fungi alleviates harmful effects of drought stress on damask rose. Saudi J Biol Sci 25(8):1772–1780. https://doi.org/10.1016/j.sjbs.2017.10.015

Säle V, Aguilera P, Laczko E, Mäder P, Berner A, Zihlmann U et al (2015) Impact of conservation tillage and organic farming on the diversity of arbuscular mycorrhizal fungi. Soil Biol Biochem 84:38–52. https://doi.org/10.1016/j.soilbio.2015.02.005

Sato T, Ezawa T, Cheng WG, Tawaraya K (2015) Release of acid phosphatase from extraradical hyphae of arbuscular mycorrhizal fungus Rhizophagus clarus. Soil Sci Plant Nutr 61:269–274

Sharma SD, Bhutani VP (2000) Leaf nutrient status of appleseedlings as in fluenced by VAM and Azotobacter and inorganicfertilizers. J Hill Res 13:63–66

Sharma SD, Sharma NC, Sharma CL, Sood R, Singh RP (2005) Studies on correlations between endomycorrhizal and Azotobacterpopulation with growth, yield and soil nutrient status of apple (Malusdomestica Borkh.) orchards in Himachal Pradesh. Acta Hortic 696:383–387

Smith SE, Read D (2008) The symbionts forming arbuscular mycorrhizas. In: Mycorrhizal symbiosis, vol 2, pp 13–41

Smith FA, Smith SE (2011a) What is the significance of the arbuscular mycorrhizal colonisation of many economically important crop plants? Plant Soil 348:63

Smith SE, Smith FA (2011b) Roles of arbuscular mycorrhizas in plant nutrition and growth: new paradigms from cellular to ecosystem scales. Annu Rev Plant Biol 62:227–250

Solanki MK, Kashyap PL, Ansari RA, Kumari B (2020) Microbiomes and plant health: panoply and their applications. Academic Press, London. https://www.elsevier.com/books/microbiomes-and-plant-health/solanki/978-0-12-819715-8

Sosa-Hernández MA, Roy J, Hempel S, Kautz T, Köpke U, Uksa M et al (2018) Subsoil arbuscular mycorrhizal fungal communities in arable soil differ from those in topsoil. Soil Biol Biochem 117:83–86. https://doi.org/10.1016/j.soilbio.2017.11.009

Spatafora JW, Chang Y, Benny GL et al (2016) A phylum-level phylogenetic classification of zygomycete fungi based on genome-scale data. Mycologia 108:1028–1046

Stirling GR, Hayden HL, Pattison AB, Stirling AM (2016) Soil health, soil biology, soil borne diseases and sustainable agriculture: a guide. CSIRO Publishing, Melbourne, VIC

Suchitra R, Kumutha K, Balachandar D (2012) Morpho-typing and molecular diversity of arbuscular mycorrhizal fungi in sub-tropical soils of Coimbatore region, Tamil Nadu, India. Indian J Microbiol 52(2):145–152. https://doi.org/10.1007/s12088-011-0206-2

Sumbul A, Mahmood I, Rizvi R, Ansari RA, Safiuddin (2017) Mycorrhiza: an alliance for the nutrient management in plants. Probiotics in Agroecosystem, 371–386

Tawaraya K (2003) Arbuscular mycorrhizal dependency of different plant species and cultivars. Soil Sci Plant Nutr 49:655–668

Tchabi A, Coyne D, Hountondji F, Lawouin L, Wiemken A, Oehl F (2008) Arbuscular mycorrhizal fungal communities in sub-Saharan Savannas of Benin, West Africa, as affected by agricultural land use intensity and ecological zone. Mycorrhiza 18(4):181–195

Thirkell TJ, Cameron DD, Hodge A (2016) Resolving the 'nitrogen paradox' of arbuscular mycorrhizas: fertilization with organic matter brings considerable benefits for plant nutrition and growth. Plant Cell Environ 39:1683–1690

Turrini A, Avio L, Giovannetti M, Agnolucci M (2018) Functional complementarity of arbuscular mycorrhizal fungi and associated microbiota: the challenge of translational research. Front Plant Sci 9:1407

Utaile YU, Van Geel M, Muys B, Cheche SS, Helsen K, Honnay O (2021) Woody encroachment of an East-African savannah ecosystem alters its arbuscular mycorrhizal fungal communities. Plant Soil 464(1):303–320

Velázquez MS, Cabello MN, Barrera M (2013) Composition and structure of arbuscular-mycorrhizal communities in El Palmar National Park, Argentina. Mycologia 105(3):509–520

Vieira LC, da Silva DKA, de Melo MAC, Escobar IEC, Oehl F, da Silva GA (2019) Edaphic factors influence the distribution of arbuscular mycorrhizal fungi along an altitudinal gradient of a tropical mountain. Microb Ecol 78(4):904–913

Walder F, Niemann H, Natarajan M, Lehmann MF, Boller T, Wiemken A (2012) Mycorrhizal networks: common goods of plants shared under unequal terms of trade. Plant Physiol 159:789–797

Wubet T, Weiss M, Kottke I, Teketay D, Oberwinkler F (2006) Phylogenetic analysis of nuclear small subunit rDNA sequences suggests that the endangered African Pencil Cedar, Juniperus procera, is associated with distinct members of Glomeraceae. Mycol Res 110(9):1059–1069

Yadav C, Pandey S (2016) Biodiversity of arbuscular mycorrhizal fungi in India: a review. Mycorrhiza News 28:7–14

Yamato M, Ikeda S, Iwase K (2009) Community of arbuscular mycorrhizal fungi in drought-resistant plants, Moringa spp., in semiarid regions in Madagascar and Uganda. Mycoscience 50(2):100–105

Zhu X, Yang W, Song F, Li X (2020) Diversity and composition of arbuscular mycorrhizal fungal communities in the cropland black soils of China. Global Ecol Conserv 22:e00964

Zou YN, Srivastava AK, Wu QS (2016) Glomalin: a potential soil conditioner for perennial fruits. Int J Agric Biol 18:293–297. https://doi.org/10.17957/IJAB/15.0085

Impact of Environmental Gases on Mycorrhizal Symbiosis and Its Influence on Ecosystem Functioning Under the Current Climate Change Scenario

3

Rizwan Ali Ansari, Aisha Sumbul, Rose Rizvi, and Irshad Mahmood

Abstract

Arbuscular mycorrhizal (AM) fungi are of paramount importance that develop a good mutual relationship with higher plants. The impact of different environmental gases on AM symbiosis has not been studied extensively. The available data suggest that elevated CO_2 increases the biomass and productivity of plants. However, the effect of CO_2 on mycorrhizal symbiosis is still a matter of debate. The impact of CO_2 on the development of mycorrhiza and spore production is a very interesting aspect to be unravelled. Assessment of the effect of SO_2 and O_3 is also not much congenial to the growth and development of mycorrhizal fungi. These gases cause enormous negative impacts on the plant's biomass and productivity, as well as mycorrhizal network. In general, the effects of these gases are influential to biomass production and mycorrhizal spore formation and thus need further attention from the researchers.

Keywords

Environmental gases · Biodiversity · Plant health · Biotic · Abiotic stressors

R. A. Ansari (✉)
Department of Plant Protection, Faculty of Agricultural Sciences, Aligarh Muslim University, Aligarh, India

A. Sumbul · R. Rizvi · I. Mahmood
Section of Plant Pathology and Nematology, Department of Botany, Aligarh Muslim University, Aligarh, India

R. A. Ansari et al. (eds.), *Mycorrhizal Symbiosis and Agroecosystem Restoration*,
https://doi.org/10.1007/978-981-99-5030-0_3

3.1 Introduction

Arbuscular mycorrhizal (AM) fungi are commonly distributed and are an indistinguishable part of the ecosystem. They possess the ability to maintain ecosystem stability and plant health. AM fungi function in different ways and help plant improve their nutrient acquisition as well as resistance abilities against biotic and abiotic stressors (Dowarah et al. 2021). More than 80% of higher plants make compatible alliances with AM fungi, making them one of the most important and common fungal symbionts (Abbott and Gazey 1994; Smith and Read 2008). AM fungi contain a unique feature that they survive on their host plants, while the plants benefit from fungus ability to access and absorb nutrients, such as phosphorous, from the soil and make them available to the plants (Smith and Smith 2012). AM fungi are also important in the maintenance of soil structure, nutrient cycling, and plant growth ameliorations and are therefore considered one of the key players in the well functioning of efficient ecosystems. AM fungal populations can be affected by various environmental factors like pollution and spatial and temporal climate change, and thus population decline of plant symbionts may cause significant effects on the host plant and ecosystem functioning (Smith and Read 2010; Bennett and Classen 2020).

Plant symbionts especially mycorrhiza develop a mutualistic relationship with higher plants, including annuals, perennials, shrubs, and trees. Fungi provide improved access to soil nutrients in exchange for photosynthetically fixed carbon from the plant (Bonfante and Genre 2010). This relationship is usually very common in cultivated and non-cultivated ecosystems and has been found to show enhanced plant biomass, survival, and ability to reduce stress. This association benefits both partners and is crucial for the survival of many plant species, especially in nutrient-poor soils (Harrison 1999). AM fungi colonize plant roots, forming structures called arbuscules that allow for the exchange of nutrients and other compounds (Luginbuehl and Oldroyd 2017). The fungus also extends its mycelium into the soil, where it can access nutrients that are not readily available to the plant roots. This allows the plant to obtain otherwise inaccessible nutrients, such as phosphorus, and improve its growth and survival. AM fungi also help in the degradation of organic compounds and improve the polluted and degraded sites of soil (Rajtor and Piotrowska-Seget 2016). The most important salient features of AM fungi include:

1. Mutualism: AM fungi develop a symbiotic alliance with the plants; in that case, both partners benefit. Fungi give access to nutrients and water, which are otherwise not available to the plants. On the other hand, plants give carbohydrates to fungi which are essentially needed for their survival (Lanfranco et al. 2016).
2. Nutrient uptake: AM fungi have an extensive network of mycelia which absorb and transport different nutrients like phosphorus, to the plant roots, making them available to the plant (George et al. 1995).
3. Soil health: AM fungi are important in maintaining soil structure, improving soil fertility, and promoting nutrient cycling. They can also improve soil water-holding capacity and reduce erosion (Gujre et al. 2021).

4. Plant growth and survival: The relationship between AM fungi and plants is crucial for the growth and survival of many plant species, especially in nutrient-poor soils. AM fungi improve plant growth, thereby enhancing the plant's ability to survive under abiotic stress (Begum et al. 2019).
5. Widespread occurrence: AM fungi can be encountered in heterogeneous environments and form symbiosis in almost all terrestrial ecosystems (Rosendahl 2008).

The total production of AM fungi is difficult to quantify because of various factors like the abundance of host plants, soil health, and environmental factors. AM fungal populations are naturally dynamic and can vary in terms of their performance greatly in different ecosystems and over time (Eom et al. 2000).

It is important to note that these estimates may not accurately reflect the actual production of AM fungi as they are based on the limited available data. Further research in this connection is needed to gain information relating to the production and dynamics of plant symbionts. It is difficult to determine the total production of AM fungi, as this type of fungus is naturally occurring and widespread in most terrestrial ecosystems. However, the diversity of AM fungi in a particular ecosystem is significantly affected by various factors like soil type, prevailing cropping pattern, and climate perturbation (Allen et al. 1995; Jeffries et al. 2003). AM fungi develop a symbiotic association with the higher plants, which are also influenced by types of mycorrhiza and the environment in which they thrive (Khaliq et al. 2022). The production of AM fungi is likely to be highest in ecosystems with abundant plant growth, such as forests and grasslands, where fungi can form mutual relationships with an array of plants (Castillo et al. 2012). However, the precise amount of AM fungal biomass produced in any given area is difficult to determine and may not necessarily be a reliable indicator of their abundance or the overall health of the ecosystem (Klironomos et al. 2000).

Meagre literature is available on important environmental gases and their influence on mycorrhizal symbiosis. An emphasis has been, therefore, given in this chapter to collect the information related to impact of elevated CO_2, SO_2, and O_3 on mycorrhizal population and mycorrhization pattern. The outcomes of this chapter will indeed help reader find updated, relevant, concrete, and authentic data on the proposed theme.

3.2 AM Fungi and Their Diversification

The current molecular studies and fossil-based analysis suggest that the association between plant and AM fungi first appeared 460 million years ago (Redecker 2002). Host receive an increased amount of nutrient and water uptake while fungal symbionts are benefitted from carbohydrates (Van Der Heijden and Horton 2009; Redecker et al. 2013; Schüßler et al. 2001). Fungi-plants symbiosis might have developed from those fungi that initially made an alliance with the earliest land plants after developing a mutual understanding with blue-green algae (Schüßler et al. 2001).

The last 100 million years have seen several emergences of such alliance that most frequently develop some specialized mycorrhiza such as ecto and ericoid mycorrhiza. It is debatable whether parasitic or saprotrophic fungi gave rise to mycorrhizal fungi because such type of frequent relationship belongs to mutualism-parasitism process that depends on the availability and accessibility of soil nutrients as well as carbohydrate production. Evidence suggests that AM fungi are developed from saprobes (Hibbett et al. 2000; James et al. 2006; Kohler et al. 2015; van der Heijden et al. 2015). There is some important regulatory modification that help in the formation of symbiosis leading to development of independent mycorrhiza are considerably important. A set of genes are also present within the host plant that governs and become very instrumental in the formation of such alliance (Wang et al. 2010). In connection to the origin and evolution of AM fungi, convergent evolution has been a convincing agency that supports the theory of ecto, ericoid, and orchid mycorrhizal fungi (Kuhn et al. 2001). It is generally considered that plant symbionts obtain the food materials like carbohydrates from the host plant. It has long been observed that such association is encountered exclusively with the plants and such plants-fungi have a close alliance. This type of association has attracted various scientists to research host–symbiosis association. Mycorrhizal fungi have a close coevolutionary relationship with their host plant. Carbohydrates from one plant may be indirectly transferred to another through mycorrhizal association (Kiers and Heijden 2006). Extraradical mycelia is an important aspect of the foraging strategy of many fungi which helps fungi acquire nutrients from the surrounding soil (Plassard et al. 2019). AM fungi produce a network of extraradical mycelia that may extend over several meters, while others may grow up to a few centimetres depending on species of fungus, nutrient and moisture availability, and the presence of competing microorganisms (Mikkelsen et al. 2008). This provides an additional surface to the plant root system to acquire more minerals and transfer them to intraradical mycelia and arbuscules which are then distributed internally within the roots. Finally, the minerals are delivered to plant roots and carbohydrates are exchanged (Nagy et al. 2009). The extraradical mycelia of AM fungi enable them to explore the soil to acquire more nutrients for a larger period, distantly located and widely arranged and making them available to the plants which are grown in nutrient-deprived soil (Allen and Shachar-Hill 2009). The functioning of an ecosystem is greatly improved due to the presence of these plant symbionts. These plant symbionts can help aggregate soil and withstands with abiotic stressors (Leifheit et al. 2014). However, physicochemical properties of the soil also affect AM fungi symbiosis and their population diversity in different environments (Li et al. 1991). These symbionts affect the physiological, biochemical status of the individual plants that help in nurturing the soil health. AM fungi also influence the microbial composition of soil that indirectly help improve plant growth characters. For example, microorganisms participating in the nitrogen cycle are markedly influenced (Veresoglou et al. 2012), and thus they exert influence on cropping pattern in a particular environment (Klironomos et al. 2011). AM fungi modulate the interaction way of plant–plant (Bever 2002), plant–pathogen (Wehner et al. 2010) and plant–herbivore (Gehring and Whitham 2002) relationship.

3.3 AM Fungi and Its Key Functions

Carbon and other minerals movement from plant to fungus in mycorrhizal plants have got good pace in the literature in addition to phosphorus (Bago et al. 2000). Phosphorus exchange from fungi to plant has better been recorded (Smith and Read 1997). An appreciable amount of carbon is clearly seen to be transported between the roots of plants through AM fungal networks (Francis and Read 1984; Grime et al. 1987; Lerat et al. 2002). The stable isotopic studies confirm that around 50% of the carbon transferred from host plant to mycorrhizal fungi can be of the root system of other plants that may subsequently be used by the plants (Graves et al. 1997). The AM symbiosis with host plants is undoubtedly very old relationship and has been the key player in the evolution of the plants. Though this assumption is very important and can unravel many hidden processes which is very important for plant–microbe association, AM fungi help plants improve plant vigour in hostile environments (Ansari and Khan 2012b; Ansari et al. 2019a, b; Sumbul et al. 2017). AM fungi alleviate the stress levels thereby enhancing crop biomass and yield (Abdel-Salam et al. 2018; Rizvi et al. 2015; Ansari et al. 2017a, b). The presence of mycorrhiza obviates the effect of many unwanted environmental factors which create stresses to the plants that include biotic and abiotic ones (Diagne et al. 2020). AM fungi-induced plant biomass enhancement is accomplished because of improved water and mineral uptake from surroundings and protects the host plant from a wide array of biotic stresses including plant pathogens (Smith and Read 2008). The mutual alliance between plants and its symbionts is of unique feature that regulates plant biomass and yield development. The mycelial cord of fungi (extraradical mycelia) is extended and reaches into the area which is otherwise inaccessible for nutrient uptake. The same fungal mycelia can create a common network between plants of different groups. This common mycelial network help improve the vigour of different crop plants (Pringle et al. 2009). The nutrients are relocated from fungi to the plants thus ameliorated plant tolerance is ensured (Plassard and Dell 2010). AM fungi improve soil health, and intensify the crop growth, morphological characters, productivity in congenial, and hostile environment (Navarro et al. 2014; Hashem et al. 2015). Plant symbionts especially mycorrhizal fungi are also considered bioinoculants, and currently receiving special attention from researchers in sustainable crop protection and production (Barrow 2012).

Glomalin is another important proteinaceous compound which is released by *Glomus* spp. that has received the attention of researchers considerably. This compound serves as glue in the adherence of soil particles. This compound sustains the water content especially in hostile environment (Wu et al. 2014). The glomalin compounds contain 30–40% of carbon which helps in protecting soil from drying (Sharma et al. 2017). Other important physiological features of the plants (efficient activity of stoma; stomatal conductance, water potential in leaf, water contents, efficiency of PSII, and carbon fixation) are imperatively improved (He et al. 2017; Chandrasekaran 2019) in glomalin-enriched soil. The glomalin can persist in the soil for many years, providing a sustained source of organic matter (Vlček and Pohanka 2020). This makes a valuable resource for maintaining soil health, especially in agroecosystem where soil is frequently disturbed or depleted of organic matter through tillage, erosion, or anthropogenic activity (Balík et al. 2022).

AM fungi also serve as biofertilizers which improve soil, plant growths, and overall biomass (Sadhana 2014). It is assumed that AM fungi may also be an alternative to inorganic fertilizers. Non-judicious application of inorganic fertilizers and pesticides has raised enormous issues for the ecosystem and human health as well. These fertilizers' non-judiciously introduction lead to significant perturbations in the terrestrial ecosystem (Dai et al. 2013). Excessive use of land may affect the biodiversity very badly and the functioning of ecosystem as well (Wagg et al. 2015). The plant symbiosis increases the accessibility of various micro and macronutrients to plants through various means. Plant symbionts colonization of the host plants boost up nutrient uptake capacity specifically phosphate (Smith et al. 2003). They help plant to uptake nutrient from nutrient-deprived soils (Kayama and Yamanaka 2014). In addition to phosphate, AM fungi are also known to transfer many micronutrients like Zn and Cu (Smith and Smith 1997). These plant symbionts may develop alliance with the plants and obtain carbohydrates for their survival. However, AM fungi give the plant various nutrients like nitrogen, phosphorus, potassium, calcium, zinc, and sulphur. Arbuscules are specialized structures and involved in the exchange of nutrients and are very instrumental in mutualistic relationship between the plant and fungus (Li et al. 2016a, b; Prasad et al. 2011). The arbuscules are highly branched, usually spherical shape that are developed within the plant cells, and provide a large surface area for the exchange of water and nutrients, primarily phosphorous, between the fungus and the plant (Smith and Read 2008). Plant symbionts inoculation progressively enhanced the nitrogen, phosphorus, and iron concentration status of *Pelagronium graveolens* under water deficient environment (Amiri et al. 2017). Application of AM fungi under salinity stress to the *Euonymus japonica* increased the phosphorus, calcium, and potassium levels in crop plants (Gómez-Bellot et al. 2015). Moreover, AM fungi and their role in nitrogen cycling have also been observed (Hodge and Fitter 2010). The extraradical hyphae may absorb the nitrogen progressively (Tanaka and Yano 2005). AM fungi usually transfer 20–75% of total nitrogen absorbed to their host plants (Tanaka and Yano 2005; Hashem et al. 2018). Plant symbionts inoculation to olive plants increased plant health and accelerate micro- and macronutrients accumulation ability (Bati et al. 2015).

AM fungi-led grown strawberry plants showed enhanced levels of biomolecule synthesis which are required for plant health (Castellanos-Morales et al. 2010). These plant symbionts also increase qualitative characters that influence the biosynthesis of several volatile compounds and also carotene-bearing compounds (Hart et al. 2015). The introduction of *Glomus versiforme* progressively enhanced important molecules (Sugars, organic acids, vitamin C, flavonoids) leading to improved quality of citrus (Zeng et al. 2014). Growth and productivity characters of *Zea mays* (Sabia et al. 2015), yam (Lu et al. 2015), and potato (Hijri 2016) were markedly found higher in AM fungi-inoculated plants. AM fungi have also been very pivotal in accelerating the biosynthesis of some important plant chemicals which make plants more fit, fine, and healthy (Sbrana et al. 2014; Rouphael et al. 2015).

AM fungi have been the subject of investigation as they may be considered as drought stress ameliorator. Drought stress has been found reducing the rate of transpiration and induces oxidatively induced stress (Impa et al. 2012). Drought stress contributes to the reduced crop growth as it affects enzymatic activity, ion uptake, and assimilation of nutrients (Ahanger and Agarwal 2017). There has been enough evidence which suggests that AM fungi can alleviate the stress of drought and improve the qualitative characters of different crops including wheat, barley, maize, soybean, strawberry, and onion (Mena-Violante et al. 2006; Ruiz-Lozano et al. 2015; Yooyongwech et al. 2016; Moradtalab et al. 2019). The plants become more tolerant to the drought and such tolerance is developed which seems to be due to the exploration of more soil by roots of mycorrhizal plants (Zhang et al. 2016). The symbiotic association of plant and AM fungi also regulates enhanced osmotic potential (Kubikova et al. 2001), activity of natural opening like stomatal opening and closure by controlled plant hormone, i.e., abscisic acid (Duan et al. 1996), increased proline accumulation (Ruiz-Sánchez et al. 2010), or enhanced glutathione level (Rani 2016). Inoculation of plant symbionts to different plants finally increases root size and uptake efficiency, leaf area index, and biomass in drought environment (Al-Karaki et al. 2004; Gholamhoseini et al. 2013). There has been a hue and cry in soil salinization which has caused a serious concern and poses a significant threat to environmental conditions. The salinity stress is well-known factor for suppressing the plant biomass and affects fixation of nitrogen (Hasanuzzaman et al. 2013). Soil salinization increases the production of oxidant molecules in an excessive amount which is also harmful to plants (Ahmad et al. 2010).

The growth rate, water potential of leaf, and the ability of plants to use water in snapdragon (*Antirrhinum majus*) were progressively challenged when grown under saline conditions (El-Nashar 2017). AM fungi inoculation progressively improved the photosynthesis rate, gaseous exchange characters, chlorophylls, and efficiency of the plants to use the water by *Ocimum brasilicum* in a salty environment (Elhindi et al. 2017). Wang et al. (2018) observed that inoculation of AM fungi alleviated the saline stress and enhanced the biomass characters of the plants. AM fungi-driven *Cucumis sativus* plants exhibited higher concentrations of total phosphorus, calcium, nitrogen, magnesium, and potassium as compared to non-AM inoculated plants growing in saline soil (Hashem et al. 2018). In another study, the chlorophyll contents, magnesium, and nitrogen uptake were markedly higher in AM fungi-inoculated plants of chilli which was grown in saline soil (Cekic et al. 2012). AM-driven plants exhibited greater biomass production, enhanced biosynthesis of proline contents, enhanced nitrogen uptake and remarkable variations in ionic exchange, and poor uptake of sodium as compared to those which are non-mycorrhizal under saline stress (Santander et al. 2019).

AM fungi also help in the regulation of plant growth hormones. For instance, AM fungi inoculation significantly improved the cytokinin concentration leading to enhanced photosynthate translocation in saline environment (Talaat and Shawky 2014). Mycorrhizal plant growth was progressively higher due to remarkable changes in polyamine pool under saline soil (Kapoor et al. 2013). Likewise, another important compound (strigolacton) in mycorrhizal plants significantly obviated the different salinity effects on lettuce. AM fungi-inoculated plants possess the characters to minimize the oxidative molecules production by reducing lipid membrane peroxidation in saline environment (Talaat and Shawky 2014). In saline soil, organic acids enhance osmoregulation even in the presence of plant symbionts. Increased level of betaine production in maize plants grown in saline soil is a traditional example (Sheng et al. 2011).

3.3.1 Heavy Metals

If heavy metals are taken for granted, the day is not far away when humankind will be facing a big challenge pertaining to ecosystem health (Liu et al. 2013; Yousaf et al. 2016). Soil containing heavy metals is yet another aspect which has obviously created chaos and unrest among biodiversity conservationists. AM fungi strengthen the defence system of plants that promote growth and development. Mycorrhizal spores infecting *Triticum aestivum* registered a significant amount of improvement in plant biomass and total yield production in aluminium-stressed soil (Aguilera et al. 2014). The heavy metal-enriched soil impairs plant growth and yield attributes causing chlorosis and finally death of the plants (Moghadam 2016). Heavy metals are immobilized in the internal and external fungal hyphae that fix them and are stored within their cells. These heavy metals are fixed in wall and vacuole that serves as storage part of the cells and are usually chelated with cytoplasm thus is a natural way to reduce the bioavailability of heavy metals in the soil and reduce their potential for harm to the ecosystem (Punamiya et al. 2010). AM fungi-led enhanced growth and development in the plants are assumed to be due to increased improvement in morpho-physiological process which consequently enhances plant biomass, and accumulation of micro- and macronutrients such as copper, zinc, and phosphorus (Miransari 2017). AM fungi also bind with Cd and Zn in the hyphae and cortical cells thereby growth and nutrients status of the plants are significantly improved (Garg and Chandel 2012). These plant symbionts also improve pH levels in the soil (Shen et al. 2006), restore cadmium in extraradical mycelia (Janouškova and Pavlíková 2010), and conjugate cadmium to glycoprotein (glomalin). Plant symbionts are highly suppressive in reducing toxic concentrations of heavy metals like cadmium by detoxifying their level on rice plants (Li et al. 2016a, b).

3.3.2 Temperature

High and low temperatures may cause significant damage to the crop yield contributing characters. Plants symbionts have always been instrumental in the obviation of such stressors and gave the researchers a convincing result. The elevated soil temperature badly affects plant growth, vegetative characters, and development which finally determines sustainability and productivity (Bunn et al. 2009). AM fungi-inoculated plants have exhibited pronounced biomass under thermal stress as compared to plants which possess no mycorrhizal spores (Gavito et al. 2005). The plant symbionts also have the ability to hold moisture for longer durations in the crop plants (Zhu et al. 2010) that enhance secondary biomolecules production leading to the improved resistance level, and enhanced protein contents that subsequently support the plant to obviate the cold stress (Abdel Latef and Chaoxing 2014). AM fungi also improve the biosynthesis of chlorophyll status when grown in a cold environment (Zhu et al. 2010).

3.4 AM Fungi in Plant Disease Management

Undoubtedly, plant symbionts always remained instrumental in the alleviation of various abiotic stressors. However, the significance of such important microorganisms cannot be overlooked in crop health improvement while grown in fungi, nematode, and bacteria-driven environment. AM fungi trigger the biosynthesis of some defensive biomolecules that makes the plants more resistant to pathogens (Sreenivasa et al. 2019; Hooker et al. 1994; Ansari et al. 2020a; Solanki et al. 2020; Ansari and Mahmood 2019a, b). AM fungi improve plant health in direct or indirect manner. The plant growth and vigour (Wang et al. 2010), phytohormone (Song et al. 2020), and food and space competition among the microorganisms (Qiao et al. 2015) are imperatively ameliorated. AM fungi also release certain phenols (Bencherif et al. 2019) that possess antimicrobial activity like polymyxin and its derivatives (Kaur and Suseela 2020; Mansfeld-Giese et al. 2002). These important plant symbionts also have the ability to release some chemicals that become lethal to the pathogens (Bruisson et al. 2016; López-Ráez et al. 2017; Wang et al. 2017). AM fungi develop a good linkage with the rhizobiome containing rhizobacteria which helps plant symbionts to be more effective in managing plant diseases and improving crop yield (Raklami et al. 2019; Ansari et al. 2020b; Ansari and Mahmood 2017a, b; Ansari and Khan 2012a). Declerck et al. (2002) found that pre-inoculation of *Glomus* sp., *G. proliferum, G. intraradices,* and *G. versiforme* not only attenuated disease severity caused by *Cylindrocladium spathiphylli* but also stimulated the growth and phosphorus contents of banana plants. The effect of AM fungi in managing different microorganisms causing disease has been demonstrated, however, inconsistency in the use of AM fungi cannot be taken for granted before commercialization of such models. Banana rhizome inoculated with *Glomus intraradices* and *Glomus* sp. isolates significantly impaired the necrosis of rhizome and the disease induced by wilt-causing fungi (*Fusarium oxysporum* f. sp. *cubense*) though, comparative analysis on the effectiveness of both AM fungi which makes the sense that both isolates were either equally effective against wilt pathogens or do not possess specificity (Jaizme-Vega 1998; Jaizme-Vega et al. 2003). The effectiveness of AM fungi was also enhanced if they are applied in mixture of other beneficial microorganisms. *Nicotiana tabacum* plants usually suffer from a major disease which is called black shank, and the disease is caused by *Phytophthora parasitica* var. *nicotianae* and damping off (*Pythium aphanidermatum*), were inoculated with a consortium of plant symbiont, *G. fasciculatum* and biocontrol agent, *T. harzianum*. It was observed that the application of these consortia effectively managed the disease leading to improved plant health (Sreeramulu et al. 1998). The effect of AM fungi on phytopathogenic bacteria and disease they cause have also been tested and showed satisfactory results that can be helpful for disease management. AM fungi, *Glomus macrocarpum,* mitigated the infection caused by *Pseudomonas lacrymans* on eggplants and cucumber (Xavier and Boyetchko 2004).

Virus–mycorrhiza interaction is perhaps the least studied aspect that also require extensive studies. It is assumed that mycorrhizae are not viral vectors as the viral particles were totally absent from plant symbiont's hyphae and arbuscules surroundings (Jabaji-Hare and Stobbs 1984). However, the effect of the virus on mycorrhiza is still a promising field and needs to be examined very skillfully. Besides, plant parasitic nematodes possess different modes of parasitism that may induce significant yield loss to crop plants across the world (Jones et al. 2013). AM fungi have been found protecting crops from nematodes through various mechanisms such as competition for space and nutrients (Parniske 2008), alterations in root morphology (Gamalero et al. 2010), induction of systemic resistance (De la Peña et al. 2006), and altered rhizosphere interaction (Hodge 2000; Ansari et al. 2020a).

3.5 Elevated CO_2 and Its Impact on Mycorrhiza and Mycorrhization

The CO_2 concentration varies depending on the location, time of day, and season. However, the average CO_2 concentration in the earth's atmosphere during 2020 was around 419 ppm (NOVA 2021). The impact of elevated CO_2 on plant mycorrhizal symbiosis has been a subject of discussion on various platforms. There are several other factors which are also influenced due to elevation of CO_2. The assumption has been that by the year 2100, the CO_2 level may arise up to 750 ppm (IPCC 2014) which may cause more perturbation to the ecosystem. The effects of elevated CO_2 on different species and wild natural ecosystem are very poorly understood. The aboveground parts of the terrestrial ecosystem have exhibited various physiological changes while growning in elevated CO_2 environment (Becklin et al. 2017; Obermeier et al. 2017). Enhanced photosynthetic activity under elevated CO_2 also increases photosynthetic transfer capacity of various plants and also influences carbon flow through soil microbial agents (Cheng et al. 2012; Drigo et al. 2010; Staddon et al. 2014).

On the other hand, few reports suggest that elevated CO_2 help plants to grow quickly than ambient CO_2 in somewhat regulated way (Poorter 1993). It is also observed that elevated levels of CO_2 may allocate an increased level of carbon in the root system as compared to shoots in a mycorrhizal plant (Rogers et al. 1995). Moreover, many researchers have attempted to show that elevated level of CO_2 below ground can stimulate the mycorrhizal colonization; however, it needs further cogent studies (Staddon and Fitter 1998; Fitter et al. 2000). In another set of experiments, Gavito et al. (2000) reported that enhanced CO_2 level registered no significant effects on colonization of AM fungus or extraradical hyphal production. Under elevated CO_2 conditions, plants exhibited a significant increase in biomass, accompanied by a concurrent enhancement in mycorrhizal symbiotic associations. Lewis et al. (1994) reported that *Pinus taeda* grown in elevated CO_2 caused significant effects on root carbohydrate levels. However, mycorrhizal colonization received no remarkable effects which reveal that there is no unique pattern of CO_2 effects on mycorrhizal symbiosis (Klironomos et al. 1998). Elevated CO_2 can impact the formation of mycorrhizal spores in the soil. Studies have shown that elevated CO_2 levels can alter the formation of mycorrhizal spores and their abundance in soil. Elevated CO_2 can increase the production of mycorrhizal spores, while others have found no effect or a decrease in their formation. The effect of elevated CO_2 on mycorrhizal spores depends on the specific plant–fungal association, as well as environmental factors such as nutrient availability, soil moisture, and temperature (Fitter et al. 2000). Overall, the effect of elevated CO_2 on mycorrhizal spores is complex and more research is needed to fully understand the mechanisms and implications of this relationship.

3.5.1 Effect of CO_2 on Plant Biomass Colonized by Mycorrhizal Fungi

Elevated CO_2 increases the total biomass production of some plant species. This is because higher CO_2 levels enhance photosynthesis in a regulated way which leads to increased growth and productivity (Morison and Lawlor 1990). This effect is known as "CO_2 fertilization effect" and has been demonstrated in different conditions. However, it is important to note that the extent of this effect may vary depending on factors such as the type of plant species, growing conditions, and soil nutrients. Elevated CO_2-exposed plants increase plant growth, alter the plant's physiology, nutrient content, and susceptibility to herbivores and diseases, which have cascading effects on ecosystems (Insam et al. 1999). The elevated CO_2 enhances the crop biomass and the presence of mycorrhizal spores pronouncedly increases plant biomass and yield characters (Gavito et al. 2000). However, substantial research work is still needed to be done to come out with any concrete conclusions.

Also, plants exposed to CO_2 inoculated with mycorrhizal spores are a good avenue for enhanced production of plant biomass and yield. Zhu et al. (2018) conducted an experiment and observed whether AM symbionts may increase the nutrient accumulations and plant biomass of wheat exposed to elevated CO_2 and saline soil. They substantially reported that elevated CO_2 enhanced the plant's symbionts' colonization and dryweight production. Elevated CO_2-exposed plants registered significant growth and biomass in non-mycorrhizal wheat. The nitrogen contents and sodium ions of shoots and roots were progressively enhanced. Overall, the plants exposed to elevated CO_2 obviated the stress and exhibited improved plant health.

3.6 Elevated O_3 and Its Impact on Mycorrhiza and Mycorrhization

AM fungi and their beneficial effects on plant are very instrumental in predicting various environmental pollutants and also their impact on sustainable crop biomass and yield (Shafer and Schoeneberger 1991). Due to obligatory nature, AM fungi are not able to produce a sufficient number of mycelia, or further development of plant symbionts is halted apparently as ramification and growth of these fungi is totally dependent on the food supply of the host plants (Smith and Read 2008). The plant health improvement depends also on AM root colonization. The association of plant symbionts with plants has improved crop health and registered a significant amount of yield in total biomass (Abbott and Robson 1979). Inoculation of AM isolate and percent root colonization has also been positively correlated (Zangaro et al. 2007). Elevated O_3 has been a major challenge ahead the environmental protectionists. It has been observed that the contribution of O_3 continues to increase in developing nations (Wang et al. 2007). O_3 also acts as an oxidant which may hamper the crop growth, physiology, and productivity. Reduced photosynthetic rate and other physiological process, leaf injury, and leaf senescence contribute markedly to yield loss. For example, elevated O_3 causes 5% and 10% of total yield reduction in maize and soybean, respectively, from 1980 to 2011 which costs around 9 billion USD in the United States (McGrath et al. 2015). A significant crop yield loss is being projected due to elevated O_3 from 2000 to 2030 (Avnery et al. 2011). The ozone effect has a negative impact on mycorrhiza which can damage the plant root cells, leading to decreased effectiveness of the mycorrhiza in providing essential nutrients and water to the host plant. This can reduce the plant's ability to resist disease and drought, and overall decrease its growth and productivity. Additionally, O_3 can directly harm the fungal partner in the symbiotic relationship, disrupting the balance of the mycorrhiza, and further reducing its benefits to the plant (Ueno et al. 2016).

Studies have shown that exposure to high levels of O_3 can disrupt this symbiotic relationship and lead to reduced plant growth and health. The O_3 can damage the plant's root cells, leading to a reduction in the production of root exudates, which are crucial for the establishment of mycorrhiza (Wilkinson and Davies 2010). Moreover, O_3 can also lead to the production of oxidative stress in mycorrhizal fungi, which can reduce their ability to form and maintain a symbiotic relationship with the plant. This, in turn, can lead to decreased plant growth and reduced tolerance to environmental stresses (Baier et al. 2005). Therefore, the ozone effect on mycorrhiza is generally considered to be negative and can lead to significant impacts on plant growth and productivity (Booker et al. 2009).

Exposure to O_3 has reduced crop health, photosynthetic variables; therefore, insufficient carbon is made available which obviously affect the AM fungi and host interaction with a plant (Parniske 2008). The rate of AM colonization in O_3 exposed environment has been negatively affected (Wang et al. 2017). It is the need of the hour to look into the effect of AM fungi on plant health and yield grown in elevated O_3 environment. A high level of O_3 also affects AM colonization, though consistency in such experiments has not been observed. Elevated O_3 cause varied effects on AM colonization. Vesicle formation and development frequency, hyphal coiling is enhanced while the frequency of arbuscules is reduced (Duckmanton and Widden 1994). For instance, increased O_3 exposure registered substantially improved vesicle formation, and reduced arbuscular and hyphal colonization in *Phseolus vulgaris* (Wang et al. 2011).

3.7 O_3 and Its Effect on Total Biomass

Elevated O_3 is well known to hamper plant growth, biomass markers, and crop yield characters in appreciable amount (Gelang et al. 2001; Feng et al. 2008; Black et al. 2012). Increased O_3 significantly harms different ionic particles of membrane, accelerates the leaf senescence and reactive oxygen species production from different enzymatic sources, and also decreases photosynthetic activity (Calatayud et al. 2011; Mills et al. 2013). However, plant symbiont's introduction induced good plant health and enhance biomass production and yield (Smith and Read 2008; Heidari and Karami 2014). Under elevated O_3 exposure, AM symbiosis may survive and trigger the enhancement of biochemical processes like stomatal conductance, transpiration, and photosynthesis (Wang et al. 2015) which suggest that AM fungi possess the ability to obviate the O_3 stress. Elevated O_3 frequently reduces stomatal conductance, CO_2 exchange rates, electron transport, PSII function, photochemical, and non-photochemical quenching (Ismail et al. 2014; Guidi et al. 2002; Pellegrini 2014).

AM symbiosis includes germination of mature spore, growing hyphae, colonization rate and pattern, formation of vesicles, functionality of arbuscules, and hyphal structure within a plant root (Bécard et al. 2004). Elevated O_3 markedly impaired the mycorrhizal colonization rate except for hyphal coiling at 41–80 ppb O_3. Higher concentration of O_3 progressively decrease the mycorrhizal colonization rate. It is speculated that elevated O_3 minimizes the rate of photosynthesis of host plants and cargo of carbohydrates from the leaf to roots of the host plants thus poor supply of carbohydrates is only left to AM fungi (Morgan et al. 2003; Feng et al. 2008). Impact of elevated O_3 on AM symbiosis is related to various important characters like exposure duration is one of the important variables that leaves substantial impact on hyphal coiling, arbuscule, and vesicle formation (Kytöviita et al. 2001; Wang et al. 2017). Elevated O_3 also determines the formation and disappearance of arbuscules which is very frequent and for shorter duration (Alexander et al. 1989; Read 1991). Elevated O_3 may enhance the biosynthesis of some biomolecules like phenols, polyphenols different plant growth regulators, and various molecules which are players in defence system of the plants (Ludwikow and Sadowski 2008; Häikiö et al. 2009). The secondary biomolecule compound synthesis is highly regarded for generation and degeneration of hyphae and arbuscules in AM symbiosis (Nagahashi and Douds 2000; Requena et al. 2007). Abbott et al. (1984) reported that in excessive phosphorus environment, vesicle formation is reduced while it is increased during the flowering period and high nitrogen environment (Gunze and Hennessy 1980; Bevege et al. 1975). Contrary to this, Yin et al. (2022) published research finding in the journal "mycorrhiza" reporting the effects of elevated O_3 on plant biomass and different nutrients. The study reported that elevated O_3 (60nml. mol[1] O_3 enrichment) reduced the net photosynthetic rate and plant's overall biomass and yield production. On the other hand, malondialdehyde was progressively increased. Moreover, Ho and Trappe (1984) conducted an experiment on the effect of O_3 on the growth and mycorrhizal symbiosis of *Festuca arundinacea*. It was observed that mycorrhiza-inoculated *F. arundinacea* and also exposed with 0.1 ppm O_3 for three months impaired the root biomass, although increase in mycorrhiza-related parameters was observed which suggests that there is no unique pattern in mycorrhization.

3.8 Elevated SO_2 and Its Impact on Mycorrhiza and Mycorrhization

Elevated SO_2 levels can acidify the soil and reduce plant growth, leading to a decrease in the benefits of mycorrhizal symbiosis. Additionally, high SO_2 levels can directly damage the fungal hyphae, reducing the efficiency of the symbiotic relationship (Clapperton and Parkinson 1990; Clapperton 1991). To mitigate the effects of elevated SO_2 on mycorrhizal symbiosis, measures can be taken to reduce SO_2 emissions and improve soil and plant health. Elevated SO_2 can have a negative impact on mycorrhizal symbiosis. Elevated SO_2 causes damage to both the plant and fungal partner, resulting in reduced plant growth and decreased efficiency of nutrient acquisition (Dighton and Jansen 1991). The mechanisms of SO_2 toxicity are not yet fully understood, but it is believed to have involved oxidative stress and cellular damage in both the plant and fungal cells (Mahoney et al. 1985). SO_2 can also interfere with the process of photosynthesis, which is the process by which plants use sunlight to produce energy and convert carbon dioxide into glucose. High levels of SO_2 can also increase acidity status in soil and in such environment, plants are failed to absorb the nutrient efficiently (Agrawal and Deepak 2003). Furthermore, continuous elevated SO_2 exposure can reduce the ability of plants to withstand against extreme environment, and fight pathogens and pests. Elevated SO_2-exposed plants exhibit reduced growth, photosynthesis, and plant health. Additionally, SO_2 can hamper the process of root development and nutrient uptake, which is a limiting factor for nutrient acquisition by plants. Elevated SO_2 may also cause acid rain, which lower soil pH and negatively reduce plant growth, health, and productivity (Gaffney et al. 1987).

Elevated SO_2-exposed plants reduce the formation and functioning of arbuscules, which can lead to reduced nutrient uptake by the plant (Clapperton et al. 1990). This can overall affect the growth and development of plant as well as efficiency of the mycorrhizal symbiosis pertaining to the acquisition of nutrients (Clapperton 1991). High levels of SO_2 can also interfere with the formation and maintenance of arbuscules, which can further reduce the effectiveness of the symbiotic relationship (Jakobsen et al. 2003; Clapperton 1991).

Elevated SO_2 is well-known gaseous pollutants which have warranted many noticeable damages to the ecosystem. The SO_2 is a highly reactive gas that harms terrestrial ecosystems when used alone or in combination with other harmful substances. Vegetation of terrestrial ecosystem has efficiently absorbed the SO_2 through stomatal absorption, water dissolution, etc. (Fowler and Unsworth 1974; Majernik and Mansfield 1970). Elevated SO_2 causes some effects on the soil microbes associated with mycorrhizosphere which is assumed to be either indirect or plant mediated. Plants exposed to low SO_2 have produced poor quality dry biomass (Bell and Clough 1973). Such degradation in plant biomass is directly linked with reduced photosynthesis and modulation in translocation of newly synthesized carbohydrates (Noyes 1980). Furthermore, biochemical and physiological changes in the host plants exposed to SO_2 have also been monitored. Rice et al. (1978) reported that increased SO_2 in open air fumigation systems showed less vesicular arbuscular mycorrhiza symbiosis with *Agropyron smithii*. Likewise, an individual field (prairie) were exposed continuously to low concentration of SO_2 which results in lower population of VAM fungi in the field condition. Clapperton et al. (1990) observed a substantial reduction in mycorrhization pattern when exposed to lower (005–007 μl) concentration of SO_2.

3.9 Conclusions and Future Outlooks

The effects of various air gases on mycorrhizal symbiosis have become a very important topic of research for plant scientists, root biologists, microbiologists, etc. Elevated CO_2, SO_2, O_3 are important environmental gases and have been instrumental in mycorrhizal symbiosis. We base on available literature suggest that elevated CO_2 exhibit varied responses in mycorrhizae formation and its development which seems to be due to various agroclimatic conditions prevailing in a particular environment. Elevated CO_2 has been somehow found helping the plants to maximize their biomass and yield perhaps due to carbon fertilization effect. Elevated CO_2 has shown inconsistent effects on mycorrhizal parameters. Moreover, elevated SO_2 has shown devastating effects on mycorrhization and overall biomass formation of plants. Important determining markers of the AM fungi are affected badly when exposed to elevated SO_2. Likewise, elevated O_3 has been found as a limiting factor that induces lethal effects on the formation of mycorrhizal networks on the plants. Elevated O_3 has negatively affected the plant growth and biomass in mycorrhiza-inoculated plants. Studies relating to effect of these gases may unravel key points which can be beneficial for environmentalists in devising different policies regarding phytoremediation.

References

Abbott LK, Gazey C (1994) An ecological view of the formation of VA mycorrhizas. Plant Soil 159(1):69–78

Abbott LK, Robson AD (1979) A quantitative study of the spores and anatomy of mycorrhizas formed by a species of Glomus, with reference to its taxonomy. Aust J Bot 27(4):363–375

Abbott LK, Robson AD, De Boer GJNP (1984) The effect of phosphorus on the formation of hyphae in soil by thevesicular-arbuscular mycorrhizal fungus, *Glomus fasciculatum*. New Phytolo 97(3):437–446
Abdel Latef AA, Chaoxing HJ (2014) Does the inoculation with *Glomus mosseae* improve salt tolerance in pepper plants? Plant Growth Regul 33:644–653. https://doi.org/10.1007/s00344-014-9414-4
Abdel-Salam E, Alatar A, El-Sheikh MA (2018) Inoculation with arbuscular mycorrhizal fungi alleviates harmful effects of drought stress on damask rose. Saudi J Biol Sci *25*(8):1772–1780
Agrawal M, Deepak SS (2003) Physiological and biochemical responses of two cultivars of wheat to elevated levels of CO2 and SO2, singly and in combination. Environ Pollut 121(2):189–197
Aguilera P, Pablo C, Fernando B, Fritz O (2014) Diversity of arbuscular mycorrhizal fungi associated with *Triticum aestivum* L. plants growing in an andosol with high aluminum level. Agric Ecosyst Environ 186:178–184. https://doi.org/10.1016/j.agee.2014.01.029
Ahanger MA, Agarwal RM (2017) Potassium up-regulates antioxidant metabolism and alleviates growth inhibition under water and osmotic stress in wheat (*Triticum aestivum* L.). Protoplasma 254(4):1471–1486. https://doi.org/10.1007/s00709-016-1037-0
Ahmad P, Jaleel CA, Salem MA, Nabi G, Sharma S (2010) Roles of enzymatic and non-enzymatic antioxidants in plants during abiotic stress. Crit Rev Biotechnol 30:161–175. https://doi.org/10.3109/07388550903524243
Alexander T, Toth R, Meier R, Weber HC (1989) Dynamics of arbuscule development and degeneration in onion, bean, and tomato with reference to vesicular–arbuscular mycorrhizae in grasses. Can J Bot 67(8):2505–2513
Al-Karaki G, Mcmichael B, Zak J (2004) Field response of wheat to arbuscular mycorrhizal fungi and drought stress. Mychorrhiza 14:263–269. https://doi.org/10.1007/s00572-003-0265-2
Allen JW, Shachar-Hill Y (2009) Sulfur transfer through an arbuscular mycorrhiza. Plant Physiol 149(1):549–560
Allen EB, Allen MF, Helm DJ, Trappe JM, Molina R, Rincon E (1995) Patterns and regulation of mycorrhizal plant and fungal diversity. Plant Soil *170*:47–62
Amiri R, Nikbakht A, Rahimmalek M, Hosseini H (2017) Variation in the essential oil composition, antioxidant capacity, and physiological characteristics of Pelargonium graveolens L. inoculated with two species of mycorrhizal fungi under water deficit conditions. J Plant Growth Regul *36*:502–515
Ansari RA, Khan TA (2012a) Parasitic association of root-knot nematode, *Meloidogyne incognita* on guava. e J Sci Technol 5(12):65–67
Ansari RA, Khan TA (2012b) Diversity and community structure of Phytonematodes associated with guava in and around Aligarh, Uttar Pradesh. Trends Biosci 5(3):202–204
Ansari RA, Mahmood I (2017a) Optimization of organic and bio-organic fertilizers on soil properties and growth of pigeonpea. Sci Hortic 226:1–9
Ansari RA, Mahmood I (2017b) Determination of disease incidence caused by *Meloidogyne* spp. and or *Fusarium udum* on pigeonpea in Aligarh district: a survey. Trends Biosci 10(24):5239–5243
Ansari RA, Mahmood I (2019a) Plant health under biotic stress. In: Ansari RA, Mahmood I (eds) Microbial interactions, vol II. Springer Nature, Singapore
Ansari RA, Mahmood I (2019b) Plant health under biotic stress. In: Ansari RA, Mahmood I (eds) Organic strategies, vol I. Springer Nature, Singapore
Ansari RA, Mahmood I, Rizvi R, Sumbul A (2017a) PGPR: current vogue in sustainable crop production. In: Kumar V (ed) Probiotics and plant health. Springer Nature, Singapore, pp 455–472
Ansari RA, Mahmood I, Rizvi R, Sumbul A, Safiuddin (2017b) Siderophores: augmentation of soil health and crop productivity. In: Kumar V (ed) Probiotics in agroecosystem. Springer Nature, Singapore, pp 291–312
Ansari RA, Sumbul A, Rizvi R, Mahmood I (2019a) Potential role of plant growth promoting Rhizobacteria in alleviation of biotic stress. In: Ansari RA, Mahmood I (eds) Plant health under biotic stress, Microbial interactions, vol II. Springer Nature, Singapore, pp 177–188

Ansari RA, Sumbul A, Rizvi R, Mahmood I (2019b) Organic soil amendments: potential tool for soil and plant health management. In: Ansari RA, Mahmood I (eds) Plant health under biotic stress, Organic strategies, vol I. Springer Nature, Singapore, pp 1–35

Ansari RA, Rizvi R, Mahmood I (eds) (2020a) Management of phytonematodes: recent advances and future challenges. Springer, Singapore

Ansari RA, Rizvi R, Sumbul A, Mahmood I (2020b) Plant-growth-promoting Rhizobacteria (PGPR)-based sustainable management of phytoparasitic nematodes: current understandings and future challenges. In: Ansari RA et al (eds) Management of phytonematodes: recent advances and future challenges. Springer Nature, Singapore. https://doi.org/10.1007/978-981-15-4087-5_3

Avnery S, Mauzerall DL, Liu JF, Horowitz LW (2011) Global crop yield reductions due to surface ozone exposure: 2. Year 2030 potential crop production losses and economic damage under two scenarios of O3 pollution. Atmos Environ 45:2297e2309

Bago B, Pfeffer PE, Shachar-Hill Y (2000) Carbon metabolism and transport in arbuscular mycorrhizas. Plant Physiol 124(3):949–958

Baier M, Kandlbinder A, Golldack D, Dietz KJ (2005) Oxidative stress and ozone: perception, signalling and response. Plant Cell Environ 28(8):1012–1020

Balík J, Kulhánek M, Černý J, Sedlář O, Suran P, Asrade DA (2022) The influence of organic and mineral fertilizers on the quality of soil organic matter and glomalin content. Agronomy 12(6):1375

Barrow CJ (2012) Biochar: potential for countering land degradation and for improving agriculture. Appl Geogr 34:21–28

Bati CB, Santilli E, Lombardo L (2015) Effect of arbuscular mycorrhizal fungi on growth and on micronutrient and macronutrient uptake and allocation in olive plantlets growing under high total Mn levels. Mycorrhiza 25:97–108. https://doi.org/10.1007/s00572-014-0589-0

Bécard G, Kosuta S, Tamasloukht M, Séjalon-Delmas N, Roux C (2004) Partner communication in the arbuscular mycorrhizal interaction. Can J Bot *82*(8):1186–1197

Becklin KM, Walker SM, Way DA, Ward JK (2017) CO 2 studies remain key to understanding a future world. New Phytol 214(1):34–40

Begum N, Qin C, Ahanger MA, Raza S, Khan MI, Ashraf M et al (2019) Role of arbuscular mycorrhizal fungi in plant growth regulation: implications in abiotic stress tolerance. Front Plant Sci 10:1068

Bell JNB, Clough WS (1973) Depression of yield in ryegrass exposed to sulphur dioxide. Nature 241(5384):47–49

Bencherif K, Djaballah Z, Brahimi F, Boutekrabt A, Dalpè Y, Sahraoui ALH (2019) Arbuscular mycorrhizal fungi affect total phenolic content and antimicrobial activity of *Tamarix gallica* in natural semi-arid Algerian areas. South Afr J Bot 125:39–45

Bennett AE, Classen AT (2020) Climate change influences mycorrhizal fungal–plant interactions, but conclusions are limited by geographical study bias. Ecology 101(4):e02978

Bevege DI, Bowen GD, Skinner MF (1975) Comparative carbohydrate physiology of ecto-and endomycorrhizas. In: Endomycorrhizas; proceedings of a symposium

Bever JD (2002) Negative feedback within a mutualism: host–specific growth of mycorrhizal fungi reduces plant benefit. Proc R Soc Lond B Biol Sci 269(1509):2595–2601

Black VJ, Stewart CA, Roberts JA, Black CR (2012) Timing of exposure to ozone affects reproductive sensitivity and compensatory ability in Brassica campestris. Environ Exp Bot 75:225–234

Bonfante P, Genre A (2010) Mechanisms underlying beneficial plant–fungus interactions in mycorrhizal symbiosis. Nat Commun 1(1):48

Booker F, Muntifering R, McGrath M, Burkey K, Decoteau D, Fiscus E et al (2009) The ozone component of global change: potential effects on agricultural and horticultural plant yield, product quality and interactions with invasive species. J Integr Plant Biol 51(4):337–351

Bruisson S, Maillot P, Schellenbaum P, Walter B, Gindro K, Deglène-Benbrahim L (2016) Arbuscular mycorrhizal symbiosis stimulates key genes of the phenylpropanoid biosynthesis and stilbenoid production in grapevine leaves in response to downy mildew and grey mould infection. Phytochemistry 131:92–99

Bunn R, Lekberg Y, Zabinski C (2009) Arbuscular mycorrhizal fungi ameliorate temperature stress in thermophilic plants. Ecology 90(5):1378–1388. https://doi.org/10.1890/07-2080.1

Calatayud V, García-Breijo FJ, Cervero J, Reig-Armiñana J, Sanz MJ (2011) Physiological, anatomical and biomass partitioning responses to ozone in the Mediterranean endemic plant Lamottea dianae. Ecotoxicol Environ Safe 74:1131–1138

Castellanos-Morales V, Villegas J, Wendelin S, Vierheilig H, Eder R, Cárdenas-Navarro R (2010) Root colonisation by the arbuscular mycorrhizal fungus *Glomus intraradices* alters the quality of strawberry fruits (*Fragaria*× *ananassa* Duch.) at different nitrogen levels. J Sci Food Agric 90(11):1774–1782

Castillo CG, Borie F, Godoy R, Rubio R, Sieverding E (2012) Diversity of mycorrhizal plant species and arbuscular mycorrhizal fungi in evergreen forest, deciduous forest and grassland ecosystems of Southern Chile.

Cekic FO, Unyayar S, Ortas I (2012) Effects of arbuscular mycorrhizal inoculation on biochemical parameters in capsicum annuum grown under long term salt stress. Turk J Bot 36:63–72. https://doi.org/10.3906/bot-1008-32

Cheng L, Booker FL, Tu C, Burkey KO, Zhou L, Shew HD et al (2012) Arbuscular mycorrhizal fungi increase organic carbon decomposition under elevated CO2. Science 337(6098):1084–1087

Chandrasekaran, M., Chanratana, M., Kim, K., Seshadri, S., & Sa, T. (2019). Impact of arbuscular mycorrhizal fungi on photosynthesis, water status, and gas exchange of plants under salt stress–a meta-analysis. Frontiers in Plant Science, 10, 457.

Clapperton MJ (1991) An Ecophysiological study of the interaction of vesicular-arbuscular mycorrhizal fungi and sulphur dioxide of the growth of phleum pratense L. University of Calgary

Clapperton MJ, Parkinson D (1990) The effect of SO2 on the vesicular–arbuscular mycorrhizae associated with a submontane mixed grass prairie in Alberta, Canada. Can J Bot 68(8):1646–1650

Clapperton MJ, Reid DM, Parkinson D (1990) Effects of sulphur dioxide fumigation on Phleum pratense and vesicular-arbuscular mycorrhizal fungi. New Phytol 115(3):465–469

Dai J, Hu J, Lin X, Yang A, Wang R, Zhang J, Wong MH (2013) Arbuscular mycorrhizal fungal diversity, external mycelium length, and glomalin-related soil protein content in response to long-term fertilizer management. J Soils Sediments 13:1–11

De La Peña E, Echeverría SR, Van Der Putten WH, Freitas H, Moens M (2006) Mechanism of control of root-feeding nematodes by mycorrhizal fungi in the dune grass Ammophila arenaria. New Phytol 169(4):829–840

Declerck S, Risède JM, Rufyikiri G, Delvaux B (2002) Effects of arbuscular mycorrhizal fungi on severity of root rot of bananas caused by Cylindrocladium spathiphylli. Plant Pathol *51*(1):109–115

Diagne N, Ngom M, Djighaly PI, Fall D, Hocher V, Svistoonoff S (2020) Roles of arbuscular mycorrhizal fungi on plant growth and performance: Importance in biotic and abiotic stressed regulation. Diversity 12(10):370

Dighton J, Jansen AE (1991) Atmospheric pollutants and ectomycorrhizae: more questions than answers? Environ Pollut 73(3-4):179–204

Dowarah B, Gill SS, Agarwala N (2021) Arbuscular mycorrhizal fungi in conferring tolerance to biotic stresses in plants. J Plant Growth Regul 41:1429–1444

Drigo B, Pijl AS, Duyts H, Kielak AM, Gamper HA, Houtekamer MJ et al (2010) Shifting carbon flow from roots into associated microbial communities in response to elevated atmospheric CO2. Proc Natl Acad Sci U S A 107(24):10938–10942

Duan X, Neuman DS, Reiber JM, Green CD, Arnold M, Saxton AM et al (1996) Mycorrhizal influence on hydraulic and hormonal factors implicated in the control of stomatal conductance during drought. J Exp Bot 47(303):1541–1550. https://doi.org/10.1093/jxb/47.10.1541

Duckmanton L, Widden P (1994) Effect of ozone on the development of vesicular-arbuscular mycorrhizae in sugar maple saplings. Mycologia 86(2):181–186

Elhindi KM, El-Din SA, Elgorban AM (2017) The impact of arbuscular mycorrhizal fungi in mitigating salt-induced adverse effects in sweet basil (*Ocimum basilicum* L.). Saudi J Biol Sci 24:170–179. https://doi.org/10.1016/j.sjbs.2016.02.010

El-Nashar YI (2017) Response of snapdragon *Antirrhinum majus* L. to blended water irrigation and arbuscular mycorrhizal fungi inoculation: uptake of minerals and leaf water relations. Photosynthetica 55(2):201–209. https://doi.org/10.1007/s11099-016-0650-7

Eom AH, Hartnett DC, Wilson GW (2000) Host plant species effects on arbuscular mycorrhizal fungal communities in tallgrass prairie. Oecologia 122:435–444

Feng Z, Kobayashi K, Ainsworth EA (2008) Impact of elevated ozone concentration on growth, physiology, and yield of wheat (Triticum aestivum L.): a meta-analysis. Glob Change Biol 14(11):2696–2708

Fitter AH, Heinemeyer A, Staddon PL (2000) The impact of elevated CO_2 and global climate change on arbuscular mycorrhizas: a mycocentric approach. New Phytol 147(1):179–187

Fowler D, Unsworth MH (1974) Dry deposition of sulphur dioxide on wheat. Nature 249(5455):389–390

Francis R, Read DJ (1984) Direct transfer of carbon between plants connected by vesicular–arbuscular mycorrhizal mycelium. Nature 307(5946):53–56

Gaffney JS, Streit GE, Spall WD, Hall JH (1987) Beyond acid rain. Do soluble oxidants and organic toxinsinteract with SO2 and NOx to increase ecosystem effects? Environ Sci Technol 21(6):519–524

Gamalero E, Pivato B, Bona E, Copetta A, Avidano L, Lingua G, Berta G (2010) Interactions between a fluorescent pseudomonad, an arbuscular mycorrhizal fungus and a hypovirulent isolate of *Rhizoctonia solani* affect plant growth and root architecture of tomato plants. Plant Biosyst 144(3):582–591

Garg N, Chandel S (2012) Role of arbuscular mycorrhizal (AM) fungi on growth, cadmium uptake, osmolyte, and phytochelatin synthesis in *Cajanus cajan* (L.) Millsp. under NaCl and Cd stresses. J Plant Growth Regul 31(3):292–308. https://doi.org/10.1007/s00344-011-9239-3

Gavito ME, Curtis PS, Mikkelsen TN, Jakobsen I (2000) Atmospheric CO2 and mycorrhiza effects on biomass allocation and nutrient uptake of nodulated pea (Pisum sativum L.) plants. J Exp Bot 51(352):1931–1938

Gavito ME, Olsson PA, Rouhier H, Medinapeñafiel A, Jakobsen I, Bago A (2005) Temperature constraints on the growth and functioning of root organ cultures with arbuscular mycorrhizal fungi. New Phytol 168:179–188. https://doi.org/10.1111/j.1469-8137.2005.01481.x

Gehring CA, Whitham TG (2002) Mycorrhizae-herbivore interactions: population and community consequences. In: Mycorrhizal ecology. Springer, Berlin, pp 295–320

Gelang J, Selldén G, Younis S, Pleijel H (2001) Effects of ozone on biomass, non-structural carbohydrates and nitrogen in spring wheat with artificially manipulated source/sink ratio. Environ Exp Bot 46:155–169

George E, Marschner H, Jakobsen I (1995) Role of arbuscular mycorrhizal fungi in uptake of phosphorus and nitrogen from soil. Crit Rev Biotechnol 15(3-4):257–270

Gholamhoseini M, Ghalavand A, Dolatabadian A, Jamshidi E, Khodaei-Joghan A (2013) Effects of arbuscular mycorrhizal inoculation on growth, yield, nutrient uptake and irrigation water productivity of sunflowers grown under drought stress. Agric Water Manag 117:106–114. https://doi.org/10.1016/j.agwat.2012.11.007

Gómez-Bellot MJ, Ortuño MF, Nortes PA, Vicente-Sánchez J, Banon S, Sánchez-Blanco MJ (2015) Mycorrhizal euonymus plants and reclaimed water: biomass, water status and nutritional responses. Sci Hortic 186:61–69

Graves JD, Watkins NK, Fitter AH, Robinson D, Scrimgeour C (1997) Intraspecific transfer of carbon between plants linked by a common mycorrhizal network. Plant Soil 192(2):153–159

Grime JP, Mackey JML, Hillier SH, Read DJ (1987) Floristic diversity in a model system using experimental microcosms. Nature 328(6129):420–422

Guidi L, Degl'Innocenti E, Soldatini GF (2002) Assimilation of CO2, enzyme activation and photosynthetic electron transport in bean leaves, as affected by high light and ozone. New Phytol 156(3):377–388

Gujre N, Soni A, Rangan L, Tsang DC, Mitra S (2021) Sustainable improvement of soil health utilizing biochar and arbuscular mycorrhizal fungi: a review. Environ Pollut 268:115549

Gunze CMB, Hennessy CMR (1980) Effect of host-applied auxin on development of endomycorrhiza in cowpeas. Trans Br Mycol Soc 74(2):247–251

Häikiö E, Freiwald V, Julkunen-Tiitto R, Beuker E, Holopainen T, Oksanen E (2009) Differences in leaf characteristics between ozone-sensitive and ozone-tolerant hybrid aspen (Populus tremula× Populus tremuloides) clones. Tree Physiol 29(1):53–66

Harrison MJ (1999) Molecular and cellular aspects of the arbuscular mycorrhizal symbiosis. Annu Rev Plant Biol 50(1):361–389

Hart M, Ehret DL, Krumbein A, Leung C, Murch S, Turi C, Franken P (2015) Inoculation with arbuscular mycorrhizal fungi improves the nutritional value of tomatoes. Mycorrhiza 25:359–376

Hasanuzzaman M, Gill SS, Fujita M (2013) Physiological role of nitric oxide in plants grown under adverse environmental conditions. In: Tuteja N, Gill SS (eds) Plant acclimation to environmental stress. Springer, New York, pp 269–322. https://doi.org/10.1007/978-1-4614-5001-6_11

Hashem A, Abd Allah EF, Alqarawi AA, Aldubise A, Egamberdieva D (2015) Arbuscular mycorrhizal fungi enhances salinity tolerance of Panicum turgidum Forssk by altering photosynthetic and antioxidant pathways. J Plant Interact 10(1):230–242

Hashem A, Alqarawi AA, Radhakrishnan R, Al-Arjani AF, Aldehaish HA, Egamberdieva D et al (2018) Arbuscular mycorrhizal fungi regulate the oxidative system, hormones and ionic equilibrium to trigger salt stress tolerance in *Cucumis sativus* L. Saudi J Biol Sci 25(6):1102–1114. https://doi.org/10.1016/j.sjbs.2018.03.009

He L, Li C, Liu R (2017) Indirect interactions between arbuscular mycorrhizal fungi and *Spodoptera exigua* alter photosynthesis and plant endogenous hormones. Mycorrhiza 27:525–535

Heidari M, Karami V (2014) Effects of different mycorrhiza species on grain yield, nutrient uptake and oil content of sunflower under water stress. J Saudi Soc Agric Sci 13(1):9–13

Hibbett DS, Gilbert LB, Donoghue MJ (2000) Evolutionary instability of ectomycorrhizal symbiosis in basidiomycetes. Nature 407(6803):506–508

Hijri M (2016) Analysis of a large dataset form field mycorrhizal inoculation trials on potato showed highly significant increase in yield. Mycorrhiza 2:209–214. https://doi.org/10.1007/s00572-015-0661-4

Hodge A (2000) Microbial ecology of the arbuscular mycorrhiza. FEMS Microbiol Ecol 32:91–96. https://doi.org/10.1111/j.1574-6941.2000.tb00702.x

Hodge A, Fitter AH (2010) Substantial nitrogen acquisition by arbuscular mycorrhizal fungi from organic material has implications for N cycling. Proc Natl Acad Sci U S A 107(31):13754–13759

Hooker JE, Jaizme-Vega M, Atkinson D (1994) Biocontrol of plant pathogens using arbuscular mycorrhizal fungi. In: Gianinazzi S, Schüepp H (eds) Impact of arbuscular mycorrhizas on sustainable agriculture and natural ecosystems. Birkhäuser Verlag, Basel, pp 191–200

Ho, I., & Trappe, J. M. (1984). Effects of ozone exposure on mycorrhiza formation and growth of Festuca arundinacea. Environmental and experimental Botany, 24(1), 71–74.

Impa SM, Nadaradjan S, Jagadish SVK (2012) Drought stress induced reactive oxygen species and anti-oxidants in plants. In: Ahmad P, Prasad MNV (eds) Abiotic stress responses in plants: metabolism, productivity and sustainability. Springer, New York, NY, pp 131–147. https://doi.org/10.1007/978-1-4614-0634-1_7

Insam H, Bååth E, Berreck M, Frostegård Å, Gerzabek MH, Kraft A et al (1999) Responses of the soil microbiota to elevated CO_2 in an artificial tropical ecosystem. J Microbiol Methods 36(1-2):45–54

IPCC (2014) Contribution of working groups I, II and III to the fifth assessment report of the intergovernmental panel on climate change. In: Core Writing Team, Pachauri RK, Meyer LA (eds) Climate change 2014: Synthesis report. IPCC, Geneva, p 151

Ismail IM, Basahi JM, Hassan IA (2014) Gas exchange and chlorophyll fluorescence of pea (Pisum sativum L.) plants in response to ambient ozone at a rural site in Egypt. Sci Total Environ 497:585–593

Jabaji-Hare, S. H., & Stobbs, L. W. (1984). Electron microscopic examination of tomato roots coinfected with Glomus sp. and tobacco mosaic virus. Phytopathology, 74(3), 277–279.

Jaizme-Vega MC (1998) He rn ᴊndez BS, Hern ᴊndez JM. Interaction of arbuscular mycorrhizal fungi and the soil pathogen Fusarium oxysporum f. sp. Cubense on the first stages of micro-propagated Grande Naine banana. Acta Hortic 490:285–295

Jaizme-Vega M, Rodríguez-Romero AS, Hermoso CM, Declerck S (2003) Growth of micro-propagated bananas colonized by root-organ culture produced arbuscular mycorrhizal fungi entrapped in Ca-alginate beads. Plant Soil 254:329–335

Jakobsen I, Smith SE, Smith FA (2003) Function and diversity of arbuscular mycorrhizae in carbon and mineral nutrition. In: Mycorrhizal ecology. Springer, pp 75–92

James TY, Kauff F, Schoch CL, Matheny PB, Hofstetter V, Cox CJ et al (2006) Reconstructing the early evolution of Fungi using a six-gene phylogeny. Nature 443(7113):818–822

Janoušková M, Pavlíková D (2010) Cadmium immobilization in the rhizosphere of arbuscular mycorrhizal plants by the fungal extraradical mycelium. Plant Soil 332:511–520. https://doi.org/10.1007/s11104-010-0317-2

Jeffries P, Gianinazzi S, Perotto S, Turnau K, Barea JM (2003) The contribution of arbuscular mycorrhizal fungi in sustainable maintenance of plant health and soil fertility. Biol Fertil Soils 37:1–16

Jones JT, Haegeman A, Danchin EG, Gaur HS, Helder J, Jones MG et al (2013) Top 10 plant-parasitic nematodes in molecular plant pathology. Mol Plant Pathol 14(9):946–961

Kapoor R, Evelin H, Mathur P, Giri B (2013) Arbuscular mycorrhiza: approaches for abiotic stress tolerance in crop plants for sustainable agriculture. In: Tuteja N, Gill SS (eds) Plant acclimation to environmental stress. Springer, New York, NY, pp 359–401. https://doi.org/10.1007/978-1-4614-5001-6_14

Kaur S, Suseela V (2020) Unraveling arbuscular mycorrhiza-induced changes in plant primary and secondary metabolome. Metabolites 10(8):335

Kayama M, Yamanaka T (2014) Growth characteristics of ectomycorrhizal seedlings of *Quercus glauca, Quercus salicina,* and *Castanopsis cuspidata* planted on acidic soil. Trees 28:569–583

Khaliq A, Perveen S, Alamer KH, Zia Ul Haq M, Rafique Z, Alsudays IM et al (2022) Arbuscular mycorrhizal fungi symbiosis to enhance plant–soil interaction. Sustainability 14(13):7840

Kiers ET, Heijden MGVD (2006) Mutualistic stability in the arbuscular mycorrhizal symbiosis: exploring hypotheses of evolutionary cooperation. Ecology 87(7):1627–1636

Klironomos JN, Ursic M, Rillig M, Allen MF (1998) Interspecific differences in the response of arbuscular mycorrhizal fungi to *Artemisia tridentata* grown under elevated atmospheric CO2. New Phytol 138(4):599–605

Klironomos JN, McCune J, Hart M, Neville J (2000) The influence of arbuscular mycorrhizae on the relationship between plant diversity and productivity. Ecol Lett 3(2):137–141

Klironomos J, Zobel M, Tibbett M, Stock WD, Rillig MC, Parrent JL et al (2011) Forces that structure plant communities: quantifying the importance of the mycorrhizal symbiosis. New Phytol 189(2):366–370

Kohler A, Kuo A, Nagy LG, Morin E, Barry KW, Buscot F et al (2015) Convergent losses of decay mechanisms and rapid turnover of symbiosis genes in mycorrhizal mutualists. Nat Genet 47(4):410–415

Kubikova E, Moore JL, Ownlew BH, Mullen MD, Augé RM (2001) Mycorrhizal impact on osmotic adjustment in *Ocimum basilicum* during a lethal drying episode. J Plant Physiol 158:1227–1230. https://doi.org/10.1078/0176-1617-00441

Kuhn G, Hijri M, Sanders IR (2001) Evidence for the evolution of multiple genomes in arbuscular mycorrhizal fungi. Nature 414(6865):745–748

Kytöviita MM, Le Thiec D, Dizengremel P (2001) Elevated CO2 and ozone reduce nitrogen acquisition by *Pinus halepensis* from its mycorrhizal symbiont. Physiol Plant 111(3):305–312

Lanfranco L, Bonfante P, Genre A (2016) The mutualistic interaction between plants and arbuscular mycorrhizal fungi. Microbiol Spectr 4(6):4–6

Leifheit EF, Veresoglou SD, Lehmann A, Morris EK, Rillig MC (2014) Multiple factors influence the role of arbuscular mycorrhizal fungi in soil aggregation—a meta-analysis. Plant Soil 374(1):523–537

Lerat S, Gauci R, Catford JG, Vierheilig H, Piché Y, Lapointe L (2002) 14C transfer between the spring ephemeral *Erythronium americanum* and sugar maple saplings via arbuscular mycorrhizal fungi in natural stands. Oecologia 132(2):181–187

Lewis J, Thomas RB, Strain BR (1994) Effect of elevated CO_2 on mycorrhizal colonization of loblolly pine (Pinus taeda L.) seedlings. *Plant and Soil 165*(1):81–88

Li XL, George E, Marschner H (1991) Extension of the phosphorus depletion zone in VA-mycorrhizal white clover in a calcareous soil. Plant Soil 136(1):41–48

Li H, Luo N, Zhang LJ, Zhao HM, Li YW, Cai QY et al (2016a) Do arbuscular mycorrhizal fungi affect cadmium uptake kinetics, subcellular distribution and chemical forms in rice? Sci Total Environ 571:1183–1190

Li X, Zeng R, Liao H (2016b) Improving crop nutrient efficiency through root architecture modifications. J Integr Plant Biol 58:193–202. https://doi.org/10.1111/jipb.12434

Liu X, Song Q, Tang Y, Li W, Xu J, Wu J et al (2013) Human health risk assessment of heavy metals in soil–vegetable system: a multi-medium analysis. Sci Total Environ 463–464:530–540. https://doi.org/10.1016/j.scitotenv.2013.06.064

López-Ráez JA, Shirasu K, Foo E (2017) Strigolactones in plant interactions with beneficial and detrimental organisms: the Yin and Yang. Trends Plant Sci 22(6):527–537

Lu F, Lee C, Wang C (2015) The influence of arbuscular mycorrhizal fungi inoculation on yam (*Dioscorea* spp.) tuber weights and secondary metabolite content. *Peer J.* 3:12–66. https://doi.org/10.7717/peerj.1266

Ludwikow A, Sadowski J (2008) Gene networks in plant ozone stress response and tolerance. *Journal of Integrative Plant Biology 50*(10):1256–1267

Luginbuehl LH, Oldroyd GE (2017) Understanding the arbuscule at the heart of endomycorrhizal symbiosis in plants. Curr Biol 27(17):R952–R963

Mahoney MJ, Chevone BI, Skelly JM, Moore LD (1985) Influence of mycorrhizae on the growth of loblolly pine seedlings exposed to ozone and sulfur dioxide. Phytopathology 75(6):679–682

Majernik O, Mansfield TA (1970) Direct effect of SO_2 pollution on the degree of opening of stomata. Nature 227(5256):377–378

Mansfeld-Giese K, Larsen J, Bødker L (2002) Bacterial populations associated with mycelium of the arbuscular mycorrhizal fungus *Glomus intraradices*. FEMS Microbiol Ecol 41(2):133–140

McGrath JM, Betzelberger AM, Wang S, Shook E, Zhu XG, Long SP, Ainsworth EA (2015) An analysis of ozone damage to historical maize and soybean yields in the United States. Proc Natl Acad Sci U S A 112(46):14390–14395

Mena-Violante HG, Ocampo-Jimenez O, Dendooven L, Martinez-Soto G, Gonzalez-Castafeda J, Davies FT et al (2006) Arbuscular mycorrhizal fungi enhance fruit growth and quality of chile ancho *Capsicum annuum* L. cv San Luis plants exposed to drought. Mycorrhiza 16:261–267. https://doi.org/10.1007/s00572-006-0043-z

Mikkelsen BL, Rosendahl S, Jakobsen I (2008) Underground resource allocation between individual networks of mycorrhizal fungi. New Phytol 180(4):890–898

Mills G, Wagg S, Harmens H (2013). Ozone pollution: impacts on ecosystem services and biodiversity. NERC/Centre for Ecology & Hydrology

Miransari M (2017) Arbuscular mycorrhizal fungi and heavy metal tolerance in plants. In: Wu QS (ed) Arbuscular mycorrhizas and stress tolerance of plants. Springer Nature, Singapore, pp 174–161. https://doi.org/10.1007/978-3-319-68867-1_4

Moghadam HRT (2016) Application of super absorbent polymer and ascorbic acid to mitigate deleterious effects of cadmium in wheat. Pesqui Agropecu Trop 6(1):9–18. https://doi.org/10.1590/1983-40632016v4638946

Moradtalab N, Roghieh H, Nasser A, Tobias EH, Günter N (2019) Silicon and the association with an arbuscular-mycorrhizal fungus (*Rhizophagus clarus*) mitigate the adverse effects of drought stress on strawberry. Agronomy 9:41. https://doi.org/10.3390/agronomy9010041

Morgan PB, Ainsworth EA, Long SP (2003) How does elevated ozone impact soybean? A meta-analysis of photosynthesis, growth and yield. Plant Cell Environ 26(8):1317–1328

Morison JIL, Lawlor DW (1999) Interactions between increasing CO2 concentration and temperature on plant growth. Plant Cell Environ 22(6):659–682

Nagahashi G, Douds DD (2000) Partial separation of root exudate components and their effects upon the growth of germinated spores of AM fungi. Mycol Res 104(12):1453–1464

Nagy R, Drissner D, Amrhein N, Jakobsen I, Bucher M (2009) Mycorrhizal phosphate uptake pathway in tomato is phosphorus-repressible and transcriptionally regulated. New Phytol 181(4):950–959

Navarro JM, Pérez-Tornero O, Morte A (2014) Alleviation of salt stress in citrus seedlings inoculated with arbuscular mycorrhizal fungi depends on the rootstock salt tolerance. J Plant Physiol 171(1):76–85

NOVA (2021) NOVA research news, Carbon dioxide peaks near 420 parts per million at Mauna Loa observatory. https://research.noaa.gov/article/ArtMID/587/ArticleID/2764/Coronavirus-response-barely-slows-rising-carbon-dioxide

Noyes RD (1980) The comparative effects of sulfur dioxide on photosynthesis and translocation in bean. Physiol Plant Pathol 16(1):73–79

Obermeier WA, Lehnert LW, Kammann CI, Müller C, Grünhage L, Luterbacher J et al (2017) Reduced CO_2 fertilization effect in temperate C3 grasslands under more extreme weather conditions. Nat Climate Change 7(2):137–141

Parniske M (2008) Arbuscular mycorrhiza: the mother of plant root endosymbiosis. Nat Rev Microbiol 6(10):763–775

Pellegrini E (2014) PSII photochemistry is the primary target of oxidative stress imposed by ozone in Tilia americana. Urban For Urban Greening 13(1):94–102

Plassard C, Dell B (2010) Phosphorus nutrition of mycorrhizal trees. Tree Physiol 30(9):1129–1139

Plassard C, Becquer A, Garcia K (2019) Phosphorus transport in mycorrhiza: how far are we? Trends Plant Sci 24(9):794–801

Poorter H (1993) Interspecific variation in the growth response of plants to an elevated ambient CO 2 concentration. In: CO_2 and biosphere. Springer, Dordrecht, pp 77–98

Prasad A, Kumar S, Khaliq A, Pandey A (2011) Heavy metals and arbuscular mycorrhizal (AM) fungi can alter the yield and chemical composition of volatile oil of sweet basil (*Ocimum basilicum* L.). Biol Fertil Soils 47:853–861

Pringle A, Bever JD, Gardes M, Parrent JL, Rillig MC, Klironomos JN (2009) Mycorrhizal symbiosis and plant invasions. Annu Rev Ecol Evol Syst 40:699–715

Punamiya P, Datta R, Sarkar D, Barber S, Patel M, Da P (2010) Symbiotic role of *Glomus mosseae* in phytoextraction of lead in vetiver grass *Chrysopogon zizanioides* L. J Hazard Mater 177:465–474. https://doi.org/10.1016/j.jhazmat.2009.12.056

Qiao, X., Bei, S., Li, C., Dong, Y., Li, H., Christie, P., ... & Zhang, J. (2015). Enhancement of faba bean competitive ability by arbuscular mycorrhizal fungi is highly correlated with dynamic nutrient acquisition by competing wheat. Scientific Reports, 5(1), 8122.

Rajtor M, Piotrowska-Seget Z (2016) Prospects for arbuscular mycorrhizal fungi (AMF) to assist in phytoremediation of soil hydrocarbon contaminants. Chemosphere 162:105–116

Raklami A, Bechtaoui N, Tahiri AI, Anli M, Meddich A, Oufdou K (2019) Use of rhizobacteria and mycorrhizae consortium in the open field as a strategy for improving crop nutrition, productivity and soil fertility. Front Microbiol 10:1106

Rani B (2016) Effect of arbuscular mycorrhiza fungi on biochemical parameters in wheat *Triticum aestivum* L. under drought conditions. *Doctoral dissertation*,. CCSHAU, Hisar

Read DJ (1991) Mycorrhizas in ecosystems. Experientia 47(4):376–391

Redecker D (2002) Molecular identification and phylogeny of arbuscular mycorrhizal fungi. In: Diversity and integration in mycorrhizas: proceedings of the 3rd international conference on mycorrhizas (ICOM3) Adelaide, Australia, 8–13 July 2001. Springer Netherlands, pp 67–73

Redecker D, Schüßler A, Stockinger H, Stürmer SL, Morton JB, Walker C (2013) An evidence-based consensus for the classification of arbuscular mycorrhizal fungi (Glomeromycota). Mycorrhiza *23*:515–531

Requena N, Serrano E, Ocón A, Breuninger M (2007) Plant signals and fungal perception during arbuscular mycorrhiza establishment. Phytochemistry 68(1):33–40

Rice PM, Pye LH, Boldi R, O'Loughlin J, Tourangeau PC, Gordon CC (1978) The effects of 'low level SO2'exposure sulphur accumulation and various plant life responses of some major grassland species on the ZAPS sites. In: Bioenvironmental impact of a coal. pp 494–591
Rizvi R, Singh G, Safiuddin, Ansari RA, Tiyagi SA, Mahmood I (2015) Sustainable management of root-knot disease of tomato by neem cake and *Glomus fasciculatum*. Cogent Food Agric 1:1008859
Rogers HH, Prior SA, Runion GB, Mitchell RJ (1995) Root to shoot ratio of crops as influenced by CO2. Plant Soil 187(2):229–248
Rosendahl S (2008) Communities, populations and individuals of arbuscular mycorrhizal fungi. New Phytol 178(2):253–266
Rouphael Y, Franken P, Schneider C, Schwarz D, Giovannetti M, Agnolucci M (2015) Arbuscular mycorrhizal fungi act as bio-stimulants in horticultural crops. Sci Hort 196:91–108. https://doi.org/10.1016/j.scienta.2015.09.002
Ruiz-Lozano JM, Aroca R, Zamarreño ÁM, Molina S, Andreo-Jiménez B, Porcel R et al (2015) Arbuscular mycorrhizal symbiosis induces *strigolactone* biosynthesis under drought and improves drought tolerance in lettuce and tomato. Plant Cell Environ 39(2):441–452. https://doi.org/10.1111/pce.12631
Ruiz-Sánchez M, Aroca R, Muñoz Y, Polón R, Ruiz-Lozano JM (2010) The arbuscular mycorrhizal symbiosis enhances the photosynthetic efficiency and the antioxidative response of rice plants subjected to drought stress. J Plant Physiol 167:862–869. https://doi.org/10.1016/j.jplph.2010.01.018
Sabia E, Claps S, Morone G, Bruno A, Sepe L, Aleandri R (2015) Field inoculation of arbuscular mycorrhiza on maize (*Zea mays* L.) under low inputs: preliminary study on quantitative and qualitative aspects. Italian J Agron 10(1):30–33
Sadhana B (2014) Arbuscular Mycorrhizal Fungi (AMF) as a biofertilizer-a review. Int J Curr Microbiol Appl Sci 3(4):384–400
Santander C, Sanhueza M, Olave J, Borie F, Valentine C, Cornejo P (2019) Arbuscular mycorrhizal colonization promotes the tolerance to salt stress in lettuce plants through an efficient modification of ionic balance. J Soil Sci Plant Nutr 19(2):321–331. https://doi.org/10.1007/s42729-019-00032-z
Sbrana C, Avio L, Giovannetti M (2014) Beneficial mycorrhizal symbionts affecting the production of health-promoting phytochemicals. Electrophoresis 35:1535–1546. https://doi.org/10.1002/elps.201300568
Schüßler A, Schwarzott D, Walker C (2001) A new fungal phylum, the Glomeromycota: phylogeny and evolution. Mycol Res 105(12):1413–1421
Shafer SR, Schoeneberger MM (1991) Mycorrhizal mediation of plant-response to atmospheric change - air-quality concepts and research considerations. Environ. Pollut. 73:163e177
Sharma S, Prasad R, Varma A, Sharma AK (2017) Glycoprotein associated with *Funneliformis coronatum, Gigaspora margarita* and *Acaulospora scrobiculata* suppress the plant pathogens in vitro. Asian J Plant Pathol 11(4):199–202. https://doi.org/10.3923/ajppaj.2017.199.202
Shen H, Christie P, Li X (2006) Uptake of zinc, cadmium and phosphorus by arbuscular mycorrhizal maize (*Zea mays*, L.) from a low available phosphorus calcareous soil spiked with zinc and cadmium. Environ Geochem Health 28:111. https://doi.org/10.1007/s10653-005-9020-2
Sheng M, Tang M, Zhang F, Huang Y (2011) Influence of arbuscular mycorrhiza on organic solutes in maize leaves under salt stress. Mycorrhiza 21:423–430. https://doi.org/10.1007/s00572-010-0353-z
Smith SE, Read DJ (1997) Mycorrhizal symbiosis. Aca-demic Press Ltd., London
Smith SE, Read DJ (2008) Mycorrhizal symbiosis. Cambridge, UK, Academic Press Ltd
Smith SE, Read DJ (2010) Mycorrhizal symbiosis. Academic press
Smith FA, Smith SE (1997) Tansley review no. 96 structural diversity in (vesicular)–arbuscular mycorrhizal symbiosis. New Phytol 137(3):373–388
Smith SE, Smith FA (2012) Fresh perspectives on the roles of arbuscular mycorrhizal fungi in plant nutrition and growth. Mycologia 104(1):1–13

Smith SE, Smith FA, Jakobsen I (2003) Mycorrhizal fungi can dominate phosphate supply to plants irrespective of growth responses. Plant Physiol 133(1):16–20
Solanki MK, Kashyap PL, Ansari RA, Kumari B (2020) Microbiomes and plant health: panoply and their applications. Academic Press, London. https://www.elsevier.com/books/microbiomes-and-plant-health/solanki/978-0-12-819715-8
Song Z, Bi Y, Zhang J, Gong Y, Yang H (2020) Arbuscular mycorrhizal fungi promote the growth of plants in the mining associated clay. Sci Rep 10(1):2663
Sreenivasa MN, Chavhan T, Kulkarni S (2019) Influence of inoculation of Efficient AM fungi (*Glomus fasciculatum*) against wilt of tomato caused by *Sclerotium rolfsii*
Sreeramulu KR, Onkarappa T, Swamy HN (1998) Biocontrol of damping off and black shank disease in tobacco nursery. Tob Res *24*:1–4
Staddon PL, Fitter AH (1998) Does elevated atmospheric carbon dioxide affect arbuscular mycorrhizas? Trends Ecol Evol 13(11):455–458
Staddon PL, Reinsch S, Olsson PA, Ambus P, Lüscher A, Jakobsen I (2014) A decade of free-air CO_2 enrichment increased the carbon throughput in a grass-clover ecosystem but did not drastically change carbon allocation patterns. Funct Ecol 28(2):538–545
Sumbul A, Mahmood I, Rizvi R, Safiuddin, Ansari RA (2017) Mycorrhiza: An alliance for the nutrient management in plants. In: Kumar V (ed) Probiotics in agroecosystem. Springer Nature, Singapore, pp 371–386
Talaat NB, Shawky BT (2014) Protective effects of arbuscular mycorrhizal fungi on wheat (*Triticum aestivum* L.) plants exposed to salinity. Environ Exp Bot 98:20–31. https://doi.org/10.1016/j.envexpbot.2013.10.005
Tanaka Y, Yano K (2005) Nitrogen delivery to maize via mycorrhizal hyphae depends on the form of N supplied. Plant Cell Environ 28(10):1247–1254
Ueno AC, Gundel PE, Omacini M, Ghersa CM, Bush LP, Martínez-Ghersa MA (2016) Mutualism effectiveness of a fungal endophyte in an annual grass is impaired by ozone. Funct Ecol 30(2):226–234
Van Der Heijden MG, Horton TR (2009) Socialism in soil? The importance of mycorrhizal fungal networks for facilitation in natural ecosystems. J Ecol 97(6):1139–1150
van Der Heijden MG, Martin FM, Selosse MA, Sanders IR (2015) Mycorrhizal ecology and evolution: the past, the present, and the future. New Phytol 205(4):1406–1423
Veresoglou SD, Chen B, Rillig MC (2012) Arbuscular mycorrhiza and soil nitrogen cycling. Soil Biol Biochem 46:53–62
Vlček V, Pohanka M (2020) Glomalin–an interesting protein part of the soil organic matter. Soil Water Res 15(2):67–74
Wagg C, Veiga R, van der Heijden MG (2015) Facilitation and antagonism in mycorrhizal networks. In: Horton TR (ed) Mycorrhizal networks. Springer, Dordrecht, pp 203–226
Wang DX, Lu YQ, He XL (2010) Effects of AM fungi on growth and physiological characters of Atractylodes Macrocephala under different P-applied levels. Acta Bot Bor Occi Sin 30(1):136–142
Wang M-Y, Ren-Xue X, Qiang-Sheng W et al (2007) Influence of arbuscular mycorrhizal fungi on microbes and enzymes of soils from different cultivated densities of red clover. Ann Microbiol 57:1–7. https://doi.org/10.1007/BF03175042
Wang B, Yeun LH, Xue JY, Liu Y, Ané JM, Qiu YL (2010) Presence of three mycorrhizal genes in the common ancestor of land plants suggests a key role of mycorrhizas in the colonization of land by plants. New Phytol 186(2):514–525
Wang S, Feng Z, Wang X, Gong W (2011) Arbuscular mycorrhizal fungi alter the response of growth and nutrient uptake of snap bean (Phaseolus vulgaris L.) to O3. J Environ Sci 23(6):968–974
Wang SG, Diao XJ, Li YW, Ma LM (2015) Effect of Glomus aggregatum on photosynthetic function of snap bean in response to elevated ozone. J Agric Sci 153(5):837–852
Wang S, Augé RM, Toler HD (2017) Arbuscular mycorrhiza formation and its function under elevated atmospheric O_3: a meta-analysis. Environ Pollut 226:104–117

Wang Y, Wang M, Li Y, Wu A, Huang J (2018) Effects of arbuscular mycorrhizal fungi on growth and nitrogen uptake of *Chrysanthemum morifolium* under salt stress. PLoS One 13(4):e0196408. https://doi.org/10.1371/journal.pone.0196408
Wehner J, Antunes PM, Powell JR, Mazukatow J, Rillig MC (2010) Plant pathogen protection by arbuscular mycorrhizas: a role for fungal diversity? Pedobiologia 53(3):197–201
Wilkinson S, Davies WJ (2010) Drought, ozone, ABA and ethylene: new insights from cell to plant to community. Plant Cell Environ 33(4):510–525
Wu QS, Cao MQ, Zou YN, He XH (2014) Direct and indirect effects of glomalin, mycorrhizal hyphae and roots on aggregate stability in rhizosphere of trifoliate orange. Sci Rep 4(1):5823
Xavier LJ, Boyetchko SM (2004) Arbuscular mycorrhizal fungi in plant disease control. In: Fungal biotechnology in agricultural, food, and environmental applications. pp 183–194
Yin R, Hao Z, Zhou X, Wu H, Feng Z, Yuan X, Chen B (2022) Ozone does not diminish the beneficial effects of arbuscular mycorrhizas on Medicago sativa L. in a low phosphorus soil. Mycorrhiza 32:33–43
Yooyongwech S, Samphumphuang T, Tisarum R, Theerawitaya C, Chaum S (2016) Arbuscular mycorrhizal fungi (AMF) improved water deficit tolerance in two different sweet potato genotypes involves osmotic adjustments *via* soluble sugar and free proline. Sci Hort 198:107–117. https://doi.org/10.1016/j.scienta.2015.11.002
Yousaf B, Liu G, Wang R, Imtiaz M, Zia-ur-Rehman M, Munir MAM et al (2016) Bioavailability evaluation, uptake of heavy metals and potential health risks *via* dietary exposure in urban-industrial areas. Environ Sci Pollut Res 23:22443–22453. https://doi.org/10.1007/s11356-016-7449-8
Zangaro W, Nishidate FR, Vandresen J, Andrade G, Nogueira MA (2007) Root mycorrhizal colonization and plant responsiveness are related to root plasticity, soil fertility and successional status of native woody species in southern Brazil. J Trop Ecol 23(1):53–62
Zeng L, Li J, Liu J, Wang M (2014) Effects of arbuscular mycorrhizal (AM) fungi on citrus fruit quality under nature conditions. Southwest China J Agric Sci 27(5):2101–2105
Zhang X, Li W, Fang M, Jixian Y, Meng S (2016) Effects of arbuscular mycorrhizal fungi inoculation on carbon and nitrogen distribution and grain yield and nutritional quality in rice (*Oryza sativa* L.). J Sci Food Agric 97:2919–2925. https://doi.org/10.1002/jsfa.8129
Zhu XC, Song FB, Xu HW (2010) Arbuscular mycorrhizae improve low temperature stress in maize *via* alterations in host water status and photosynthesis. Plant Soil 331:129–137. https://doi.org/10.1007/s11104-009-0239-z
Zhu X, Cao Q, Sun L, Yang X, Yang W, Zhang H (2018) Stomatal conductance and morphology of arbuscular mycorrhizal wheat plants response to elevated CO2 and NaCl stress. Front Plant Sci 9:1363

Propagation of Mycorrhizal Fungi and Their Strength in Plant Health Ameliorations

4

Suresh Patil, Rizwan Ali Ansari, Chennappa Gurikar, S. Bhavan, and K. Hariprasad

Abstract

Arbuscular mycorrhizae (AM) fungi interact with plant roots in a symbiotic manner. These fungi serve well in the maintenance of soil fertility and crop health improvement. The current chapter entails soil sample collection, isolation, characterization, and enumeration of AM fungi and their mass multiplication. These procedures necessitate a skilled hand and high level of knowledge, and thus not much common. The current state of the art in inoculation, root colonization assessment, and massive propagation have been discussed. The techniques that are used for AM fungi production are slowly gaining attention and being popularized nowadays. Molecular approaches for the identification of any microorganisms, including AM fungi, are also gaining significant attention from the researchers. Some techniques like fluorescent antibody methodology, enzyme-linked immunosorbent assay, and molecular quantification through PCR, qPCR technique, and Illumina MiSeq high-throughput DNA sequence-based analysis, have also been explained.

S. Patil (✉) · S. Bhavan · K. Hariprasad
Department of Plant Pathology, University of Agricultural and Horticultural Sciences, Shivamogga, Karnataka, India

R. A. Ansari
Department of Plant Protection, Faculty of Agricultural Sciences, Aligarh Muslim University, Aligarh, India

C. Gurikar
Department of Food Technology, MS Ramaiah University of Applied Sciences, Bangalore, Karnataka, India

R. A. Ansari et al. (eds.), *Mycorrhizal Symbiosis and Agroecosystem Restoration*,
https://doi.org/10.1007/978-981-99-5030-0_4

Keywords
AM fungi · Plant health maintenance · Organic agriculture · Induced systemic resistance

4.1 Introduction

A mutual alliance of host plants and a special group of fungi form mycorrhizae. Mycorrhizal plants are divided into two key groups as ectomycorrhiza (ECM) and endomycorrhiza (ENM), and are bifurcated based on their form and anatomy (Subba Rao 1977). ECM is mostly found in coniferous trees, and these trees cannot survive without the help of mycorrhizal fungi (Policelli et al. 2020). ENM are symbiotic fungi that prevail in plant and soils of cultivated crops and some important weeds (Santos et al. 2013; Hayman 1980; Trappe 1982; Maitra et al. 2020).

AM fungi form endomycorrhizae and have vesicles, a food storage organ, and arbuscules, a nutrient transfer site (Sylvia 1998). AM fungi are long-lived structures that persist in the soil as spores or in connection with some roots. They provide several benefits to their host, like nutrient absorption and enhanced stress tolerance (Smith and Read 2008; Ansari et al. 2020a; Solanki et al. 2020; Ansari and Mahmood 2019a). The quantitative and qualitative occurrence of AM fungi under different agroclimatic conditions aids in root colonization, spore survival, and their judicious dissemination (Daniels 1984). A better understanding of these aspects is required to maximize crop production and productivity through the use of AM fungi. Different mechanisms implicated in the biosynthesis of plant growth enhancer molecules are still poorly understood, which contribute significantly for the maximization of crop biomass (Ansari and Mahmood 2017a, b, 2019b; Ansari et al. 2020b). AM fungi enhance the movement and uptake of major and micronutrients and water, which reduces water stress (Zhao et al. 2015). Plants' ability to withstand diseases is developed due to the involvement of plant symbionts that enhance crop biomass and development process; also, nutrient absorption and fertilizer use efficiency of the plants are improved (Sood et al. 2020). AM fungi increase drought resistance levels, plants' ability against other stress, and also resistance level against plant pathogens (Begum et al. 2019; Nemec and Myhre 1984; Pozo and Azcón-Aguilar 2007). Various plant metabolites such as plant growth regulators, enzymes, siderophores, and antibiotics are used to stimulate crop growth and biomass (Sood et al. 2020). The application of AM fungi in sustainable cropping systems has been an emerging spot of research. Researchers are putting more stress on finding out the suitability of AM fungi that can withstand harsh environments and produce more yields in a nutrient-deprived ecosystem. Our understanding of AM fungi as bioprotectants or symbionts, how they work, and why they are important in plant health management in sustainable agriculture needs rigorous studies (Kumar 2013). The current section emphasizes on the basic work of AM fungi such as isolation, identification, and their propagation. The chapter also appraises the readers on different techniques being used in the maximization of AM fungal spore production and their consistent supply.

4.2 Morphological Features of AM Fungi

Mycelium is coarse, coenocytic, and grows intracellularly, which produces varying sizes of spores (Drijber and McPherson 2021). Mycorrhizal fungi are usually hyaline to dark in color and texture mostly smooth to highly echinulate (Vaingankar 2012). After penetration, the hyphae produce branches constantly to form a dichotomous tree-like assembly known as arbuscules, having a short life span of 15 days that later degenerate forming new arbuscules in new areas (Hata et al. 2010; Davison et al. 2015; Roth et al. 2019).

4.3 Collection of Soil Samples

The collection techniques of soil samples are also crucial in the isolation of AM fungi for their application. The population and beneficial aspects of AM fungi are directly related to soil type and geographical conditions. Other factors such as microbial activities, location, and sampling depth are also important in isolating the AM fungi to determine their efficiency. Many reports state that almost all microbial activities take place near the rhizosphere (Chennappa et al. 2014a, b).

4.4 Isolation and Characterization

The selection and production of inoculants begin with the proper isolation of AM fungi from the various ecosystem. For the isolation of AM fungi, various approaches are utilized to produce it on a large-scale using soil and inert substrate. Efficient AM fungi can be found in natural conditions, which are used in agroecosystem restoration. However, soil inoculum can be erroneous unless specific information on propagule volume, variety, and infectivity is provided. Spores collected from the earth are generally used as starters for making crude inoculum (Habte and Osorio 2001).

4.4.1 AM Fungi Sampling

The sampling approach varies with a natural and an agricultural ecosystem.

1. Depending on the site and varied soil usage, 3–5 sampling zones are chosen.
2. Five to eight sampling spots with a distance of 50 m, randomly in each sampling region, are selected.
3. At each sampling site, two concentric circles with a radius of 3 and 6 m are drawn.
4. A total of 12 soil subsamples are taken out with the help of soil core sampler.
5. The dirt from the 12 subsamples are manually, homogenously shredded and combined.
6. Approximately 2 kg of sample is collected and kept in a labeled plastic bag at 4 °C until further processing.

4.5 Enumeration of AM Fungal Spore Population

Generally, a commonly used method (wet sieving) is practiced to sum up the AM fungi, as per standard protocol (Gardemann and Nicolson 1963). In a nutshell, 50 g of dirt soil in 200 ml of water is dissolved. To remove the larger organic matter particles included in the samples, the suspension is decanted using a 710-micron sieve. Then after, the residue is re-suspended in more water and sieved once more. The suspension is allowed to pass through the filter paper, agitate, and keep it for a while. Heavier particles are settled down, and then the supernatant is decanted through a 250-micron mesh. A 45-micron nylon mesh is used to catch spores suspended in the solution, which is stored in a labelled Petri plate, and the quantity of spores is counted using a binocular stereo microscope. The spores that have become shriveled and dried are removed. Only living spores are counted based on surface conditions, shape, and spore content assessment (Eom et al. 2001).

4.5.1 Spore Extraction

By following the wet sieving and decanting method, the spores are extracted (Gardemann and Nicolson 1963; Sieverding and Howeler 1983). The following techniques are used to find out and isolate the mycorrhizal fungi from soil.

1. Wet sieving and decanting technique (Varma et al. 1991).
2. Sucrose centrifugation technique (Daniels and Skipper 1982).
3. Adhesion-flotation technique (Sutton and Barron 1972).
4. Extraction of AM fungi spores by using capillary rise method (Shamini and Amutha 2014).

4.6 Morphological Identification of AM Fungi

More than twenty distinct unique features (morphological) are employed for the characterization of plant symbionts (INVAM 2019). Extraradical (out-root) spore color, transparency, size, and shape; the shape of spore attachment to subtending hypha; number, thickness, density, elasticity or fragility, and color in Melzer's reagent of layers of spore walls and subtending hypha; presence or absence of spore wall, etc., are some of the parameters used to determine the parameters of mycorrhization and the type of mycorrhization (Kariman et al. 2005).

Biotechnological approaches in the characterization and identification of AM fungi have improved the qualitative approach of mycorrhizological studies.

4.7 Spore Characterization

Spores are collected from rhizosphere samples collected, mounted with polyvinyl lacto-glycerol, and examined under a compound microscope (100–1000). Bibliographies can be used for the morphological identification of the spores (Almeida and Schenck 1990; Morton 1996; Weickel et al. 1997; Walker and Vestberg 1998; Rodrigues and Muthkumar 2009; Oehl et al. 2010), and also the culture database established by INVAM is frequently used.

Trappe (1982) used synoptic keys of zygomycetes to identify spores of common AM fungi, while Hall (1988) and Abbott and Robson (1982) used a photographic slide collection to illustrate some aspects of endogonaceae (Rodrigues and Muthkumar 2009).

4.7.1 Visualization of Mycorrhizal Fungi

Staining is a relatively easy and inexpensive method for visualizing mycorrhizal fungi in roots (Phillips and Hayman 1970).

4.7.2 Protocol as Described by Phillips and Hayman (1970)

- Pick roots that are smaller than 2 mm in diameter and rinse them.
- In a beaker, combine the roots and 10 mL of 10% KOH. Incubate it in an 80 °C water bath for 15–120 min.
- Take a beaker out of the water bath, add a drop of 30% H_2O_2, and then incubate for 10 minutes at room temperature.
- In a Petri dish, place the roots. Rinse the roots for 15 s with tap water.
- Combine the roots and 10 mL of 10% HCl in a beaker (enough to cover roots).
- Incubate it for 5 min at room temperature.
- Directly from the 10% HCl solution, place roots in a beaker containing 10 mL 0.05 percent aniline blue/lactic acid solution. After that, incubate the beaker for 30 min in an 80 °C water bath.
- Place the roots in a 10-mL lactic acid solution containing 85 percent lactic acid after removing them from the aniline blue/lactic acid solution. At room temperature, incubate for 5 min.
- Store the roots in lactic acid at an 85 percent concentration.
- Examine the dyed roots under a microscope after they have been wet mounted.
- Although the fungus appears blue, the root tissue is not.

4.8 Molecular Approaches in the Identification of Mycorrhizal Fungi

The appropriate identification of mycorrhizal fungi through molecular approaches has revolutionized the researchers ability to study and understand these essential plant symbionts. Molecular techniques provide higher resolution, accuracy, and speed as compared to traditional methods. Here are some common molecular techniques used in the identification of mycorrhizal fungi:

4.8.1 Advancement in Identification of Mycorrhizal Fungi

Morphological identification is still considered an important tool, but some recent advanced techniques have augmented the identification of different AM fungi from the heterogeneous environment (Declerck et al. 1998).

4.8.2 Fluorescent Antibody Technique (FAT)

Aldwell (1983) used FAT techniques to treat mycorrhizal fungi, but the antisera had low specificity because the antigens were made from crushed spores, which contained a lot of non-specific chitinaceous spore wall material, and were given as a single intramuscular injection with no adjuvant or boosters. Mampaso and Wilson (1983) grounded, centrifuged, and sonicated germ tube hyphae and auxiliary cells gathered from spores germinated between millipore filters before producing antisera by subcutaneous injection with Freund's Complete Adjuvant (FCA) and intravenous boosters. They were able to identify between *Eucalyptus decipiens*, *Acaulospora laevis*, and other non-mycorrhizal mucoraceous fungi using similar techniques.

Aldwell (1983) devised an antisera technique for washing mycorrhizal hyphae off, which later became target antigens. He further narrated important characters that are used nowadays to differentiate between *Glomus, Gigaspora, Acaulospora,* and *Sclerocystis*, as well as the non-mycorrhizal fungi Pezizella ericae, *Mortierella wolfii, Mucor hiemalis*, and *Rhizopus oryzae*, through these approaches.

4.8.3 Enzyme-Linked Immunosorbent Assay (ELISA)

Serological techniques, particularly the ELISA, hold great promise as taxonomic tools for the mycorrhizal fungi. Several immunochemical approaches to characterizing and detecting endo- and ectomycorrhizal fungi have recently been developed (Wright et al. 1987; Cleyet-Marel et al. 1990; Perotto et al. 1992).

Serological techniques are dependent on the production of highly specific antibodies in animals when they are inoculated with the chemicals contained in or produced by microorganisms. If antibody and antigen that elicited its formation are incubated together, they will react. A wide range of assays have been established to detect the presence of either the antibody or the antigen. Of these tests, the radioimmunoassay (RIA) and ELISA are important.

Following Aldwell's failure to distinguish between species within a genus using the FAT, he chose to study VAM using the ELISA rather than the radioimmunoassay (RIA). The RIA has not yet been used to identify mycorrhizal fungi; it is unlikely to be widely used due to a lack of equipment. Aldwell (1983) used an ELISA and antisera made using processes similar to those used to make antisera for the FAT.

The development of more specific antisera against ectomycorrhizal fungi than those raised by Seviour et al. (1973) and the use of the ELISA could provide a more rapid and more reliable technique for identifying the fungal symbiont in mycorrhizas collected than those currently used (Schenck 1982). These techniques could also be useful for confirming the presence of a desired fungal symbiont on seedlings raised in the nursery.

4.8.4 Recent Advances in the Identification of AM Fungi

Morphological, biochemical, physiological, and many more markers are importantly used for the identification of AM fungi (Diédhiou et al. 2014; Edgar (2010), Edgar (2013), Edgar (2018). However, we shall only discuss here the molecular techniques as these are the latest, robust, and trustable techniques that are being used by different workers. In comparison with other approaches, the technology behind genetic markers has improved very quickly. Markers that are based on morphology and biochemistry are being phased out in favor of DNA-based approaches of increasing complexity.

4.8.5 RAPD: Random Amplified Polymorphic DNA

This technique depends on the synthesis of the DNA containing nine or ten bases that hybridize to chromosomal DNA at a very low annealing temperature (Demeke et al. 1992). The amplification is closely followed by agarose gel electrophoresis that produces a band pattern that is unique to the AM fungus. RAPD is an effective diagnostic technique for assessing isolates of certain species and potentially distinguishing different taxa (Kruger et al. 2009; Wyss and Bonfante 1993). This technique has been a popular way to discern *Glomus, Gigaspora,* and *Acaulospora* from a mixed population (Gomez-Leyva et al. 2008).

RAPD offers all the advantages of a PCR-based marker, in addition to the fact that primers are commercially available and conceptual comprehension of the sequence of the target DNA is not required. Besides, screening a large number of loci is facilitated by RAPD. The use of RAPD markers is influenced by the presence of paralogous PCR products and by the reduced reproductive potential resulting from a lower temperature, which is required for annealing employed in PCR amplification.

4.8.6 Restriction Fragment Length Polymorphism (PCR RFLP)

Botstein et al. (1980) used for the first the RFLP markers. The RFLP probe employs endonucleases to break DNA into parts of varying sizes, which are then separated by electrophoresis. The technique takes advantage of naturally occurring genetic variations in an organism's DNA sequence that can result in alterations in DNA pattern in those fragments that are produced by restriction enzyme digestion (Marshall et al. 1999).

PCR-RFLP was utilized efficiently to differentiate AM fungi species from spore-isolated DNA (Sanders et al. 1996). However, this approach may induce polymorphism in non-target organisms when applied to field samples. Using *Hind*I endonuclease, AM fungi permitted discrimination between *F. mosseae* and other related species.

4.8.7 T-RFLP: Terminal Restriction Fragment Length Polymorphism

T-RFLP is a unique technique that is used to analyze and compare microbial communities. It is a fingerprinting method that involves cutting of DNA from the microbial sample using restriction enzymes. The restriction enzymes break the DNA at a very specific site. The electrophoresis is then used to separate the length and size of DNA fragments. T-RFLP investigation rests on the digestion of restriction enzymes that are fluorescently labelled products of PCR. Electrophoresis is a method for separating the digests, which are then detected by a sequencer that is automatically operated. Outcomes reflect a picture of some microorganisms in the given communities (Mummey and Rillig 2006). For T-RFLP analysis, there are some important aspects that need to be used meticulously such as primers, target site, and restriction endonucleases. For the identification of specific fungi from the microbial community, internal transcribed spacer (ITS), small subunit (SSU) or large subunit (LSU) is utilized. In the case of AM fungi, ITS4/ITS5 are utilized for ITS amplification (Lekberg et al. 2018); AM1/NS31 are specifically utilized for the amplification of SSU (Uibopuu et al. 2009). Likewise, FLR3/FLR4 are usually accepted to amplify LSU of AM fungi (Koch et al. 2011). Restriction enzymes are also another crucial things that are chosen wisely. A restriction enzyme of the highest quality will generate a wide range of fragment sizes with little variance between the species. *AluI* and *MboI* are available in the market and have been utilized in many cases (Barto et al. 2011).

4.8.8 DGGE: Denaturing Gradient Gel Electrophoresis and TGGE: Temperature Gradient Gel Electrophoresis

The DGGE is a molecular technique used to separate DNA fragments based on their difference in the sequence rather than sizes (Hovig et al. 1991). DGGE exploits such difference by creating a gradient of chemical substances (urea and formamide) across a polyacrylamide gel. The gradient is created by mixing different concentrations of denaturants into the gel before polymerization. Sequences of organisms are segregated based just on G + C content of PCR-amplified DNA using mixtures of formamide and urea (Kirk et al. 2004). Triple hydrogen bonds between G and C are harder to break than the two bonds of hydrogen between A and T. As DNA breaks, movement in the gel starts to slow down. Attachment of GC clamp to PCR primer prevents complete strand separation, producing a fine band on gel (Sheffield et al. 1989). The heterogeneous amplified PCR is separated during electrophoresis. The results that are depicted in the form of various bands give a glimpse of the approximate population and abundance of a community. Similar patterns and principles are also operated in TGGE except the gradient where in this case temperature instead of chemical is the final gradient. Nested PCR in which AM1-NS31 is a primer gave a picture of high quality of AM fungi collected from polluted soil (Krishnamoorthy et al. 2014, 2015).

4.8.9 Illumina MiSeq for the Identification of AM Fungi

Morphological identification is frequently unreliable due to the vast number of cryptic species (Kryukov et al. 2020). The number of AM fungal biodiversity research using modern NGS-based technologies, especially the Illumina MiSeq, has increased in recent years. Illumina MiSeq is a next-generation sequencing (NGS) platform developed by Illumina, Inc. It is designed for targeted resequencing, metagenomics, small genome sequencing, and amplicon sequencing applications. MiSeq uses Illumina's sequencing by synthesis (SBS) technology, which is based on reversible terminators and fluorescent detection. This technique helps us identify eight out of nine isolates of AM fungi (Kryukov et al. 2020). It permits the identification of a great amount of AM fungi from a fungal community, particularly when the percentage of genes of interest is very low in the given sample. MiSeq can give longer reads and fewer mistakes than other techniques such as HiSeq and Ion Torrent, correspondingly (Salipante et al. 2014; Razzauti et al. 2015).

Fungi with the letters AM selecting a marker for barcoding, the most important factors to consider are whether to use conservative or flexible sequences, as well as whether to use universal or AM fungi-specific primers. Using universal primers ITS3 and ITS4, the ITS 5.8S rDNA–ITS2 area of the 35S rRNA was successfully sequenced in Illumina MiSeq, which included both a conservative and aggressive sequence.

4.9 AM Fungi Propagation

Some approaches for AM fungi propagation have been found in recent decades. The obligatory nature of this biotrophic fungus is the chief hurdle, so AM fungi cannot be cultured. For bulk production of AM fungi, in vitro growing methods such as hydroponics (Das et al. 2020) and root organ culture (Mathur and Vyas 2007) are used. These techniques are used to produce cost-effective AM fungal propagules keeping in mind that their quality is retained. Other techniques such as substrate-based production (Khaliq et al. 2010), substrate-free production (Hawkins and George 1997), and in vitro production (Lee and George 2005; De Boulois et al. 2006) are the commonly used techniques for AM fungi.

4.9.1 Methodology for Mass Propagation

AM fungi and their mass production is still a great challenge; however, substrate-based inoculum, nutrient film technique, aeroponic culture, root organ culture method are some of the techniques that may be used for this purpose.

4.9.2 Substrate-Based Inoculum Propagation

This is a widely accepted normal method of propagating AM culture. Sterile soil treated with a pure culture of a particular AM fungi containing a plant roots is a commonly utilized media for this purpose (Khaliq et al. 2010). For the inoculation, inoculum containing spores, hyphal fragments, and infected root portions are used. Despite of having good popularity among the workers, it has some disadvantages. To begin with, this approach can only create a little amount of inoculum. The product is big and heavy, making it difficult to maintain and move to the application site. There are some inert substances like vermiculite, perlite, sand, and mixture of all substances that are used and replaced with soil. Substrate-based inoculum is frequently used for bulk generation of AM fungus (Schenck 1982). The major goal of this technology is to create as many infective propagules as possible in a short time while minimizing impurities. Large plastic troughs or cement cisterns can be filled with perlite: soil rite mix (1:1), inoculated with the chosen AM fungal starter culture and seeded with Rhodes grass as the host (Khaliq et al. 2010). Guinea grass or any other suitable grass can be substituted, if Rhodes grass is unavailable (Khaliq et al. 2010). As a nitrogen and phosphorus source, calcium ammonium nitrate (80 ppm N) and rock phosphate (10 ppm phosphorus) can be also added (Daniel and Arya 2012). The fungicide, acaricide, and insecticide can be added with the substrate at half of the recommended dosage to minimize pests and pathogens population in the inoculum while not affecting AM fungus. Plants are harvested after 75 days. The roots are cut into small pieces, mixed with substrate, air-dried, and packaged in polythene bag (Coelho et al. 2015).

4.9.3 Nutrient Film Technique

Warner et al. (1985) devised a peat moss-based technique for mass-producing AM fungi. Peat moss is combined with three times to its weight in water and a tiny amount of lime to maintain a pH that is congenial for AM fungi multiplication. Pot culture mycorrhizal inoculum is put into the peat moss and crushed into 4 × 4 × 4 cm blocks. The lettuces are then planted and allowed to germinate and grow in such a way that the root system reaches throughout the peat (Warner et al. 1985). These root peat segments are included in the nutrition film where a nutritious solution is passed through the channels. These plants are allowed to grow up to 8–10 weeks. AM fungus grows throughout the root and peat block during this period. The lettuce is harvested and eaten as a vegetable, while the peat blocks are dried. The entire peat block is then crushed and used as source of AM fungi (Warner et al. 1985).

4.9.4 Aeroponic Culture

Aeroponics is a way of cultivating plants that do not require soil. This method was developed by David Sylvia of the University of Florida, the United States. The host roots containing AM fungi are dipped in nutrient solution. Spraying micro-droplets improves medium aeration and enables gas exchange through the liquid layer surrounding the roots. With the mist sprayer, a nutrient solution is put into the bottom of a plastic container. Seedlings of mycorrhizal grass are placed in a container with a perforated lid. The nutrient solution is sprayed using a mist sprayer. After a few weeks, the roots are sliced into 1 cm pieces, shade dried, and used as a source of AM fungi (Jarstfer and Sylvia 1994; Mohammad et al. 2020).

4.9.5 Root Organ Culture

Mugnier and Mosse (1987) were the first to use Ri-plasmid-modified root cultures. Yve Piche and co-workers in Canada established the full potential of this culture method (Becard and Piche 1989; Chabot and Becard 1992). The most extensively utilized plant symbionts in this technique are *Rhizophagus intraradices* and *Gigaspora margarita* in axenic cultures.

4.9.6 Substrate-Free Production System

The substrate-free growing systems or nutrient flow approaches are already accessible. The mechanism of aeration and application of the nutritional solution may change across these different strategies. In brief, the nutrition solution is aerated by an aeration pump in a static type of system to prevent oxygen deprivation of the roots. The pumps must be switched on and off constantly to avoid the rush of nutritional solutions and gas bubbles from injuring the fragile extraradical mycelium (Hawkins and George 1997).

4.10 Significance of Plant Symbionts (AM Fungi) in Plant Health Improvement in Pathogen-Induced Environment

Mycorrhizal plants are able to withstand diseased plants and improve the status of plants grown in a pathogen-induced environment (Ansari and Khan 2012a, b; Ansari et al. 2019a, b).

4.10.1 Competition

AM fungi do not compete for nutrients in soil; they require spore reserves till the root connectivity is achieved. After penetration into roots, further processes such as infection site competition, photosynthetically produced materials, and root space for their survival begin (Rodrigo Mendes et al. 2013). Pathogens can be physically excluded by taking advantage of AM fungi-pathogen competition. Physiological and biochemical alteration in plants takes place in mycorrhizal plants. Phosphorus levels in the host root tissue are frequently elevated, and the phospholipid profile and permeability of the membrane are altered, leading to the generation of less net sugar, carboxylic acid, and amino acid leakage into the rhizosphere (Mendes et al. 2013).

4.10.2 Systemic-Induced Resistance (SIR), Phytoalexins, and Phytoanticipins

SIR is the process of inducing disease resistance in plants over time through inoculation with a pathogen.

Phytoalexins are synthesized by plants in reaction to microbes, while phytoanticipins are maintained before the pathogen infection (Paxton 1981). The levels of phytoalexin elicited by causal agent are found substantially greater than those triggered by symbiotic species. Based on the molecules' cellular location, they are lipophilic, allowing them to pass through the plasma membrane and perform their actions inside the cell (Braga et al. 1991). Mycorrhizal plant roots enhanced secondary metabolites such as coumaric acids, lignin, ferulic, syringic phenolics, isoflavonoids, or flavonoids. Plants have an improved defense system against *F. oxysporum* with high beta-glucosidase and phenylalanine activity, as well as total phenol concentration in their roots (Dehne and Schönbeck 1979). Furthermore, phytoalexins are necessary to neutralize the anti-pathogen effect in AM plants compared to control plants (Caron et al. 1986). Phytoalexin and phytoanticipin are isoflavonoid compounds produced by AM fungi (Morandi 1996).

4.10.3 Organic molecules behaviour

The growth media of a *Glomus intraradices* produce some organic molecules that are responsible for either the stimulation or inhibition of microorganism. The extraradical mycelia of *G. intraradices* release molecules that impact microorganisms' equilibrium. It is vital to understand and implement AM fungi significance in plant disease control (Lisette et al. 2003). Filion et al. (1999) reported that some antimicrobial substances synthesized by *G. intraradices* regulate the conidial germination of *F. oxysporum* f. sp. *chrysanthemi* under the non-influential effect of pH. However, further studies are to be ascertained in order to get any concrete findings over it as this kind of study are scanty.

4.11 AM Fungi in Disease Management Caused by Phytopathogenic Fungi

Plant symbionts that inhabit the plant's root living in rhizosphere are thought advantageous to the growth of the plant. Plant symbionts improve overall crop growth and productivity (Linderman 2000; Begum et al. 2019). Plants do have certain tolerance mechanisms that assist them in avoiding the negative repercussions of diverse environmental circumstances (Table 4.1; Graham 1981; Rouhier et al. 2008; Xavier and Boyetchko 2003; Xavier and Boyetchko 2004; Sumbul et al. 2017; Rizvi et al. 2015; Ansari et al. 2017a, b).

AM fungi promotes plant growth and metabolisms, leading to enhanced crop yield and productivity (Fall et al. 2022). AM fungi in stress reduction is self-evident, as they improve the essential defensive systems of host plants. AM fungi stimulate some metabolites such as indole-3-acetaldehyde, indole-3-carboxyaldehyde, and indole-3-ethanol, volatile compounds, and peptides (Lorito and Woo 2015; Colla et al. 2015; Rouphael et al. 2017).

4.11.1 AM Fungi in Bacterial Disease Management

Glomus mosseae that lowered the pathogen's population density in the rhizosphere prevented *P. syringae* infection of soybean plants (Dreischhoff et al. 2020; Mingqin et al. 2004; Shalaby and Hanna 1998). Despite no growth or production, *G. macrocarpum* reduced *P. lacrymans* infection in eggplant and cucumber, indicating resistance to the pathogen as a potential causal agent (Li et al. 1997). The occurrence of *P. syringae pv. mori* bacterial blight was reduced when mulberry trees were inoculated with plant symbionts (*G. fasciculatum* and *G. mosseae*) along with 60–90 kg of phosphorus/ha/year.

Table 4.1 AM fungi-mediated plant disease control in different crops

Sl. no.	Mycorrhizal fungi	Targeted pathogenic fungi	Host	Mechanism(s)	References
1	*Glomus fasciculatum*	*Fusarium oxysporum* f. sp. *lycopersici*	*Solanum lycopersicum*	AM fungi inoculated plants increased the amino acid contents, leading to improved resistance level against pathogens	Manila and Nelson (2017)
		Fusarium oxysporum f. sp. *ciceris*	*Cicer arietinum*	AM fungi induce the changes in the host root exudation pattern and inhibit the multiplication of the pathogens. Inoculation of AM fungi helps plants absorb more nutrient and improves plant biomass	Siddiqui and Singh (2004)
2	*Glomus mosseae*	*Phytophthora capsici*	*Piper nigrum*	AM fungi significantly increases capsidiol level in pepper	Ozgonen and Erkilic (2007)
		P. syringae	*Lycopersicon esculentum*	AM fungi-inoculated crop improves biomass and reduces the pathogen activity	Garcia-Garrido and Ocampo (1989)
3	*Funneliformis mosseae*	*Gaeeumanomyces gramanis*	*Triticum aestivum*	AM fungi-inoculated plants significantly increase phytoalexins, enzymes of phenyl propanoid pathway, chitinase peroxidases, PR-related proteins, phenolics, etc.	Falahian et al. (2007)
		Tomato yellow leaf curl virus and cucumber mosaic virus	*Lycopersicon esculentum and Cucumis sativus*	AM fungi-inoculated crop advances the resistance levels against these viruses	Maffei et al. (2014)
4	*Rhizophagus irregularis*	Potato virus Y Potato virus Y	*Solanum tuberosum*	AM fungi inoculation improves the resistance levels against these virus	Sipahioglu et al. (2009)
		Tobacco mosaic virus	*Nicotiana tabacum*		

4.11.2 AM Fungi for the Management of Phytoparasitic Nematodes

The plant parasitic nematode is a diverse group of species with different life behaviors that can destroy some key crops around the world (Schouteden et al. 2015).

4.11.3 Major Challenges in the Application of AM Fungi in Plant Disease Management

AM fungi's effectiveness against plant diseases has only been studied in controlled circumstances that suggest that AM fungi has a high potential for managing plant infections. AM fungi research on the biological control of plant diseases in field situations has two drawbacks.

1. Due to obligatory nature, mass production of inoculum is a difficult task (Budi et al. 1999).
2. After introduction into the field, there is negative reciprocity/interrelation between the introduced AM fungi and the indigenous AM fungi, as well as other microbial populations (Bever et al. 1996).

The efficacy of AM fungi inoculum in disease management under field conditions is often determined by the challenges given by interactions between AM fungi and the indigenous microbial community, as well as edaphic factors. Understanding the characteristics that determine plant symbiont efficacy as bioagents can help them survive, compete, and be more effective (Xavier 1999).

4.12 Conclusions and Future Outlooks

Plant symbionts are beneficial microorganisms that develop mutual relationships with higher plant roots. These fungi are phenomenal in long run soil fertility, health maintenance, and sustainable plant production systems.

Some robust techniques for culturing AM fungi have recently come out. The obligatory nature of these biotrophic fungi is the main impediment to mass production. Though, AM fungi products are based on substrate-based inoculum, root organ culture, substrate-free production systems. Application of AM fungi in disease management and maintaining sustainability in agroecosystem has a crucial role. The rhizosphere is home to AM fungi's diverse services, which include improved plant nutrients, pathogen resistance, qualitative and quantitative changes in pathogen biomass, competition impacts on rhizosphere microbial populations, and systemic induction resistance. These are some mechanisms that help plants cope with any biotic stressors. Some efficient techniques for propagating AM fungi are the need of the hour, and it must be realized that these microorganisms if used judiciously may help improve crop health.

References

Abbott LK, Robson AD (1982) The role of vesicular arbuscular mycorrhizal fungi in agriculture and the selection of fungi for inoculation. Aust J Agric Res 33(2):389–408

Aldwell FEB (1983) Immunological techniques- to identify VAM fungi. Ph.D. Thesis

Almeida RT, Schenck NC (1990) A revision of the genus Sclerocystis (Glomaceae, Glomales). Mycologia 82(6):703–714

Ansari, R.A. and Khan, T.A. (2012a). Parasitic association of root-knot nematode, *Meloidogyne incognita* on guava, e J Sci Technol 5 (12) 65–67

Ansari RA, Khan TA (2012b) Diversity and community structure of Phytonematodes associated with guava in and around Aligarh, Uttar Pradesh. Trends Biosci 5(3):202–204

Ansari RA, Mahmood I (2017a) Optimization of organic and bio-organic fertilizers on soil properties and growth of pigeonpea. Sci Hortic 226:1–9

Ansari RA, Mahmood I (2017b) Determination of disease incidence caused by *Meloidogyne* spp. and or *Fusarium udum* on pigeonpea in Aligarh district: a survey. Trends. Bioscience 10(24):5239–5243

Ansari RA, Mahmood I (2019a) In: Ansari RA, Mahmood I (eds) Plant health under biotic stress, Microbial interactions, vol II. Springer Nature, Singapore. https://doi.org/10.1007/978-981-13-6040-4

Ansari RA, Mahmood I (2019b) In: Ansari RA, Mahmood I (eds) Plant health under biotic stress, Organic strategies, vol I. Springer Nature, Singapore. https://doi.org/10.1007/978-981-13-6043-5

Ansari RA, Mahmood I, Rizvi R, Sumbul A (2017a) PGPR: current vogue in sustainable crop production. In: Kumar V (ed) Probiotics and plant health. Springer Nature, Singapore, pp 455–472

Ansari RA, Mahmood I, Rizvi R, Sumbul A, Safiuddin (2017b) Siderophores: augmentation of soil health and crop productivity. In: Kumar V (ed) Probiotics in agroecosystem. Springer Nature, Singapore, pp 291–312

Ansari RA, Sumbul A, Rizvi R, Mahmood I (2019a) Potential role of plant growth promoting Rhizobacteria in alleviation of biotic stress. In: Ansari RA, Mahmood I (eds) Plant health under biotic stress, Microbial interactions, vol II. Springer Nature, Singapore, pp 177–188

Ansari RA, Sumbul A, Rizvi R, Mahmood I (2019b) Organic soil amendments: potential tool for soil and plant health management. In: Ansari RA, Mahmood I (eds) Plant health under biotic stress, Organic strategies, vol I. Springer Nature, Singapore, pp 1–35

Ansari RA, Rizvi R, Mahmood I (2020a) Management of phytonematodes: management of phytonematodes: recent advances and future challenges. Springer Nature, Singapore. https://doi.org/10.1007/978-981-15-4087-5

Ansari RA, Rizvi R, Sumbul A, Mahmood I (2020b) Plant-growth-promoting Rhizobacteria (PGPR)-based sustainable Management of Phytoparasitic Nematodes: current understandings and future challenges. In: Ansari RA et al (eds) Management of phytonematodes: recent advances and future challenges. Springer Nature, Singapore. https://doi.org/10.1007/978-981-15-4087-5_3

Barto EK, Antunes PM, Stinson K, Koch AM, Klironomos JN, Cipollini D (2011) Differences in arbuscular mycorrhizal fungal communities associated with sugar maple seedlings in and outside of invaded garlic mustard forest patches. Biol Invasions 13:2755–2762

Becard G, Piche Y (1989) New aspects of the acquisition of biotrophic status by a vesicular-arbuscular mycorrhizal fungus, Gigaspora margarita. New Phytol 112:77–83

Begum N, Qin C, Ahanger MA, Raza S, Khan MI, Ashraf M, Ahmed N, Zhang L (2019) Role of arbuscular mycorrhizal fungi in plant growth regulation: implications in abiotic stress tolerance. Front Plant Sci 10:1068

Bever JD, Morton JB, Antonovics J, Schultz PA (1996) Host dependent sporulation and species diversity of arbuscular mycorrhizal fungi in a mown grassland. J Ecol 84:71–82

Botstein D, White RL, Skolnick M, Davis RW (1980) Construction of genetic linkage map in man using restriction length polymorphisms. Am J Hum Genet 32:314–331

Braga MR, Claudia M, Young M, Dietrich SMC, Gottlieb OR (1991) Phytoalexin induction in Rubiaceae. J Chem Ecol 17:1079–1090

Budi SW, Van TD, Martinotti G, Gianinazzi S (1999) Isolation from the Sorghum bicolor mycorrhizosphere of a bacterium compatible with arbuscular mycorrhiza development and antagonistic towards soilborne fungal pathogens. Appl Environ Microbiol 65:5148–5150

Caron M, Fortin JA, Richard C (1986) Effect of phosphorus concentration and *Glomus intraradices*. Phytopathology 76:942–946

Chabot S, Becard G (1992) Life cycle of Glomus intra-radix in root organ culture. Mycologia 84:315–321

Chennappa G, Adkar-Purushothama CR, Suraj U, Tamilvendan K, Sreenivasa MY (2014a) Pesticide tolerant Azotobacter isolates from paddy growing areas of northern Karnataka, India. World J Microbiol Biotechnol 30(1):1–7

Chennappa G, Adkar-Purushothama CR, Naik MK, Suraj U, Sreenivasa MY (2014b) Impact of pesticides on PGPR activity of Azotobacter sp. isolated from pesticide flooded paddy soils. Greener J Biol Sci 4(4):117–129

Cleyet-Marel JC, Bousquet N, Mousain D (1990) The immunochemical approach for the characterization of ectomycorhizal fungi. Agric Ecosyst Environ 28(1–4):79–83

Coelho IR, Pedone-Bonfim MV, Silva FS, Maia LC (2015) Optimization of the production of mycorrhizal inoculum on substrate with organic fertilizer. Braz J Microbiol 45(4):1173–1178

Colla G, Rouphael Y, Bonini P, Cardarelli M (2015) Coating seeds with endophytic fungi enhances growth, nutrient uptake, yield and grain quality of winter wheat. Int J Plant Prod 9:171–189

Daniel M, Arya A (2012) Phytotechnology: emerging trends. Scientific Publishers

Daniels, B. A. 1984. Ecology of VA mycorrhizal hngi. In: VA mycorrhiza. (Eds. C. LI. Powell and D .J. Bagyaraj), CRC Press, Boca Raton, FL, pp,205–222

Daniels BA, Skipper HD (1982) Methods for the recovery and quantitative estimation ofpropagules from soil. In: Schneck NC (ed) Methods and principles of mycorrhizalResearch. American Phytopathological Society, St Paul, MN, p 244

Das D, Torabi S, Chapman P, Gutjahr C (2020) A flexible, low-cost hydroponic co-cultivation system for studying arbuscular Mycorrhiza Symbiosis. Front Plant Sci 11:63

Davison J, Moora M, Öpik M, Adholeya A, Ainsaar L, Bâ A, Burla S, Diedhiou AG, Hiiesalu I, Jairus T, Johnson NC (2015) Global assessment of arbuscular mycorrhizal fungus diversity reveals very low endemism. Science 349(6251):970–973

De Boulois HD, Voets L, Delvaux B, Jakobsen I, Declerck S (2006) Transport of radiocaesium by arbuscular mycorrhizal fungi to Medicago truncatula under in vitro conditions. Environ Microbiol 8(11):1926–1934

Declerck S, Strullu DG, Plenchette C (1998) Monoxenic culture of the intraradical forms of glomus sp. isolated from a tropical ecosystem: a proposed methodology for germplasm collection. Mycologia 90:579–585

Dehne, H.W. and Schönbeck, F. 1979. Untersuchungen zum einfluss der endotrophen Mycorrhiza auf Pflanzenkrankheiten: II. Phenolstoffwechsel und lignifizierung. J Phytopathol, 95, 210–21216

Demeke T, Adams RP, Chibbar R (1992) Potential taxonomic use of random amplified polymorphic DNA (RAPD): a case study in brassica. Theor Appl Genet 84:990–994

Diédhiou AG, Borges WL, Sadio O, Faria SM (2014) Assessment of DNA extraction methods for detection of arbuscular mycorrhizal fungi in plant roots by nested-PCR *Acta Scientiarum*. Biol Sci 36(4):433–441

Dreischhoff S, Das IS, Jakobi M, Kasper K, Polle A (2020) Local responses and systemic induced resistance mediated by ectomycorrhizal fungi. Front Plant Sci 11:14

Drijber RA, McPherson MR (2021) Mycorrhizal symbioses. In: Principles and applications of soil microbiology. Elsevier, pp 303–325

Edgar RC (2010) Search and clustering orders of magnitude faster than BLAST. Bioinformatics (Oxford, England) 26(19):2460–2461. https://doi.org/10.1093/bioinformatics/btq461

Edgar RC (2013) UPARSE: highly accurate OTU sequences from microbial amplicon reads. Nat Methods 10(10):996–998

Edgar RC (2018) Updating the 97% identity threshold for 16S ribosomal RNA OTUs. Bioinformatics 34(14):2371–2375
Eom AH, Wilson GWT, Hartnett DC (2001) Effect of ungulate grazers on arbuscular mycorrhizal symbiosis and fungal community structure in tall grass prairie. Mycologia 93:233–242
Falahian F, Ardebili ZO, Fahimi F, Khavarinejad R (2007) Effect of mycorrhizal fungi on some defense enzymes against Gaeumannomyces gaminis in wheat. Pak J Biol Sci 10:2418–2422
Fall AF, Nakabonge G, Ssekandi J, Founoune-Mboup H, Apori SO, Ndiaye A, Badji A, Ngom K (2022) Roles of arbuscular mycorrhizal fungi on soil fertility: contribution in the improvement of physical, chemical, and biological properties of the soil. Front Fungal Biol 3:723892
Filion M, St-Arnaud M, Fortin JA (1999) Direct interaction between the arbuscular mycorrhizal fungus Glomus intraradices and different rhizosphere microorganisms. New Phytol 141:525–533
Garcia-Garrido JM, Ocampo JA (1989) Effect of VA mycorrhizal infection of tomato on damage caused by Pseudomonas syringae. Soil Biol Biochem 21:165–167
Gardemann JW, Nicolson TH (1963) Spores of mycorrhizal Endogone species extracted from soil by wet sieving and decanting. Trans Br Mycol Soc 46:235–244
Gomez-Leyva JF, Lara-Reyna J, Hernandez-Cuecas LV, Martinez-Soriano JP (2008) Specific polymerase chain reaction based assay for the identification of the arbuscular mycorrhizal fungus *Glomus intraradices*. J Biol Sci 8:563–569
Graham JH (1981) Interactions of mycorrhizal fungi with soil-borne plant pathogens and other organisms. Phytopathology 78:365–374
Habte M, Osorio NW (2001) Arbuscular mycorrhizas: producing and applying arbuscular mycorrhizal inoculum. University of Hawaii
Hall IR (1988) Potential for exploiting vesicular-arbuscular mycorrhizas in agriculture. Adv Biotechnol Process 9:141–174
Hata S, Kobae Y, Banba M (2010) Interactions between plants and arbuscular mycorrhizal fungi. Int Rev Cell Mol Biol 281:1–48
Hawkins HJ, George E (1997) Hydroponic culture of the mycorrhizal fungus *glomus mosseae* with *Linumusitatissimum* L., *Sorghum bicolor* L. and *Triticum aestivum* L. Plant Soil 196:143–149
Hayman DS (1980) Mycorrhiza and crop production. Nature 287:487–488
Hovig E, Smithsorensen B, Brogger A, Borresen AL (1991) Constant denaturant gel-electrophoresis, a modification of denaturing gradient gel-electrophoresis, in mutation detection. Mutat Res 262 :63–71. https://doi.org/10.1016/0165-7992(91)90108-G
INVAM (2019) Species descriptions from reference cultures. http://fungi.invam.wvu.edu/the-fungi/species-descriptions.html
Jarstfer AG, Sylvia DM (1994) Aeroponic culture of VAM fungi. In: Varma AK, Hock B (eds) Mycorrhiza: structure, function, molecular biology and biotechnology. Springer-Verlag, Berlin, pp 427–441
Kariman KH, Goltapeh EM, Minassian V (2005) Arbuscular mycorrhizal fungi from Iran. J Agric Technol 1(2):301–313
Khaliq A, Bagyaraj DJ, Alam M (2010) Advances in mass production technology of arbuscular mycorrhiza. In: Thangadurai D, Busso CA, Hijri M (eds) Mycorrhizal biotechnology. Capital Pub. Co., New Delhi, pp 43–55
Kirk JL, Beaudette LA, Hart M (2004) Methods of studying soil microbial diversity," J Microbiol Methods 58(2): 169–188
Koch AM, Antunes PM, Barto EK, Cipollini D, Mummey DL, Klironomos JN (2011) The effects of arbuscular mycorrhizal (AM) fungal and garlic mustard introductions on native AM fungal diversity. Biol Invasions 13:1627–1639
Krishnamoorthy R, Kim KY, Kim CG, Sa TM (2014) Changes of arbuscular mycorrhizal traits and com-munity structure with respect to soil salinity in a coastal reclamation land. Soil Biol Biochem 72:1–10
Krishnamoorthy R, Kim CG, Subramanian P, Kim KY, Selvakumar G, Sa TM (2015) Arbuscular mycor-rhizal fungi community structure, abundance and species richness changes in soil by

different levels of heavy metal and metalloid concentration. PLoS One 10(6):e0128784. https://doi.org/10.1371/journal.pone.0128784
Kruger M, Stockinger H, Kruger C, Schuler A (2009) DNA-based species-level detection of Glomeromycota: one PCR primer set for all arbuscular mycorrhizal fungi. New Phytol 183:212–223
Kryukov AA, Gorbunova AO, Machs EM, Mikhaylova YV, Rodionov AV, Zhurbenko PM, Yurkov AP (2020) Perspectives of using Illumina MiSeq for identification of arbuscular mycorrhizal fungi. Vavilovskii Zhurnal Genet Selektsii 24(2):158–167. https://doi.org/10.18699/VJ19.38-o
Kumar S (2013) Trichoderma: a biological weapon for managing plant diseases and promoting sustainability. Int J Agric Sci Med Vet 1(3):106–121
Lee YJ, George E (2005) Development of a nutrient film technique culture system for arbuscular mycorrhizal plants. Hort Sci 40:378–380
Lekberg Y, Vasar M, Bullington LS (2018) More bang for the buck, can arbuscular mycorrhizal fungal communities be characterized adequately alongside other fungi using general fungal primers. New Phytol 220:971–976
Li SL, Zhao SJ, Zhao LZ (1997) Effects of VA mycorrhizae on the growth of eggplant and cucumber and control of diseases. Acta Phytophylac Sin 24:117–120
Linderman RG (2000) Effects of mycorrhizas on plant tolerance to diseases. In: Kapulnik Y, Douds DD (eds) Arbuscular mycorrhizas: physiology and function. Springer, Dordrecht. https://doi.org/10.1007/978-94-017-0776-3_15
Lisette J, Xavier C, Germida JJ (2003) Selective interactions between arbuscular mycorrhizal fungi and Rhizobium leguminosarum bv. viceae enhance pea yield and nutrition. Biol Fertil Soils 37:261–267
Lorito M, Woo SL (2015) Discussion agronomic. In: Lugtenberg B (ed) Principles of plant-microbe interactions. Springer, Berlin, pp 345–353. https://doi.org/10.1007/978-3-319-08575-3_36
Maffei G, Miozzi L, Fiorilli V, Novero M, Lanfranco L, Accotto GP (2014) The arbuscular mycorrhizal symbiosis attenuates symptom severity and reduces virus concentration in tomato infected by tomato yellow leaf curl Sardinia virus (TYLCSV). Mycorrhiza 24:179–186
Maitra D, Dhara D, Sultana S, Chakraborty R, Mitra A, Choudhury S (2020) Mycorrhizal association in agriculture and their role in preventing climate change. In: Precision agriculture and sustainable crop production. Today & Tomorrow's Printers and Publishers
Mampaso FM, Wilson CB (1983) Characterization of inflammatory cells in autoimmune tubulointerstitial nephritis in rats. Kidney Int 23(3):448–457
Manila R, Nelson R (2017) Nutrient uptake and promotion of growth by arbuscular mycorrhizal fungi in tomato and their role in bio-protection against the tomato wilt pathogen. J Microbiol Biotechnol, Res 3(4):4246
Marshall SM, Melito PL, Woodward DL, Johnson WM, Rodgers FG, Mulvey MR (1999) Rapid identification of campylobacter, Arcobacter, and Helicobacter isolates by PCR-restriction fragment length polymorphism analysis of the 16S rRNA gene. J Clin Microbiol 37(12):4158–4160
Mathur N, Vyas A (2007) Arbuscular Mycorrhiza on root-organ cultures. Am J Plant Physiol 2:122–138
Mendes R, Garbeva P, Raaijmakers JM (2013) The rhizosphere microbiome: significance of plant beneficial, plant pathogenic, and human pathogenic microorganisms. FEMS Microbiol Rev 37(5):634–663
Mingqin G, Yu C, Fengzhen W (2004) Resistance of the AM fungus eucalyptus seedlings against Pseudomonas solanacearum. For Res 17(4):441–446
Mohammad A, Khan AG, Kuek C (2020) Improved aeroponic culture of inocula of arbuscular mycorrhizal fungi. Mycorrhiza 9(6):337–339
Morandi D (1996) Occurrence of phytoalexins and phenolic compounds in endomycorrhizal interactions, and their potential role in biological control. Plant Soil 2:241–251
Morton JB (1996) Redescription of glomus caledonium based on correspondence of spore morphological characters in type specimens and a living reference culture. Mycorrhiza 6(3):161–166
Mugnier J, Mosse B (1987) Vesicular-arbuscular mycorrhizal infection in transformed root-inducing T-DNA roots grown axenically. Phytopathology 77:1045–1050

Mummey DL, Rillig MC (2006) The invasive plant species Centaurea maculosa alters arbuscular mycorrhizal fungal communities in the field. Plant Soil 288:81–90

Nemec S, Myhre D (1984) Virus-Glomus etunicatum interactions in citrus rootstocks. Plant Dis 68:311–314

Oehl F, Laczko E, Bogenrieder A, Stahr K, Bösch R, van der Heijden M, Sieverding E (2010) Soil type and land use intensity determine the composition of arbuscular mycorrhizal fungal communities. Soil Biol Biochem 42(5):724–738

Ozgonen H, Erkilic A (2007) Growth enhancement and Phytophthora blight (Phytophthora capsici L.) control by arbuscular mycorrhizal fungal inoculation in pepper. Crop Protec 26:1682–1688

Paxton JD (1981) Phytoalexins-a working redefinition. Phytopathology 10:106–109

Perotto S, Malavasi F, Butcher GW (1992) 12 Use of monoclonal antibodies to study mycorrhiza: present applications and perspectives. In: Methods in microbiology, vol 24. Academic Press, pp 221–248

Phillips JM, Hayman DS (1970) Improved procedures for clearing roots and staining parasitic and vesicular-arbuscular mycorrhizal fungi for rapid assessment of infection. Trans Br Mycol Soc 55:157–160

Policelli N, Horton TR, Hudon AT, Patterson TR, Bhatnagar JM (2020) Back to roots: the role of ectomycorrhizal fungi in boreal and temperate forest restoration. Front For Global Change 3:97

Pozo MJ, Azcón-Aguilar C (2007) Unravelling mycorrhiza induced resistance. Curr Opin Plant Biol 10:393–398

Razzauti M, Galan M, Bernard M, Maman S, Klopp C, Charbonnel N et al (2015) A comparison between transcriptome sequencing and 16S metagenomics for detection of bacterial pathogens in wildlife. PLoS Negl Trop Dis 9(8):e0003929

Rizvi R, Singh G, Safiuddin ARA, Tiyagi SA, Mahmood I (2015) Sustainable management of root-knot disease of tomato by neem cake and *Glomus fasciculatum*. Cogent Food Agric 1:1008859

Rodrigues BF, Muthkumar T (2009) Arbuscular mycorrhizae of Goa—a manual of identification protocols. Goa University, Goa

Roth R, Hillmer S, Funaya C, Chiapello M, Schumacher K, Presti LL, Kahmann R, Paszkowski U (2019) Arbuscular cell invasion coincides with extracellular vesicles and membrane tubules. Nat Plants 5(2):204–211

Rouhier N, San Koh C, Gelhaye E, Corbier C, Favier F, Didierjean C, Jacquot JP (2008) Redox based anti-oxidant systems in plants: biochemical and structural analyses. Biochim Biophys Acta 1780:1249–1260

Rouphael Y, Cardarelli M, Bonini P, Colla G (2017) Synergistic action of a microbial-based biostimulant and a plant derived-protein hydrolysate enhances lettuce tolerance to alkalinity and salinity. Front Plant Sci 8:131. https://doi.org/10.3389/fpls.2017.00131

Salipante SJ, Kawashima T, Rosenthal C, Hoogestraat DR, Cummings LA, Sengupta DJ et al (2014) Performance comparison of Illumina and ion torrent next-generation sequencing platforms for 16S rRNA-based bacterial community profiling. Appl Environ Microbiol 80(24):7583–7591

Sanders IR, Clapp JP, Wiemken A (1996) The genetic diversity of arbuscular mycorrhizal fungi in natural ecosystems–a key to understanding the ecology and functioning of the mycorrhizal symbiosis. New Phytol 133(1):123–134

Santos EAD, Ferreira LR, Costa MD, Silva MDCSD, Reis MRD, França AC (2013) Occurrence of symbiotic fungi and rhizospheric phosphate solubilization in weeds. Acta Sci Agron 35:49–55

Schenck NC (1982) Methods and principles of mycorrhizal research. The American Phytopathological Society

Schouteden N, De Waele D, Panis B, Christine M (2015) Arbuscular mycorrhizal fungi for the biocontrol of plant-parasitic nematodes: a review of the mechanisms involved. Front Microbiol 6:1280

Seviour RJ, Willing RR, Chilvers CA (1973) ES! Phytol ZE:381–385

Shalaby AM, Hanna MM (1998) Preliminary studies on interactions between VA mycorrhizal fungus *Glomus mosseae, Bradyrhizobium japonicum* and *Pseudomonas syringae* in soybean plants. Acta Microbiol Pol 47(4)

Shamini S, Amutha K (2014) Techniques for extraction of arbuscular mycorrhizal fungi spores. Int J Front Sci Technol 2(5):1–6

Sheffield VC, Cox DR, Lerman LS, Myers RM (1989) Attachment of a 40-base-pair G+C-rich sequence (GC-clamp) to genomic DNA fragments by the polymerase chain reaction results in improved detection of single-base changes. P Natl Acad Sci USA86:232–236

Siddiqui ZA, Singh LP (2004) Effects of soil inoculants on the growth, transpiration and wilt disease of chickpea. J Plant Dis Protec 111:151–157

Sieverding E, Howeler RH (1983) Potentials and limitations of mycorrhizal inoculation illustrated by experiments with field-grown cassava. Plant Soil 75(2):245–261

Sipahioglu MH, Demir S, Usta M, Akkopru A (2009) Biological relationship of potato virus Y and arbuscular mycorrhizal fungus Glomus intraradices in potato. Pest Technol 3:63–66

Smith S, Read D (2008) Mycorrhiza symbiosis, 3rd edn. Academic Press, San Diego, CA

Solanki MK, Kashyap PL, Ansari RA, Kumari B (2020) Microbiomes and plant health: panoply and their applications. Academic Press, London. https://www.elsevier.com/books/microbiomes-and-plant-health/solanki/978-0-12-819715-8

Sood M, Kapoor D, Kumar V, Sheteiwy MS, Ramakrishnan M, Landi M, Araniti F, Sharma A (2020) Trichoderma: the "secrets" of a multitalented biocontrol agent. Plan Theory 9(6):762

Subba Rao NS (1977) Soil microorganisms and plant growth. Oxford & IBH Publishing Co., New Delhi, p 289

Sumbul A, Mahmood I, Rizvi R, Safiuddin and Ansari RA (2017). Mycorrhiza: an alliance for the nutrient management in plants In: Probiotics in agroecosystem (Kumar V. Ed.), Springer Nature, Singapore Pp. 371–386

Sutton JC, Barron GL (1972) Population dynamics of endogone spores in soil. Can J Bot 50:1909–1914

Sylvia DM (1998) Mycorrhizal symbioses. In: Sylvia DM, Furmann JJ, Hartel PG, Zuberar DA (eds) Principles and applications of soil microbiology. Prentice Hall, Upper Saddle River, NJ, pp 408–426

Trappe JM (1982) Synoptic key to the genera and species of Zygomycetous myconhizal fungi. Phytopathology 72:297–304

Uibopuu A, Moora M, Saks Ü, Daniell TJ, Zobel M, Öpik M (2009) Differential effect of arbuscular mycorrhizal fungal communities from ecosystems along management gradient on the growth of forest understory plant species. Soil Biol Biochem 41:2141–2146

Vaingankar JD (2012) Studies on arbuscular mycorrhizal (am) fungi in ornamental flowering plants commonly found in Goa. Doctoral dissertation, Goa University

Varma AK, Trappe JM, Fortin JA, Norris JR (1991) Wet-sieving and decanting technique for spores extraction. In: Methods in microbiology, Experiments with Mycorrhizae, vol 23. Academic Press, London

Walker C, Vestberg M (1998) Synonymy amongst the arbuscular mycorrhizal fungi: Glomus claroideum, G. maculosum, G. multisubstenum and G. fistulosum. Ann Bot 82(5):601–624

Warner A, Mosse B, Dingemann L (1985) The nutrient film technique for inoculum production. In: Molina R (ed) Proceedings of the 6th NACOM. Oregon State University, Corvallis, pp 85–86

Weickel B, Dodd JC, Dehne HW (1997) Isolation and identification of arbuscular mycorrhizal fungi (AMF). In: Diagnosis and identification of plant pathogens. Springer, Dordrecht, pp 353–356

Wright SF, Morton JB, Sworobuk JE (1987) Identification of a vesicular-arbuscular mycorrhizal fungus by using monoclonal antibodies in an enzyme-linked immunosorbent assay. Appl Environ Microbiol 53(9):2222–2225

Wyss P, Bonfante P (1993) Amplification of genomic DNA of arbuscular-mycorrhizal (AM) fungi by peR using short arbitrary primers. Mycol Res 97:1351–1357

Xavier LJC (1999) Effects of interactions between arbuscular mycorrhizal fungi and Rhizobium leguminosarum on pea and lentil PhD dissertation, Saskatoon, Saskatchewan, Canada: University of Saskatchewan

Xavier LJC, Boyetchko SM (2003) Arbuscular mycorrhizal fungi in plant disease control. https://doi.org/10.1201/9780203913369.ch16

Xavier LJ, Boyetchko SM (2004) Arbuscular mycorrhizal fungi in plant disease control. In: Arora DK, Bridge PD, Bhatnagar D (eds) Fungal biotechnology in agricultural, food, and environmental applications. Marcel Dekker, New York, pp 183–194
Zhao R, Guo W, Bi N, Guo J, Wang L, Zhao JI, Zhang J (2015) Arbuscular mycorrhizal fungi affect the growth, nutrient uptake andwater status of maize (*Zea mays* L.) grown in two types of coal minespoils under drought stress. Appl Soil Ecol 88:41–49

Part II

Applications of Mycorrhizal Fungi

Root Exudates and Their Importance in Arbuscular Mycorrhizal Symbiosis and Nutrients Navigation from Inaccessible Soil: An Efficient Mediator of Mineral Acquisition in Nutrient Deprived Soil

5

Ravichandran Koshila Ravi and Thangavelu Muthukumar

Abstract

The growth and development of plants happen through internal molecular communications that rely on adequate nutrient supplements to roots from the soil. Plants take mineral nutrients from the rhizosphere, where the microbes interact with plant root exudates. The root exudates secreted by the plants into the rhizosphere have a pivotal function in mediating the relationship between plants and soil microbes. These metabolites from plant roots have different responses toward soil organisms and modify their composition and activity. Plants utilize root exudates to balance the role of mineral nutrient transporters in recognizing the availability of nutrients and further the direct supply of nutrients depending on the demand. The root exudates are affected by fungal communities associated with the roots depending upon their abundance and composition. Root exudation commences and regulates the communications channel between plant roots and the associated soil microbes like arbuscular mycorrhizal (AM) fungi. In addition, the root exudates act as signals that instigate the establishment of AM symbiosis. During the pre-symbiotic phase, some metabolites secreted by roots are necessary for AM fungal spore germination, hyphal growth, and root colonization. The metabolites in the root exudates are sensed by AM fungi, and these metabolites may be stimulatory or inhibitory to the symbiotic establishment. In this chapter, we highlight the role of root exudates in plant nutrient availability, the establishment of AM symbiosis, and also the influence of AM fungi on plant root exudates.

R. Koshila Ravi · T. Muthukumar (✉)
Root and Soil Biology Laboratory, Department of Botany, Bharathiar University, Coimbatore, Tamilnadu, India

R. A. Ansari et al. (eds.), *Mycorrhizal Symbiosis and Agroecosystem Restoration*, https://doi.org/10.1007/978-981-99-5030-0_5

Keywords

Hyphosphere · Mycorrhizosphere · Nutrient cycling · Nutrient stress · Rhizodeposits · Rhizosphere · Symbiosis

5.1 Introduction

The lack of mobility has rendered plants to adopt strategies to reach for resources that are heterogeneously distributed in the environment. This has resulted in the indefinite growth of shoots and roots from seed germination to senescence. In addition to reaching toward the resources, plants also adopt strategies that would enable them to make resources available in their vicinity, thereby reducing the cost of resource acquisition. One such mechanism is the creation of the rhizosphere, the region of the soil that is influenced by plant roots. The term rhizosphere, coined by Hiltner (1904), is a complicated ecosystem and a hotspot for millions of diverse microorganisms. According to Pinton et al. (2007), the rhizosphere is the soil region influenced by the exudation from plant roots and colonized by microorganisms. For example, the populations of the bacteria have been estimated to be 10 to 100 times more in the rhizosphere than in the bulk soil. However, the population and diversity of microorganisms in the rhizosphere vary with plant genotypes, developmental stages of plants, and the soil environment (Liu et al. 2020). The plant rhizosphere can be broadly differentiated into three different zones (Fig. 5.1). The ectorhizosphere or exorhizosphere is the soil outside the plant roots that are affected by rhizodeposits, root surface, rhizoplane, and the endorhizosphere: region of root tissues flanked by the epidermis and the endodermis, i.e., the root cortex. Based on these, the rhizosphere microbiome is classified into endophytic (residing within roots), ectophytic (rhizoplane), and rhizospheric (present in soil affected by roots) (Edwards et al. 2015). The microorganisms coexisting in the different regions of the rhizosphere are involved in different types of dynamic interactions, and fluctuating metabolites in root and rhizodeposits originating from plant roots shape the rhizosphere microbiota. The interactions among organisms in the rhizosphere can be neutral, positive, or negative (Hernandez et al. 2015; Ansari and Mahmood 2017, 2019a, b). Although the role of the rhizosphere microbiome on plant health is well recognized, the various interactions among these microorganisms are rather obscure (Ansari et al. 2017a, b). However, some progress has been made in understanding the interactions among the rhizosphere microbiome and their effect on plant health through the use of sequencing technologies along with the application of proteomic, transcriptomic, metagenomic, metatranscriptomic, and metabolomic approaches (Schlaeppi and Bulgarelli 2015; Olanrewaju et al. 2019).

Of the various microorganisms inhabiting different regions of the rhizosphere, the most influential ones are the endophytic microorganisms that reside in the endorhizosphere of plant roots. Nevertheless, some microorganisms like mycorrhizal fungi extend their presence in the ecto- and endorhizosphere, thereby directly bridging the root to the soil environment (Ansari et al. 2019a, b). Among the different

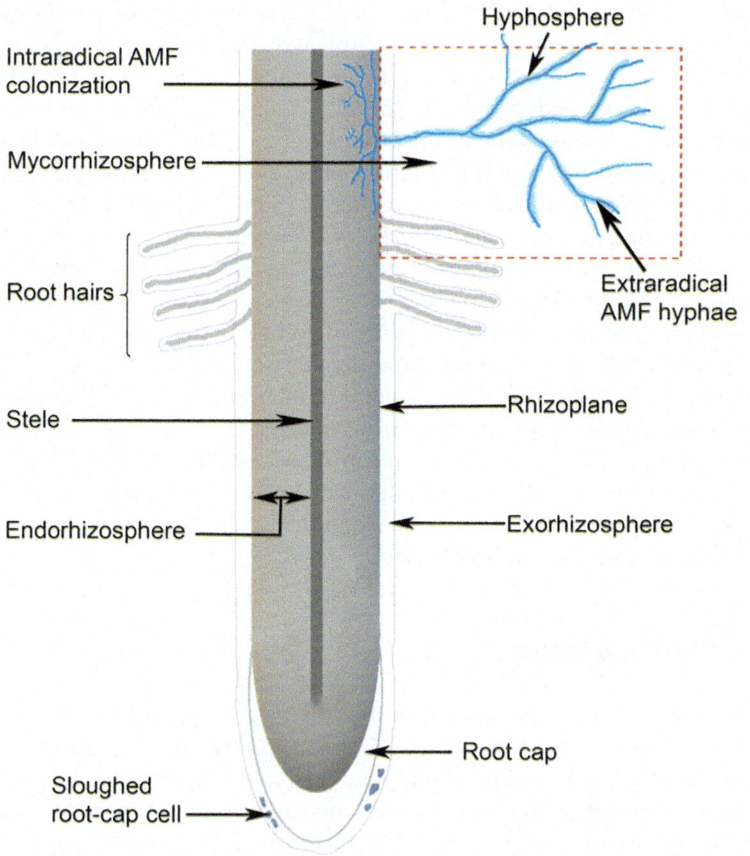

Fig. 5.1 Structure of rhizosphere and mycorrhizosphere. *AMF* arbuscular mycorrhizal fungi

types of mycorrhizal fungi, the most common and widespread is the arbuscular mycorrhizal (AM) fungi which occur in more than 70% of the terrestrial plant species (Brundrett and Tedersoo 2018). The AM fungi are obligate symbionts, and the establishment of symbiosis is mandatory to complete the fungal life cycle (Smith and Read 2008). Indeed, different AM fungi may colonize the same host root system and functionally complement in rendering host benefit (Jansa et al. 2008).

The establishment of the symbiosis between the AM fungus and plant roots happens with the germination of spores or soil hyphae originating from mycorrhizal roots. Studies have shown that a large number of molecules produced by fungi and plants are involved in the chemical dialogue before any physical contact in the symbiotic establishment (Sun et al. 2015). The AM fungus develops an asymbiotic phase involving the germination of spores, the development of germ tubes, and limited presymbiotic mycelium prior to the establishment of the symbiosis. The energy for this asymbiotic phase is obtained from the metabolism of triglycerides stored in the fungal structures (Giovannetti et al. 2010). Once established, the AM fungi produce the extraradical hyphae, which extend from the roots and explore the surrounding soil (Sumbul et al. 2017). Like plant roots, AM fungi also exude metabolites into the surrounding soil resulting in the formation of a hyphosphere (Fig. 5.1). For example, the extraradical hyphae of *Rhizophagus clarus* originating from roots of *Allium fistulosum* and *Linum usitatissimum* release acid phosphatases that hydrolyze organic phosphatic sources and enhances the availability of inorganic phosphorus in the soil solution under in vitro conditions (Sato et al. 2015). The extraradical hyphae of AM fungi *Gigaspora margarita* or *Claroideoglomus etunicatum* colonizing roots if *Allium cepa* seedlings secrete citric acid capable of solubilizing iron phosphate ($FePO_4$) in the soil solution (Tawaraya et al. 2006). In addition to these, the hyphosphere also maintains a distinct microbiome that is specific for an AM fungal taxon (Zhou et al. 2020). These demonstrate that AM fungi can directly mediate the availability of nutrient elements in the soil. The rhizosphere along with the hyphosphere is known as the mycorrhizosphere. The formation and functioning of the mycorrhizosphere are dealt with in detail elsewhere (Priyadharsini et al. 2016). In this chapter, the role and importance of root exudates are discussed in detail. Root exudates and their importance in the establishment and functioning of AM symbiosis have also been unraveled.

5.2 Root Exudates

Roots perform several functions in addition to anchoring plants to the substrates. Apart from acquiring nutrients and water from the soil or other substrates, roots also modify the soil structure and function (Fig. 5.2). A large number of microorganisms reside in the rhizosphere surrounding the roots, and several of these plays a central role in plant growth and development. The rhizospheric microbiota is different from those that inhabit the bulk soil. For instance, the majority (~80%) of the rhizospheric bacteria is larger >0.3 μm than those in the bulk soil (~37%) (Olanrewaju et al. 2019). This rhizosphere microbiota plays an important role in facilitating nutrient availability to plants and protecting the roots against pathogenic invasion (Rizvi et al. 2015; Solanki et al. 2020). The interaction among soil microbes inhabiting the rhizosphere and plants is mostly mediated by root exudates. These exudates monitor several crucial ecosystem processes, including soil biogeochemical cycles (Meier et al. 2017), modify the structure of soil, solubilize/mobilize nutrients (Pantigoso et al. 2020), and liberate allelochemicals which may promote or suppress plant growth (Bouhaouel et al. 2019) and also regulate the establishment of interaction among different soil microbial communities and plant–microbe associations (Olanrewaju et al. 2019). Normally plant roots exude a diverse range of plant metabolites (Table 5.1), and it has been estimated to involve up to 10–50% of the carbon fixed by plants (Korenblum et al. 2020). Primary metabolites such as amino acids, carbohydrates, membrane lipids, and organic acids exuded by roots into the soil provide nutrients and energy to the soil microorganisms (Canarini et al. 2019). Root exudates also contain secondary metabolites such as flavonoids,

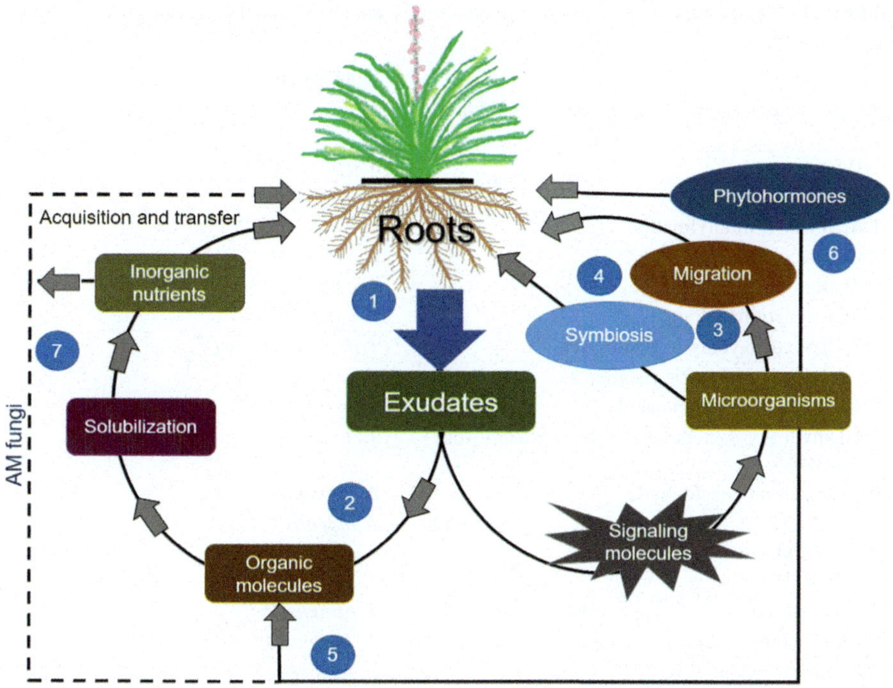

Fig. 5.2 Influence of root exudates on the nutrient availability in the rhizosphere. (1) Nutrient stressed plants exude metabolites into the rhizosphere; (2) Some exudates (e.g., organic acids) directly solubilize the organic molecules releasing the inorganic nutrients that are directly taken up by plant roots; (3) Certain metabolites in the root exudates act as signaling molecules (e.g., flavonoids) chemotactically attracting the microorganisms resulting in changes in their diversity and abundance; (4) Signal molecules (e.g., strigolactones) also play an important role in the reorganization and establishment of symbiosis between plant and microorganisms; (5) Inorganic nutrients are made available indirectly to the plants through microbial mediated nutrient solubilization and fixation; (6) Phytohormones produced by microorganisms increase root proliferation; (7) Available inorganic nutrients are acquired from the soil and transported to plant roots through arbuscular mycorrhizal fungi (AMF) (shown in broken lines)

phenolics, and terpenoids; inorganic molecules such as water and carbon dioxide, enzymes, nucleosides, and vitamins (Olanrewaju et al. 2019).

Plants release root exudates into the rhizosphere through active and passive transports. Generally, the process of root exudation secretion is passive and is regulated by various pathways including transport via vesicles and ionic channels in the root membrane (Maurer et al. 2021). The positive interaction involves a mutual relationship with beneficial soil organisms, including mycorrhizal fungi, plant growth-promoting rhizobacteria, etc. Likewise, negative interactions with plants comprise association with pathogenic microorganisms and invertebrate herbivores (Haldar and Sengupta 2015). The constituents of root exudates differ according to environmental conditions, plant species, root characters, developmental stages of the plant, and availability of nutrients (Herz et al. 2018). The composition of the

Table 5.1 Metabolites identified in root exudates of some terrestrial plant species under different types of stresses

Compounds	Plant species
1,2-Benzenedicarboxylic acid, mono (2-ethylhexyl) ester	*Flaveria bidentis*[14]
1-Hexacosene	*Scirpus triqueter*[9]
1-Hexadecanol, 2-methyl	*S. triqueter*[9]
1-Methoxy-3-indolylmethyl	*Brassica rapa*[13]
1-Octadeccane	*F. bidentis*[14]
2-(2-Butoxyethoxy)ethoxy	*Arachis hypogaea*[1]
2(3H)-Benzothiazolone	*F. bidentis*[14]
2,3,4 Trihydroxybutyric (isomer 1)	*Trifolium pratense*[12]
2,3,4 Trihydroxybutyric (isomer 2)	*T. pratense*[12]
2,3,4-Trihydroxybutyric acid	*A. hypogaea*[1]
2,4-Dihydroxy-7-methoxy-1,4-benzoxazin-3-one (DIMBOA)	*Triticum aestivum*[3]
2,4-Dimethyl benzaldehyde	*A. hypogaea*[1]
2,6,10,14,18,22-Tetracosahexaene, 2,6,10,1519,23-hexamethyl-(all-E)-	*F. bidentis*[14]
2-Aminoadipic	*T. pratense*[12]
2-Butenedioic	*T. pratense*[12]
2-Ethylhexanoic acid	*A. hypogaea*[1]
2-Furancarboxylic acid	*A. hypogaea*[1]
2-Hydroxy-3-butenyl	*B. rapa*[13]
2-Hydroxy-4-pentenyl	*B. rapa*[13]
2-Hydroxyglutaric	*T. pratense*[12]
2-Ketoglutaric acid	*S. triqueter*[9]
2-Mercaptobenzothiazole	*F. bidentis*[14]
2-methyl butyric acid	*S. triqueter*[9]
2-Monopalmitoylglycerol	*A. hypogaea*[1]
2-Monostearin	*A. hypogaea*[1]
2″-O-glucosyl-8-C-glucosylapigenin	*Desmodium incanum*[8], *Desmodium intortum*[8], *Desmodium uncinatum*[8]
2″-O-glucosylvitexin	*D. uncinatum*[8]
2-Picolinic acid	*A. hypogaea*[1]
2-Pyrrolidone carboxylic acid	*A. hypogaea*[1]
3- Butenyl	*B. rapa*[13]
3-Phenylpropanoic acid	*T. pretense*[12]
3,4-Dihydroxybutanoic acid	*A. hypogaea*[1]
3-Furoic acid	*A. hypogaea*[1]
3-Hydroxybutyric acid	*A. hypogaea*[1]
3-Indolylmethyl	*B. rapa*[13]
3-Pyridinecarboxylic acid	*A. hypogaea*[1]
4-Hydroxy-3,5-dimethoxy-, hydrazide	*S. triqueter*[9]
4-Hydroxy-3-indolylmethyl	*B. rapa*[13]

(continued)

Table 5.1 (continued)

Compounds	Plant species
4-Hydroxybenzoic acid	*T. pratense*[12]
4-Hydroxybutanoic acid	*T. pratense*[12]
4-Methoxy-3-indolylmethyl	*B. rapa*[13]
4-Methylbenzoate	*A. hypogaea*[1]
4-Pentenyl	*B. rapa*[13]
5-Oxo-proline	*A. hypogaea*[1]
5-Tetradecene	*S. triqueter*[9]
6,8-di-C-hexosylapigenin ferulate	*D. incanum*[8], *D. intortum*[8], *D. uncinatum*[8]
6,8-di-C-hexosylapigenin sinapinate	*D. incanum*[8], *D. intortum*[8], *D. uncinatum*[8]
6-C-arabinosyl-8-C- galactosylapigen	*D. incanum*[8], *D. intortum*[8], *D. uncinatum*[8]
6-C-arabinosyl-8-C-glucosylapigenin (isoschaftoside)	*D. incanum*[8], *D. intortum*[8], *D. uncinatum*[8]
6-C-galactosyl-8-C-arabinosylapigenin	*D. incanum*[8], *D. intortum*[8], *D. uncinatum*[8]
6-C-galactosyl-8-C-glucosylapigenin	*D. incanum*[8], *D. intortum*[8], *D. uncinatum*[8]
6-C-glucosyl-8-C- galactosylapigen	*D. incanum*[8], *D. intortum*[8], *D. uncinatum*[8]
6-C-glucosyl-8-C-glucosylapigenin (Vicenin-2)	*D. incanum*[8], *D. intortum*[8], *D. uncinatum*[8]
8-C-glucosylapigenin (vitexin)	*D. incanum*[8], *D. intortum*[8], *D. uncinatum*[8]
9,12-Octadecadienoic acid	*A. hypogaea*[1]
9-Octadecenamide, N,N-dimethyl-	*F. bidentis*[14]
9-Octadecenamide,	*F. bidentis*[14]
9-Octadecenamide	*S. triqueter*[9]
a-Amino-3-hydroxy-4-methoxyacetophenone	*S. triqueter*[9]
Acetic acid	*Oryza sativa* [2]
Aconitic acid	*T. pratense*[12]
Adipic acid	*A. hypogaea*[1]
Alanine	*A. hypogaea*[10], *T. pretense*[12]
Amine (3TMS)	*A. hypogaea*[1]
Arginine	*A. hypogaea*[10]
Aspartic acid	*A. hypogaea*[10], *T. pratense*[12]
Azelaic acid	*T. pratense*[12]
Azelaic acid (2TMS)	*A. hypogaea*[1]
Azulene	*A. hypogaea*[1]
Benzo[1,3]dioxolecarbonitriles	*Secale cereale* [4]
Benzoic acid	*A. hypogaea*[10], *S. triqueter*[9], *T. pretense*[12], *Citrullus lanatus*[7], *O. sativa*[7]
Butanedioic acid	*A. hypogaea*[1]
Butanoic acid	*A. hypogaea*[1]

(continued)

Table 5.1 (continued)

Compounds	Plant species
Butanol (4TMS)	*A. hypogaea*[1]
Butylated hydroxytoluene	*F. bidentis*[14]
C-hexosyl-C-pentosylapigenin ferulate	*D. incanum*[8], *D. intortum*[8], *D. uncinatum*[8]
C-hexosyl-C-pentosylapigenin sinapinate	*D. incanum*[8], *D. intortum*[8], *D. uncinatum*[8]
Cholestan-3-ol, 2-methylene-,(3 beta., 5alpha.)-(Methylthio)-acetonitrile)	*F. bidentis*[14]
Cinnamic acid	*Musa paradisiaca* AAA Cavendish cv. Brazil[15], *C. lanatus*[7], *O. sativa*[7]
Citric acid	*O. sativa*[2], *S. triqueter*[9], *T. pratense*[12]
Coumaric acid	*A. hypogaea*[10]
Crotonic acid	*S. triqueter*[9]
Cyanato- and carbonitrile-benzo[1,3]dioxoles	*S. cereale*[4]
Cyanatophenol	*S. cereale*[4]
Cycloeicosane	*F. bidentis*[14]
Decane, 2,4,6-trimethyl-	*F. bidentis*[14]
Decanoic acid	*S. triqueter*[9]
Deoxytetronic acid	*T. pratense*[12]
Di-n-octyl phthalate	*F. bidentis*[14]
D-pinitol	*T. pratense*[12]
Eicosane	*F. bidentis*[14]
Eicosanoic acid	*A. hypogaea*[1]
Ethanol, 2-methoxy- carbonate	*A. hypogaea*[1]
Ethyl citrate	*F. bidentis*[14]
Ethylmalonic/methylsuccinic acid	*T. pratense*[12]
Ferulic acid	*C. lanatus*[7]
Fructose	*T. pratense*[12]
Galactaric acid	*T. pratense*[12]
Galactopyranose	*A. hypogaea*[1]
Galactose	*T. pratense*[12]
Gallic acid	*O. sativa*[7], *C. lanatus*[7]
Gluconic acid (isomer 1)	*T. pratense*[12]
Gluconic acid (isomer 2)	*T. pratense*[12]
Glucose	*T. pratense*[12]
Glutamic acid	*A. hypogaea*[10], *S. triqueter*[9], *T. pretense*[12]
Glutaric acid (pentanedioic)	*T. pratense*[12]
Glyceric acid	*T. pratense*[12]
Glycerol	*T. pratense*[12]
Glycine	*A. hypogaea*[10], *T. pretense*[12]
Glycolic acid	*T. pratense*[12]
Heptacosane	*A. hypogaea*[1]
Heptacosane	*F. bidentis*[14]
Heptadecanoic acid	*A. hypogaea*[1]

(continued)

Table 5.1 (continued)

Compounds	Plant species
Hexadecane	*A. hypogaea*[1]
Hexadecanoic acid	*S. triqueter*[9], *A. hypogaea*[1]
Hexagecane 2,6,10,14-tetramethyl	*F. bidentis*[14]
Histidine	*A. hypogaea*[10]
Homoserine	*T. pratense*[12]
Isoleucine	*T. pratense*[12]
Isoschaftoside	*D. intortum*[8], *D. uncinatum*[8]
Lactic acid	*T. pratense*[12]
Lauric acid	*A. hypogaea*[1]
L-Indole-3-lactic acid	*T. pratense*[12]
Lysine	*A. hypogaea*[10]
Malic acid	*O. sativa*[2], *T. pretense*[12]
Malonic acid (propanedioic)	*T. pratense*[12]
Maltose	*A. hypogaea*[1], *T. pretense*[12]
Methylmalonic	*T. pratense*[12]
Myoinositol	*A. hypogaea*[1], *T. pretense*[12]
Myristic acid	*A. hypogaea*[1], *S. triqueter*[9]
Nicotinic acid	*A. hypogaea*[1]
N-Methacryloylglycine	*T. pratense*[12]
Nonadecane	*F. bidentis*[14]
Nonahexacontanoic acid	*F. bidentis*[14]
Octacosane	*F. bidentis*[14]
Octadecane	*A. hypogaea*[1]
Octadecanoic acid	*S. triqueter*[9]
Octanedioic acid	*A. hypogaea*[1]
Octanoic acid	*S. triqueter*[9]
Oleic acid	*A. hypogaea*[1], *S. triqueter*[9]
O-phthalic acid	*S. triqueter*[9]
Oxalic acid	*T. pratense*[12]
Oxoproline	*T. pratense*[12]
p-coumaric acid	*C. lanatus*[7], *O. sativa*[7]
Peagol	*Pisum sativum*[5]
Peagoldione	*P. sativum*[5]
Pentadecane	*F. bidentis*[14]
Pentadecanoic acid	*S. triqueter*[9]
Pentanedioic acid	*A. hypogaea*[1]
Pentanoic acid	*S. triqueter*[9]
Phen-1,4-diol, 2,3-dimethyl-5-trifluoromethyl-	*S. triqueter*[9]
Phthalic acid	*M. paradisiaca*[15], *C. lanatus*[7], *O. sativa*[7]
P-hydroxybenzoic acid	*C. lanatus*[7], *O. sativa*[7], *A. hypogaea*[10]
P-phthalic acid	*S. triqueter*[9]
Proline	*A. hypogaea*[10], *T. pretense*[12]
Propachlor	*A. hypogaea*[1]
Propanedioic acid	*S. triqueter*[9], *A. hypogaea*[10]

(continued)

Table 5.1 (continued)

Compounds	Plant species
Propanoic acid	*A. hypogaea*[1]
Protocatechuic acid	*T. pratense*[12]
Rhamnose	*T. pratense*[12]
Ribitol	*A. hypogaea*[1]
Ribonic acid	*A. hypogaea*[1]
Ribose	*T. pratense*[12]
Ryecarbonitriline A	*S. cereal*[4]
Ryecyanatine A	*S. cereal*[4]
Ryecyanatine B	*S. cereal*[4]
Saccharose	*T. pretense*[12]
Salicylic acid	*M. paradisiaca*[15], *C. lanatus*[7], *O. sativa*[7], *A. hypogaea*[1]
Serine	*T. pratense*[12]
Shikimic acid	*T. pratense*[12]
Silanamine	*A. hypogaea*[1]
Silane	*A. hypogaea*[1]
Silanol, trimethyl-, phosphate (3:1)	*A. hypogaea*[1]
Soyasapogenol B [olean-12-ene-3,22,24-triol(3β,4β,22β)]	*Vicia sativa*[6]
Succinic acid	*S. triqueter*[9], *T. pretense*[12]
Sulfurous acid, butyl heptadecyl ester	*F. bidentis*[14]
Syringic acid	*C. lanatus*[7], *O. sativa*[7]
Tartaric acid	*O. sativa*[2], *A. hypogaea*[1], *T. pretense*[12]
Terephthalic acid	*T. pratense*[12]
Tetradecanoic acid (1TMS)	*A. hypogaea*[1]
Tetradecanoic acid	*S. triqueter*[9]
Thieno[3,2-e]benzofuran	*F. bidentis*[14]
Threonine	*A. hypogaea*[10], *T. pretense*[12]
Trans-22-dehydrocampesterol [(ergosta-5,22-dien-3-ol, (3β,22E,24S)]	*V. sativa*[6]
Tridecane	*A. hypogaea*[1]
Tridecane,5-propyl-	*F. bidentis*[14]
Tris borate (1TMS)	*A. hypogaea*[1]
Tris(hydroxymethyl)aminomethane (3TMS)	*A. hypogaea*[1]
Uracil	*T. pratense*[12]
Urea (3TMS)	*A. hypogaea*[1], *T. pratense*[12]
Valine	*A. hypogaea*[10], *T. pretense*[12]
Vanillic acid	*S. triqueter*[9], *O. sativa*[7], *C. lanatus*[7]
Vicenin-2	*D. intortum*[8]
Vitexin	*D. uncinatum*[8]

[1]Ankati et al. (2019); [2]Bhattacharyya et al. (2013); [3]Chen et al. (2010); [4]Cimmino et al. (2015); [5]Evidente et al. (2009); [6]Evidente et al. (2011); [7]Hao et al. (2010); [8]Hooper et al. (2015); [9]Hou et al. (2015); [10]Li et al. (2013); [11]Liu et al. (2015); [12]Molina et al. (2021); [13]Schreiner et al. (2011); [14]Xing et al. (2014); [15]Yuan et al. (2018)

root exudate is also determined by the photosynthetic modes of the plants. As erythritol, inositol, and ribitol are dominant sugars in root exudates of C4 plants, ribose, maltose, and mannose are the prominent sugars in C3 plant root exudates. Moreover, C4 plants release greater numbers of amino acids and organic acids compared to the greater exudation of organic carbons and carbohydrates by the C3 plants (Olanrewaju et al. 2019). The developmental stage of the plants also affects the nature and extent of root exudation. Young plants excrete more exudates than older plants due to the asymmetrical portioning of carbon to shoots and roots during the early stages of plant development (Pausch and Kuzyakov 2018). Continuous exudation of metabolites and antimicrobial substances (phytoanticipins and phytoalexins) incurs a significant carbon cost to the plant (Bamji and Corbitt 2017). Therefore, strict regulation on the release of root deposits and other plant metabolites is essential to reduce the plant's energy cost. Most of our understanding of the cost–benefit ratio of root exudates on plant health has arisen from the studies on the role of root exudates in plant-phytopathogen interactions (Olanrewaju et al. 2019).

5.3 Plant Nutrient Stress and Root Exudation

Plants growing in stressed environments release a wide range of metabolites into the rhizosphere (Table 1), and the exudation of various inorganic and organic molecules into the rhizosphere mostly depends on the nutritional status of the plant (Vives-Peris et al. 2020). For example, certain plant species release more organic acids into the rhizosphere during iron and phosphorus deficiency and phytosiderophores during zinc and iron deficiency (Olanrewaju et al. 2019). The release of these compounds increases the availability of these nutrient elements in the soil to be acquired by roots (Fig. 5.2). Plants like *Nicotina tabacum*, *Beta vulgaris*, *Hyoscyamus albus*, and *Medicago truncatula* secretes riboflavin or its derivatives under iron-deficient conditions (Chen et al. 2017). Likewise, *Arabidopsis thaliana* and *Brassica napus* excrete the phenolic compound coumarin to acquire iron from substrates that are highly alkaline (Clemens and Weber 2016). Carvalhais et al. (2011) studied the influence of nitrogen, phosphorus, potassium, and iron deficiencies on the exudation patterns of amino acids, sugars, and organic acids in *Zea mays* plants. The results of the study indicated that root exudates of iron-deficient *Z. mays* plants contained higher concentrations of citrate, glucose, glutamate, and ribitol, whereas plants deficient in phosphorus exuded more carbohydrates and *c*-aminobutyric acid. Potassium deficiency reduced the excretion of sugars, especially fructose, glycerol, maltose, and ribitol, while nitrogen deficiency decreased the quantities of amino acids released (Carvalhais et al. 2011). Further specified concentrations of nitrogen or phosphorus upregulate the production of secondary metabolites like the isoflavonoids in plants, and these metabolites, when secreted into the soil, act as signal molecules to soil microbiota (White et al. 2017). Reciprocally, the presence of microbes in the rhizosphere has a powerful influence on the rate of root exudation, facilitating the uptake of some nutrient elements by plant roots. This clearly shows that the quantity and quality of metabolites in root exudates are altered with plants' exposure to nutrient limitations and can act as a distinct signaling factor in the rhizosphere. Once within the rhizosphere, several bacteria undergo rapid multiplication in response to the flavonoid molecules released by roots. This in turn triggers plants to exude more new or existing flavonoids in the rhizosphere (Olanrewaju et al. 2019). Several organic molecules released into the rhizosphere, especially carbohydrates, amino acids, proteins, vitamins, and organic acids, promote ethylene synthesis by microorganisms, which is a powerful chemical signal in controlling plant development. Finally, these components play various roles in nutrient element acquisition by plants. Strigolactone levels in root exudates of maize cultivars with varied susceptibility to *Striga* infestation were low or below detectable limits under normal nutrient conditions (Yoneyama et al. 2015). But, a deficiency of nitrogen and phosphorus increased the presence of strigolactone in root exudates of both cultivars (Yoneyama et al. 2015). Phosphorus limitation also increases the exudation of citrate by tobacco roots (Del-Saz et al. 2017). All these clearly show that plant nutrient stress could qualitatively and quantitatively modify root exudates to increase the availability of nutrients in the soil.

5.4 Propagules of AM Fungi

The propagule composition of AM fungi can have a strong influence on the establishment of symbiosis with plants. The different kinds of AM fungal propagules include mycorrhizal roots, soil hyphae, and spores (Smith and Read 2008; Varela-Cervero et al. 2016). Of these, specific types of propagules dominate the mycorrhization of plant roots in various habitats. For example, spores appear to be the dominant form of propagules in seasonal and disturbed vegetations, whereas soil hyphae and mycorrhizal roots are the chief propagules in vegetations where new roots are formed throughout the year (Paz et al. 2021). Moreover, AM fungal taxa vary in their strategies of colonization, and these differences are often associated with the type of propagules involved. Mycorrhizal roots are the major propagules for taxa in Claroideoglomeraceae and Glomeraceae rather than the soil hyphae or spores (Varela-Cervero et al. 2016). Contrarily, spores are the primary propagules for initiating colonization in roots by taxa in Gigasporaceae. Taxa in Diversisporaceae and Pacisporaceae adopt an intermediate strategy of initiating colonization from both root and soil-based propagules (Varela-Cervero et al. 2016). Nonetheless, exceptions tend to persist like certain members of Glomeraceae where effective colonization of roots is initiated through spores as well as soil hyphae (Varela-Cervero et al. 2016). Previous studies have shown the variation in the longevity of different types of AM fungal propagules. Spores and mycorrhizal roots are considered to be more resistant to soil disturbance than soil hyphae (Varela-Cervero et al. 2015). Several factors influence the initiation of mycorrhization by different types of propagules. These involve a large number of interacting host-fungus-soil factors (Fig. 5.3).

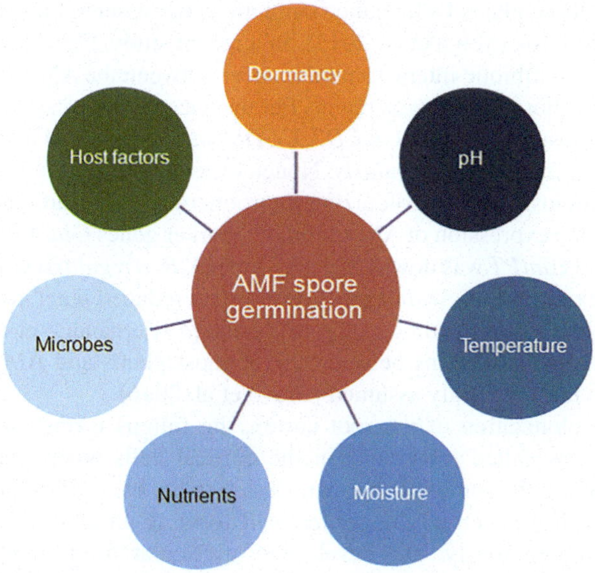

Fig. 5.3 Factors affecting arbuscular mycorrhizal fungi (AMF) spore germination. Soil factors are indicated in shades of blue

5.5 Establishment of the Symbiosis

The process of mycorrhization begins with the reciprocal exchange of signaling molecules before any physical contact between the engaging symbiotic partners (Bonfante and Genre 2015). Plant roots exudate strigolactones, the signal molecules that are recognized by the fungal symbiont, which eventually induce substantial AM fungal hyphal branching (Mori et al. 2016; Lanfranco et al. 2018a, b). Increasing hyphal branching enhances the chances of the fungus coming into contact with the host root (Besserer et al. 2006). In turn, the fungus releases signal molecules termed "Myc factors" into the rhizosphere, which are recognized by the host roots (Maillet et al. 2011). Recognition of Myc factors by the host roots activates unique calcium spiking in the root cells. This initiates a symbiotic program called the "Sym pathway," which comprises a morphological, physiological, and transcriptional change in the host roots to harbor the fungal symbiont (Gough and Cullimore 2011). After the recognition of the molecular signals between the symbionts, the fungus enters the root and colonizes the root cortex. For the successful establishment of functional symbiosis, the fungus has to modify plants' defense signaling to decrease defense reactions in the root (Fernández et al. 2019). To overcome the host defense, AM fungus, *Rhizophagus irregularis*, secretes effector proteins like SP7 to suppress host defense and allows the fungus to establish itself in the root cortex of the host plants (Kloppholz et al. 2011). Nevertheless, several studies have reported limited colonization of nonhost plant roots by AM fungi when raised along with mycorrhizal host plants despite these plants lacking the symbiotic genes essential for establishing the functional symbiosis (Veiga et al. 2013). In a recent study, Fernández et al. (2019) showed that presymbiotic interactions can occur between the AM fungus *R. irregularis* and the nonhost *Arabidopsis* plants. For instance, the expression of the specific strigolactone biosynthesis genes (*CCD7*, *CCD8*) was upregulated in the *Arabidopsis* roots in the presence of *R. irregularis*. Nonetheless, the fungus failed to establish a functional symbiosis despite colonizing the root cortex. In addition to the absence of arbuscular, the expression of AM symbiotic marker genes *GintAMT2*, *GintMST2*, *GintMST4*, and *GintPT* was downregulated. Indeed, *R. irregularis* colonization triggered defense response in *Arabidopsis,* resulting in reduced plant growth and resistance against the fungal pathogen *Botrytis cinerea* (Fernández et al. 2019). This indicates that the interactions between the nonhost plants and AM fungi may be more intricate than generally assumed (Bravo et al. 2016).

After the colonization of the root cortex, the fungus forms highly branching hyphal structures called arbuscules in the cortical cells where the exchange of resources between the symbiotic partners occurs (Choi et al. 2018; Voß et al. 2018). The fungal hyphae penetrate the cortical cell forming an arbuscular trunk before elaborating dichotomously. The developing arbuscular trunk and the arbuscular branches are enveloped by the periarbuscular membrane, which is continuous with the cortical cell membrane (Choi et al. 2018). Despite their intricate structure and functional significance, arbuscules are ephemeral, with a limited lifespan of a few days. The mechanism that initiates arbuscule senescence is obscure. Normally the arbuscular degeneration commences in the fine branches and slowly progresses toward the arbuscular trunk. The collapse of the arbuscules is preceded by the retraction of the cytoplasm and septation (Choi et al. 2018).

5.6 Role of Root Exudates in Mycorrhization

Root exudates, in addition to structuring the soil microbiota, play a pivotal role in regulating AM symbiosis. Available evidence does suggest that plant root exudates can positively influence the rate of AM spore germination and proliferation of hyphae (Nagata et al. 2016). The addition of activated carbon to the soil reduces mycorrhization and alters the AM fungal community composition in *Solidago canadensis* (Yuan et al. 2014). These changes in AM status of *S. canadensis* were attributed to the absorption of molecules released by the roots by activated carbon and confirm the view that root exudates play an important role in plant–AM fungal interactions (Yuan et al. 2014). Tian et al. (2021) investigated the presence of flavonoids quercitrin and quercetin in root exudates of different populations (introduced vs. native) of the invasive *Triadica sebifera* and examined its impact on AM fungi. The results of the study indicated the presence of more quercetin in root exudates and higher AM colonization in the introduced *T. sebifera* population. Further, the application of quercetin or root exudates from introduced population enhanced AM colonization in target *T. sebifera* plants and germination of *Funneliformis mosseae* spores (Tian et al. 2021). These observations suggest variations in flavonoid content in root exudate may play a vital role in increasing AM association and performance of invasive plants.

The occurrence of factors of unknown chemical nature in root exudates can influence the establishment of AM symbiosis. For example, exudates produced by roots of tomato cultivar Micro-Tom M161, a pre-mycorrhizal colonization mutant, failed to induce hyphal growth and branching of the AM fungus, *Rhizophagus intraradices* (Sun et al. 2012). This failure of M161 tomato root exudates to influence the presymbiotic development of *R. intraradices* was attributed to the absence of an active factor in the root exudates. The active factor isolated from root exudates of wild tomatoes was shown to promote fungal development, hyphal branching, and the development of viable spores in dual culture conditions (Sun et al. 2012). The phytohormone jasmonic acid that regulates plant growth and development, in addition to modulating plants' response to various types of stresses, also affects AM symbiosis. Nagata et al. (2016) studied the influence of Red/Far Red light conditions on strigolactone and jasmonic acid signaling mediated AM (*R. irregularis*) symbiosis in legume (*Lotus japonicus*) and nonlegume (*Solanum lycopersicum*) plants. The results of the study indicated that plants grown in high Red/Far Red light accumulated and excreted more jasmonic acid in their root exudates resulting in increased hyphal length, enhanced colonization, and mycorrhization of plants (Nagata et al. 2016).

A qualitative and quantitative variation in the composition of strigolactone composition significantly affects the interactions between AM fungi and the plant host. Mutants of tomato and pea deficient in strigolactone production were less susceptible to AM fungi and exhibited low levels of colonization compared to their respective wild types (Gomez-Roldan et al. 2008; Koltai et al. 2010). Moreover, strigolactone plays a significant role in stimulating AM fungal hyphal branching in *Gigaspora margarita* colonizing crop plants. Nevertheless, the stimulatory activity of the nonhydroxy-strigolactone (5-deoxystrigol) was thirty times greater than the hydroxy-strigolactone sorgomol (Akiyama et al. 2010). In addition to exudates released from roots, seed exudates also influence the germination of AM fungal spores to a greater extent. For example, crude seed exudates of *Sesbania virgata* stimulated the spore germination and mycelial development of a symbiotic phase of *Gigaspora albida* (Coelho et al. 2019). Nevertheless, the stimulatory effect of seed exudate appears to be concentration dependent, as increasing concentrations of *S. virgata* root exudates inhibited spore germination of *G. albida* (Coelho et al. 2019).

5.6.1 AM Induced Changes in Root Exudates

Apart from being influenced by root exudates, AM fungi can also modulate the composition of the root exudates produced by plant roots. Inoculation of AM fungus, *Acauloapora scrobiculata* decreased the relative abundances of aldehydes, alkanes, olefins, and sterols but increased the relative abundances of nitriles and ketones in root exudates of peach seedlings (Lǚ et al. 2019). The presence of *A. scrobiculata* in peach roots also reduced the presence of allelochemicals like benzoic acid, diisooctyl phthalate, n-hexadecanoic acid, phenols, and sterols (Lǚ et al. 2019). Root exudates of maize had reduced citric, fumaric, malic, malonic, oxalic, and T-aconitic acids when colonized by diverse AM fugal taxa like *C. etunicatum*, *Diversispora versiformis, F. mosseae,* and *R. intraradices* (Hussain et al. 2021). Although it is generally presumed that AM fungi are nonhost specific, a certain level of partner preference has been reported (Davison et al. 2016; Põlme et al. 2018). Evidence suggests that exudates from mycorrhizal roots could affect the germination of AM fungal taxa differently. An examination of five chickpea genotypes colonized by *R. intraradices* indicated variations in the composition of the methanolic fractions of the root exudates (Ellouze et al. 2012). Moreover, some of these methanolic fractions inhibited the germination of *Claroideoglomus* and *Gigaspora rosea* spores under in vitro conditions. Certain proteins in the root exudate bioactive fractions were identified which are involved in plant defense responses indicating that these chemicals could play some role in the patterns of partner preferences in AM symbiosis (Ellouze et al. 2012).

Changes in the composition of root exudates by AM fungi depend on nutrient availability in the soil. Colonization by *R. intraradices* generally decreased the exudation of carboxylates by tobacco roots in both phosphorus-deficient and sufficient soils (Del-Saz et al. 2017). More specifically, *R. intraradices* colonization reduced the citrate concentration by 72% and malate concentration by 50% under phosphorus-limiting conditions. However, no significant changes were noted in lactate, fumarate, and oxalate concentrations in root exudates of nonmycorrhizal and mycorrhizal tobacco plants growing under phosphorus-sufficient and deficient conditions (Del-Saz et al. 2017). A gas chromatography–mass spectrometry analysis of root exudates of tomato roots colonized by *F. mosseae* indicated an AM-dependent hike in sugars like malate and glucose and a decrease in amino acids and organic acids (Hage-Ahmed et al. 2013). These AM-mediated changes in root exudates are shown to reduce the incidence of pathogenicity in plants. Inoculation with AM fungus *F. mosseae* and *Fusarium oxysporum* decreased the incidence of wilt disease in *Citrullus lanatus* (Ren et al. 2015). The exudation of free amino acid from *C. lanatus* roots decreased by 50–85%, but malic acid and *p*-coumaric acid exudation increased by 142% and 62%, respectively (Ren et al. 2015).

5.6.2 AM-Mediated Changes in Rhizosphere Microflora

It is now well known that AM-induced changes in root exudates can modify microbial diversity, thereby indirectly affecting nutrient availability in the rhizosphere (Fig. 5.2). Field inoculation of four autochthonous shrub species *Lavandula dentata, Salvia officinalis, Santolina chamaecyparissus,* and *Thymus vulgaris* growing in a semiarid soil with the AM fungus, *R. intraradices* affected changes in the rhizosphere bacterial community structure and composition as well as enhanced plant growth (Rodríguez-Caballero et al. 2017). Generally, AM symbiosis promotes or represses certain groups of microorganisms in the rhizosphere. For example, different studies have shown an increase in the abundance of anaerobic bacteria belonging to the family Anaerolineaceae in response to AM fungal inoculation (Rodríguez-Caballero et al. 2017; Cao et al. 2016; Qin et al. 2016). Taxa in Anaerolineaceae are shown to degrade microbial products like sugars and other cellular components (Miura and Okabe 2008) like the chitin that constitute the fungus cell wall besides plant and microbial deposits, thereby aiding nutrient cycling in the rhizosphere (Rodríguez-Caballero et al. 2017). Similarly, changes in the abundance of bacteria (*Agrococcus*, *Bacillus*, *Skermanella*, *Microbacterium*, *Nitrospira*, *Planomicrobium*, *Gemmatimonas*, *Lysobacter*, *Streptomyces*) and fungi (*Penicillium*) in the rhizosphere of mycorrhizal plants are often correlated to increased nutrient availability and plant growth (Cao et al. 2016; Rodríguez-Caballero et al. 2017; Hao et al. 2021). Moreover, an increase in the abundance of plant growth-promoting rhizobacteria in the rhizosphere of AM plants could also increase root proliferation and modify root architecture, thereby increasing nutrient acquisition by plants (Nanjundappa et al. 2019).

Many bacteria associated with the rhizosphere or AM fungal propagules can stimulate plant growth as well as the establishment of symbiosis and therefore termed as mycorrhizal helper bacteria (MHB). For instance, *Azospirillum*, *Rhizobium*, *Bacillus*, *Pseudomonas*, and *Paenibacillus* are shown to enhance plant growth and can also act as MHB by increasing the germination of AM fungal propagules or markedly increase hyphal growth (Bidondo et al. 2016). Experimental studies have shown that dual inoculation of AM fungi and MHB can significantly increase growth and nutrient uptake in plants like onion, *Medicago sativa*, *Lactuca sativa*, *Tagetes erecta*, *Artemisia annua*, *Geranium* sp., *Stevia rebaudiana*, *Withania somnifera*, *Cucumis sativus*, *S. lycopersicum*, *Musa* sp., *Cymbopogon citratus*, and *Capsicum* sp. (Carina et al. 2016; Nanjundappa et al. 2019; da Cruz et al. 2020; Saia et al. 2020). Based on these evidence, it is clear that AM fungi-mediated changes in root exudates may significantly influence the availability of nutrients through alteration in the microbial community.

5.7 Conclusions and Future Prospects

Plants sustain a stable rhizobiome through the release of fixed carbon into the rhizosphere. These microorganisms in turn help plants in their nutrient and water acquisition and protection against various abiotic and biotic stresses. Thus, the rhizosphere plays a central role in the growth and wellbeing of plants. The quality and quantity of root exudates are influenced by several factors including the plant factor. Root exudates influence the establishment of symbiotic interactions in plants. Ample evidence now exists that root exudates can influence the germination of AM fungal propagules, development of the asymbiotic phase, and symbiosis establishment in roots. The AM fungi, in turn, could modify the composition of the root exudates enabling changes in microbiome structure and nutrient availability in the rhizosphere. Moreover, the AM fungi-mediated changes in the plant root exudates also protect plants against different stresses, thereby enabling their survival and establishment in harsh environments. However, most studies examining the role of root exudates on AM symbiosis are conducted under controlled conditions involving a few plants host and fungal species. AM fungi are presumed to be nonhost specific; therefore, it would be interesting to see how different fungi colonizing the same root system modifies the chemical composition of the mycorrhizal roots and those that are released from the root. Like roots, AM fungal hyphae also exude carbon into the soil, and information on which is very limited. Although the nutritional aspects of AM fungi–plant interaction are well worked out, the intricacies of symbiosis establishment are not well resolved. Untangling the secrets of AM fungal–plant chemical dialogues will enable us to better understand and exploit this symbiosis for improving crop growth. However, the collaboration of the symbiosis and the influence of various external factors can make this a daunting task but worth enough to work out.

References

Akiyama K, Ogasawara S, Ito S, Hayashi H (2010) Structural requirements of strigolactones for hyphal branching in AM fungi. Plant Cell Physiol 51:1104–1117

Ankati S, Rani TS, Podile AR (2019) Changes in root exudates and root proteins in groundnut–*Pseudomonas* sp. interaction contribute to root colonization by bacteria and defense response of the host. J Plant Growth Regul 38:523–538

Ansari RA, Mahmood I (2017) Optimization of organic and bio-organic fertilizers on soil properties and growth of pigeonpea. Sci Hortic 226:1–9

Ansari RA, Mahmood I (2019a) Plant health under biotic stress. In: Ansari RA, Mahmood I (eds) Microbial interactions, vol II. Springer Nature, Singapore. https://doi.org/10.1007/978-981-13-6040-4

Ansari RA, Mahmood I (2019b) Plant health under biotic stress. In: Ansari RA, Mahmood I (eds) Organic strategies, vol I. Springer Nature, Singapore. https://doi.org/10.1007/978-981-13-6043-5

Ansari RA, Mahmood I, Rizvi R, Sumbul A (2017a) PGPR: current vogue in sustainable crop production. In: Kumar V (ed) Probiotics and plant health. Springer Nature, Singapore, pp 455–472

Ansari RA, Mahmood I, Rizvi R, Sumbul A, Safiuddin (2017b) Siderophores: augmentation of soil health and crop productivity. In: Kumar V (ed) Probiotics in agroecosystem. Springer Nature, Singapore, pp 291–312

Ansari RA, Sumbul A, Rizvi R, Mahmood I (2019a) Potential role of plant growth promoting rhizobacteria in alleviation of biotic stress. In: Ansari RA, Mahmood I (eds) Plant health under biotic stress, Microbial interactions, vol II. Springer Nature, Singapore, pp 177–188

Ansari RA, Sumbul A, Rizvi R, Mahmood I (2019b) Organic soil amendments: potential tool for soil and plant health management. In: Ansari RA, Mahmood I (eds) Plant health under biotic stress, Organic strategies, vol I. Springer Nature, Singapore, pp 1–35

Ansari RA, Rizvi R, Sumbul A, Mahmood I (2020) Plant-growth-promoting Rhizobacteria (PGPR)-based sustainable management of phytoparasitic nematodes: current understandings and future challenges. In: Ansari RA et al (eds) Management of phytonematodes: recent advances and future challenges. Springer Nature, Singapore. https://doi.org/10.1007/978-981-15-4087-5_3

Bamji SF, Corbitt C (2017) Glyceollins: soybean phytoalexins that exhibit a wide range of health-promoting effects. J Funct Foods 34:98–105

Besserer A, Puech-Pagès V, Kiefer P, Gomez-Roldan V, Jauneau A, Roy S, Portais JC, Roux C, Bécard G, Séjalon-Delmas N (2006) Strigolactones stimulate arbuscular mycorrhizal fungi by activating mitochondria. PLoS Biol 4:e226. https://doi.org/10.1371/journal.pbio.0040226

Bhattacharyya P, Das S, Adhya TK (2013) Root exudates of rice cultivars affect rhizospheric phosphorus dynamics in soils with different phosphorus statuses. Commun Soil Sci Plant Anal 44:1643–1658

Bidondo LF, Colombo R, Bompadre J, Benavides M, Scorza V, Silvani V, Pérgola M, Godeas A (2016) Cultivable bacteria associated with infective propagules of arbuscular mycorrhizal fungi. Implications for mycorrhizal activity. Appl Soil Ecol 105:86–90

Bonfante P, Genre A (2015) Arbuscular mycorrhizal dialogues: do you speak 'plantish'or 'fungish'? Trends Plant Sci 20:150–154

Bouhaouel I, Richard G, Fauconnier ML, Ongena M, Franzil L, Gfeller A, Slim Amara H, du Jardin P (2019) Identification of barley (*Hordeum vulgare* L. subsp. *vulgare*) root exudates allelochemicals, their autoallelopathic activity and against *Bromus diandrus* Roth. Germination. Agronomy 9:345. https://doi.org/10.3390/agronomy9070345

Bravo A, York T, Pumplin N, Mueller LA, Harrison MJ (2016) Genes conserved for arbuscular mycorrhizal symbiosis identified through phylogenomics. Nat Plant 2:15208. https://doi.org/10.1038/nplants.2015.208

Brundrett MC, Tedersoo L (2018) Evolutionary history of mycorrhizal symbioses and global host plant diversity. New Phytol 220:1108–1115

Canarini A, Kaiser C, Merchant A, Richter A, Wanek W (2019) Root exudation of primary metabolites: mechanisms and their roles in plant responses to environmental stimuli. Front Plant Sci 10:157. https://doi.org/10.3389/fpls.2019.00157

Cao J, Feng Y, Lin X, Wang J (2016) Arbuscular mycorrhizal fungi alleviate the negative effects of iron oxide nanoparticles on bacterial community in rhizospheric soils. Front Environ Sci 4:10. https://doi.org/10.3389/fenvs.2016.00010

Carina VT, Emilia MM, Miguel US, Ortiz DR, Manjunatha B, Thangaswamy S, Mulla SI (2016) Effect of arbuscular mycorrhizal fungi (AMF) and *Azospirillum* on growth and nutrition of banana plantlets during acclimatization phase. J Appl Pharma Sci 6:131–138

Carvalhais LC, Dennis PG, Fedoseyenko D, Hajirezaei MR, Borriss R, von Wirén N (2011) Root exudation of sugars, amino acids, and organic acids by maize as affected by nitrogen, phosphorus, potassium, and iron deficiency. J Plant Nutr Soil Sci 174:3–11

Chen KJ, Zheng YQ, Kong CH, Zhang SZ, Li J, Liu XG (2010) 2, 4-Dihydroxy-7-methoxy-1, 4-benzoxazin-3-one (DIMBOA) and 6-methoxy-benzoxazolin-2-one (MBOA) levels in the wheat rhizosphere and their effect on the soil microbial community structure. J Agric Food Chem 58:12710–12716

Chen YT, Wang Y, Yeh KC (2017) Role of root exudates in metal acquisition and tolerance. Curr Opin Plant Biol 39:66–72

Choi J, Summers W, Paszkowski U (2018) Mechanisms underlying establishment of arbuscular mycorrhizal symbioses. Annu Rev Phytopathol 56:135–160

Cimmino A, Fernández-Aparicio M, Avolio F, Yoneyama K, Rubiales D, Evidente A (2015) Ryecyanatines A and B and ryecarbonitrilines A and B, substituted cyanatophenol, cyanatobenzo [1, 3] dioxole, and benzo [1, 3] dioxolecarbonitriles from rye (*Secale cereale* L.) root exudates: novel metabolites with allelopathic activity on *Orobanche* seed germination and radicle growth. Phytochemistry 109:57–65

Clemens S, Weber M (2016) The essential role of coumarin secretion for Fe acquisition from alkaline soil. Plant Signal Behav 11:e1114197. https://doi.org/10.1080/15592324.2015.1114197

Coelho L, Mignoni DS, Silva FS, Braga MR (2019) Seed exudates of *Sesbania virgata* (Cav.) Pers. stimulate the asymbiotic phase of the arbuscular mycorrhizal fungus *Gigaspora albida* Becker & Hall. Hoehnea 46:e272018. https://doi.org/10.1590/2236-8906-27/2018

da Cruz RM, Alberton O, da Silva Lorencete M, da Cruz GL, Gasparotto-Junior A, Cardozo-Filho L, de Souza SG (2020) Phytochemistry of *Cymbopogon citratus* (DC) Stapf inoculated with arbuscular mycorrhizal fungi and plant growth promoting bacteria. Ind Crop Prod 149:112340. https://doi.org/10.1016/j.indcrop.2020.112340

Davison J, Moora M, Jairus T, Vasar M, Öpik M, Zobel M (2016) Hierarchical assembly rules in arbuscular mycorrhizal (AM) fungal communities. Soil Biol Biochem 97:63–70

Del-Saz NF, Romero-Munar A, Cawthray GR, Aroca R, Baraza E, Flexas J, Lambers H, Ribas-Carbó M (2017) Arbuscular mycorrhizal fungus colonization in *Nicotiana tabacum* decreases the rate of both carboxylate exudation and root respiration and increases plant growth under phosphorus limitation. Plant Soil 416:97–106

Edwards J, Johnson C, Santos-Medellin C, Lurie E, Podishetty NK, Bhatnagar S, Eisen JA, Sundaresan V (2015) Structure, variation, and assembly of the root-associated microbiomes of rice. Proc Natl Acad Sci U S A 112:E911–E920

Ellouze W, Hamel C, Cruz AF, Ishii T, Gan Y, Bouzid S, St-Arnaud M (2012) Phytochemicals and spore germination: at the root of AMF host preference? Appl Soil Ecol 60:98–104

Evidente A, Fernández-Aparicio M, Cimmino A, Rubiales D, Andolfi A, Motta A (2009) Peagol and peagoldione, two new strigolactone-like metabolites isolated from pea root exudates. Tetrahedron Lett 50:6955–6958

Evidente A, Cimmino A, Fernández-Aparicio M, Rubiales D, Andolfi A, Melck D (2011) Soyasapogenol B and trans-22-dehydrocam-pesterol from common vetch (*Vicia sativa* L.) root exudates stimulate broomrape seed germination. Pest Manag Sci 67:1015–1022

Fernández I, Cosme M, Stringlis IA, Yu K, de Jonge R, van Wees SM, Pozo MJ, Pieterse CM, van der Heijden MG (2019) Molecular dialogue between arbuscular mycorrhizal fungi and the nonhost plant *Arabidopsis thaliana* switches from initial detection to antagonism. New Phytol 223:867–881

Giovannetti M, Avio L, Sbrana C (2010) Fungal spore germination and pre-symbiotic mycelial growth – physiological and genetic aspects. In: Koltai H, Kapulnik Y (eds) Arbuscular mycorrhizas: physiology and function. Springer, Dordrecht, pp 3–32

Gomez-Roldan V, Fermas S, Brewer PB, Puech-Pagès V, Dun EA, Pillot JP, Letisse F, Matusova R, Danoun S, Portais JC, Bouwmeester H (2008) Strigolactone inhibition of shoot branching. Nature 455:189–194

Gough C, Cullimore J (2011) Lipo-chitooligosaccharide signaling in endosymbiotic plant-microbe interactions. MPMI 24:867–878

Hage-Ahmed K, Moyses A, Voglgruber A, Hadacek F, Steinkellner S (2013) Alterations in root exudation of intercropped tomato mediated by the arbuscular mycorrhizal fungus *Glomus mosseae* and the soilborne pathogen *Fusarium oxysporum* f. sp. *lycopersici*. J Phytopathol 161:763–773

Haldar S, Sengupta S (2015) Plant-microbe cross-talk in the rhizosphere: insight and biotechnological potential. Open Microbiol J 9:1–7. https://doi.org/10.2174/1874285801509010001

Hao WY, Ren LX, Ran W, Shen QR (2010) Allelopathic effects of root exudates from watermelon and rice plants on *Fusarium oxysporum* f. sp. *niveum*. Plant Soil 336:485–497

Hao L, Zhang Z, Hao B, Diao F, Zhang J, Bao Z, Guo W (2021) Arbuscular mycorrhizal fungi alter microbiome structure of rhizosphere soil to enhance maize tolerance to La. Ecotoxicol Environ Saf 212:111996. https://doi.org/10.1016/j.ecoenv.2021.111996

Hernandez M, Dumont MG, Yuan Q, Conrad R (2015) Different bacterial populations associated with the roots and rhizosphere of rice incorporate plant-derived carbon. Appl Environ Microbiol 81:2244–2253

Herz K, Dietz S, Gorzolka K, Haider S, Jandt U, Scheel D, Bruelheide H (2018) Linking root exudates to functional plant traits. PLoS One 13:e0204128. https://doi.org/10.1371/journal.pone.0204128

Hiltner L (1904) Uber neuere Erfahrungen und Probleme auf dem Gebiete der Bodenbakteriologie unter besonderden berucksichtigung und Brache. Arb Dtsch Landwirtsch Gesellschaft 98:59–78

Hooper AM, Caulfield JC, Hao B, Pickett JA, Midega CAO, Khan ZR (2015) Isolation and identification of *Desmodium* root exudates from drought tolerant species used as intercrops against *Striga hermonthica*. Phytochemistry 117:380–387

Hou Y, Liu X, Zhang X, Chen X, Tao K, Chen X, Liang X, He C (2015) Identification of *Scirpus triqueter* root exudates and the effects of organic acids on desorption and bioavailability of pyrene and lead in co-contaminated wetland soils. Environ Sci Pollut Res 22:17780–17788

Hussain HA, Qingwen Z, Hussain S, Hongbo L, Waqqas A, Li Z (2021) Effects of arbuscular mycorrhizal fungi on maize growth, root colonization, and root exudates varied with inoculum and application method. J Soil Sci Plant Nutr 21:1577–1590

Jansa J, Smith FA, Smith SE (2008) Are there benefits of simultaneous root colonization by different arbuscular mycorrhizal fungi? New Phytol 177:779–789

Kloppholz S, Kuhn H, Requena N (2011) A secreted fungal effector of *Glomus intraradices* promotes symbiotic biotrophy. Curr Biol 21:1204–1209

Koltai H, LekKala SP, Bhattacharya C, Mayzlish-Gati E, Resnick N, Wininger S, Dor E, Yoneyama K, Yoneyama K, Hershenhorn J, Joel DM, Kapulnik Y (2010) A tomato strigolactone-impaired mutant displays aberrant shoot morphology and plant interactions. J Exp Bot 61:1739–1749

Korenblum E, Dong Y, Szymanski J, Panda S, Jozwiak A, Massalha H, Meir S, Rogachev I, Aharoni A (2020) Rhizosphere microbiome mediates systemic root metabolite exudation by root-to-root signaling. Proc Natl Acad Sci U S A 117:3874–3883

Lanfranco L, Fiorilli V, Gutjahr C (2018a) Partner communication and role of nutrients in the arbuscular mycorrhizal symbiosis. New Phytol 220:1031–1046

Lanfranco L, Fiorilli V, Venice F, Bonfante P (2018b) Strigolactones cross the kingdoms: plants, fungi, and bacteria in the arbuscular mycorrhizal symbiosis. J Exp Bot 69:2175–2188

Li XG, Zhang TL, Wang XX, Hua K, Zhao L, Han ZM (2013) The composition of root exudates from two different resistant peanut cultivars and their effects on the growth of soil-borne pathogen. Int J Biol Sci 9:164–173

Liu Y, Li X, Cai K, Cai L, Lu N, Shi J (2015) Identification of benzoic acid and 3-phenylpropanoic acid in tobacco root exudates and their role in the growth of rhizosphere microorganisms. Appl Soil Ecol 93:78–87

Liu L, Huang X, Zhang J, Cai Z, Jiang K, Chang Y (2020) Deciphering the relative importance of soil and plant traits on the development of rhizosphere microbial communities. Soil Biol Biochem 148:107909. https://doi.org/10.1016/j.soilbio.2020.107909

Lǚ LH, Zou YN, Wu QS (2019) Mycorrhizas mitigate soil replant disease of peach through regulating root exudates, soil microbial population, and soil aggregate stability. Commun Soil Sci Plant Anal 50:909–921

Maillet F, Poinsot V, André O, Puech-Pagès V, Haouy A, Gueunier M, Cromer L, Giraudet D, Formey D, Niebel A, Martinez EA (2011) Fungal lipochitooligosaccharide symbiotic signals in arbuscular mycorrhiza. Nature 469:58–63

Maurer D, Malique F, Alfarraj S, Albasher G, Horn MA, Butterbach-Bahl K, Dannenmann M, Rennenberg H (2021) Interactive regulation of root exudation and rhizosphere denitrification by plant metabolite content and soil properties. Plant Soil 467:107–127. https://doi.org/10.1007/s11104-021-05069-7

Meier IC, Finzi AC, Phillips RP (2017) Root exudates increase N availability by stimulating microbial turnover of fast-cycling N pools. Soil Biol Biochem 106:119–128

Miura Y, Okabe S (2008) Quantification of cell specific uptake activity of microbial products by uncultured *Chloroflexi* by microautoradiography combined with fluorescence in situ hybridization. Environ Sci Technol 42:7380–7386

Molina L, Udaondo Z, Montero-Curiel M, Wittich RM, García-Puente A, Segura A (2021) Clover root exudates favor *Novosphingobium* sp. hr1a establishment in the rhizosphere and promote phenanthrene rhizoremediation. Msphere 6:e00412–e00421. https://doi.org/10.1128/mSphere.00412-21

Mori N, Nishiuma K, Sugiyama T, Hayashi H, Akiyama K (2016) Carlactone-type strigolactones and their synthetic analogues as inducers of hyphal branching in arbuscular mycorrhizal fungi. Phytochemistry 130:90–98

Nagata M, Yamamoto N, Miyamoto T, Shimomura A, Arima S, Hirsch AM, Suzuki A (2016) Enhanced hyphal growth of arbuscular mycorrhizae by root exudates derived from high R/FR treated *Lotus japonicus*. Plant Signal Behav 11:e1187356. https://doi.org/10.1080/15592324.2016.1187356

Nanjundappa A, Bagyaraj DJ, Saxena AK, Kumar M, Chakdar H (2019) Interaction between arbuscular mycorrhizal fungi and *Bacillus* spp. in soil enhancing growth of crop plants. Fungal Biol Biotechnol 6:1–10

Olanrewaju OS, Ayangbenro AS, Glick BR, Babalola OO (2019) Plant health: feedback effect of root exudates-rhizobiome interactions. Appl Microbiol Biotechnol 103:1155–1166

Pantigoso HA, Yuan J, He Y, Guo Q, Vollmer C, Vivanco JM (2020) Role of root exudates on assimilation of phosphorus in young and old *Arabidopsis thaliana* plants. PLoS One 15:e0234216. https://doi.org/10.1371/journal.pone.0234216

Pausch J, Kuzyakov Y (2018) Carbon input by roots into the soil: quantification of rhizodeposition from root to ecosystem scale. Glob Chang Biol 24:1–12

Paz C, Öpik M, Bulascoschi L, Bueno CG, Galetti M (2021) Dispersal of arbuscular mycorrhizal fungi: evidence and insights for ecological studies. Microb Ecol 81:283–292

Pinton R, Varanini Z, Nannipieri P (2007) The rhizosphere. Biochemistry and organic substances at the soil-plant interface, 2nd edn. CRC Press, Boca Raton, FL

Põlme S, Bahram M, Jacquemyn H, Kennedy P, Kohout P, Moora M, Oja J, Öpik M, Pecoraro L, Tedersoo L (2018) Host preference and network properties in biotrophic plant–fungal associations. New Phytol 217:1230–1129

Priyadharsini P, Rojamala K, Koshila Ravi R, Muthuraja R, Nagaraj P, Muthukumar T (2016) Mycorrhizosphere: the extended rhizosphere and its significance. In: Choudhary DK, Varma A, Tuteja N (eds) Plant-microbe interaction: an approach to sustainable agriculture. Springer Nature, Singapore, pp 97–124

Qin H, Brookes P, Xu J (2016) Arbuscular mycorrhizal fungal hyphae alter soil bacterial community and enhance polychlorinated biphenyls dissipation. Front Microbiol 7:939. https://doi.org/10.3389/fmicb.2016.00939

Ren L, Zhang N, Wu P, Huo H, Xu G, Wu G (2015) Arbuscular mycorrhizal colonization alleviates *Fusarium* wilt in watermelon and modulates the composition of root exudates. Plant Growth Regul 77:77–85

Rizvi R, Singh G, Safiuddin ARA, Tiyagi SA, Mahmood I (2015) Sustainable management of root-knot disease of tomato by neem cake and *Glomus fasciculatum*. Cogent Food Agric 1:1008859. https://doi.org/10.1080/23311932.2015.1008859

Rodríguez-Caballero G, Caravaca F, Fernández-González AJ, Alguacil MM, Fernández-López M, Roldán A (2017) Arbuscular mycorrhizal fungi inoculation mediated changes in rhizosphere bacterial community structure while promoting revegetation in a semiarid ecosystem. Sci Total Environ 584:838–848

Saia S, Aissa E, Luziatelli F, Ruzzi M, Colla G, Ficca AG, Cardarelli M, Rouphael Y (2020) Growth-promoting bacteria and arbuscular mycorrhizal fungi differentially benefit tomato and corn depending upon the supplied form of phosphorus. Mycorrhiza 30:133–147

Sato T, Ezawa T, Cheng W, Tawaraya K (2015) Release of acid phosphatase from extraradical hyphae of arbuscular mycorrhizal fungus *Rhizophagus clarus*. Soil Sci Plant Nutr 61:269–274

Schlaeppi K, Bulgarelli D (2015) The plant microbiome at work. MPMI 28:212–217

Schreiner M, Krumbein A, Knorr D, Smetanska I (2011) Enhanced glucosinolates in root exudates of *Brassica rapa* ssp. *rapa* mediated by salicylic acid and methyl jasmonate. J Agric Food Chem 59:1400–1405

Smith SE, Read DJ (2008) Mycorrhizal Symbiosis, 3rd edn. Academic Press, London

Solanki MK, Kashyap PL, Ansari RA, Kumari B (2020) Microbiomes and plant health: panoply and their applications. Academic Press, London. https://www.elsevier.com/books/microbiomes-and-plant-health/solanki/978-0-12-819715-8

Sumbul A, Mahmood I, Rizvi R, Safiuddin, Ansari RA (2017) Mycorrhiza: an alliance for the nutrient management in plants. In: Kumar V (ed) Probiotics in agroecosystem. Springer Nature, Singapore, pp 371–386

Sun S, Wang J, Zhu L, Liao D, Gu M, Ren L, Kapulnik Y, Xu G (2012) An active factor from tomato root exudates plays an important role in efficient establishment of mycorrhizal symbiosis. PLoS One 7:e43385. https://doi.org/10.1371/journal.pone.0043385

Sun XG, Bonfante P, Tang M (2015) Effect of volatiles versus exudates released by germinating spores of *Gigaspora margarita* on lateral root formation. Plant Physiol Biochem 97:1–10

Tawaraya K, Naito M, Wagatsuma T (2006) Solubilization of insoluble inorganic phosphate by hyphal exudates of arbuscular mycorrhizal fungi. J Plant Nutr 29:657–665

Tian B, Pei Y, Huang W, Ding J, Siemann E (2021) Increasing flavonoid concentrations in root exudates enhance associations between arbuscular mycorrhizal fungi and an invasive plant. ISME J 15:1919–1930

Varela-Cervero S, Vasar M, Davison J, Barea JM, Öpik M, Azcón-Aguilar C (2015) The composition of arbuscular mycorrhizal fungal communities differs among the roots, spores and extraradical mycelia associated with five Mediterranean plant species. Environ Microbiol 17:2882–2895

Varela-Cervero S, López-García Á, Barea JM, Azcón-Aguilar C (2016) Differences in the composition of arbuscular mycorrhizal fungal communities promoted by different propagule forms from a Mediterranean shrubland. Mycorrhiza 26:489–496

Veiga RSL, Faccio A, Genre A, Pieterse CMJ, Bonfante P, Van der Heijden MGA (2013) Arbuscular mycorrhizal fungi reduce growth and infect roots of the non-host plant Arabidopsis thaliana. Plant Cell Environ 36:1926–1937

Vives-Peris V, de Ollas C, Gómez-Cadenas A, Pérez-Clemente RM (2020) Root exudates: from plant to rhizosphere and beyond. Plant Cell Rep 39:3–17

Voß S, Betz R, Heidt S, Corradi N, Requena N (2018) RiCRN1, a crinkler effector from the arbuscular mycorrhizal fungus *Rhizophagus irregularis*, functions in arbuscule development. Front Microbiol 9:2068. https://doi.org/10.3389/fmicb.2018.02068

White LJ, Ge X, Brözel VS, Subramanian S (2017) Root isoflavonoids and hairy root transformation influence key bacterial taxa in the soybean rhizosphere. Environ Microbiol 19:1391–1406

Xing YANG, Zhang LH, Shi CP, Shang Y, Zhang JL, Han JM, Dong JG (2014) The extraction, isolation and identification of exudates from the roots of *Flaveria bidentis*. J Integr Agric 13:105–114

Yoneyama K, Arakawa R, Ishimoto K, Kim HI, Kisugi T, Xie X, Nomura T, Kanampiu F, Yokota T, Ezawa T, Yoneyama K (2015) Difference in *Striga*-susceptibility is reflected in strigolactone secretion profile, but not in compatibility and host preference in arbuscular mycorrhizal symbiosis in two maize cultivars. New Phytol 206:983–989

Yuan YG, Tang JJ, Leng D, Hu SJ, Yong JWH, Chen X (2014) An invasive plant promotes its arbuscular mycorrhizal symbioses and competitiveness through its secondary metabolites: indirect evidence from activated carbon. PLoS One 9:e97163. https://doi.org/10.1371/journal.pone.0097163

Yuan J, Wu Y, Zhao M, Wen T, Huang Q, Shen Q (2018) Effect of phenolic acids from banana root exudates on root colonization and pathogen suppressive properties of *Bacillus amyloliquefaciens* NJN-6. Biol Control 125:131–137

Zhou J, Chai X, Zhang L, George TS, Wang F, Feng G (2020) Different arbuscular mycorrhizal fungi cocolonizing on a single plant root system recruit distinct microbiomes. Msystems 5:e00929-20. https://doi.org/10.1128/mSystems.00929-20

6 Advancement in Mycorrhizal Fungi-Based Sustainable Plant Disease Management

Tesleem Taye Bello and Oluwatoyin Adenike Fabiyi

Abstract

Good agricultural soil requires a decent nutrient retention ability which could be largely enhanced by the collective activities of soil microorganisms. However, in practice, agricultural soils are largely exposed to disturbance as a result of tillage activities which break up the soil and make it liable to nutrient leaching in soils with low nutrient retention ability. In the past decades, there has been increased awareness about the inherent potential of microorganisms in sustainable agriculture. This is largely attributed to expanding bodies of research exposing the eco-friendly nature of microorganisms like the mycorrhizal fungi, and the roles they play in soil fertility, crop health improvement and agricultural disease management, which ultimately provide the needed support for crops to achieve optimal productivity. Since the importance of mycorrhizal fungi has become an increasingly popular discussion in sustainable crop production, emphasis should be placed on how to manage our crop production systems in such a way to be able to harness the potential of mycorrhizal fungi while sustaining their continued propagation in order to retain the ecological and production benefits.

T. T. Bello
Department of Plant Soil and Microbial Sciences, Michigan State University, East Lansing, MI, USA

Department of Agricultural Science Education, Federal College of Education, Abeokuta, Ogun, Nigeria

O. A. Fabiyi (✉)
Department of Crop Protection, Faculty of Agriculture, University of Ilorin, Ilorin, Kwara, Nigeria

R. A. Ansari et al. (eds.), *Mycorrhizal Symbiosis and Agroecosystem Restoration*,
https://doi.org/10.1007/978-981-99-5030-0_6

Keywords

Plant health · Disease severity · Plant pathogens · Management strategies · Rhizosphere

6.1 Introduction

Arbuscular mycorrhizal fungi (AM fungi) are described from almost known habitats and ecosystems throughout the world (Öpik et al. 2006, 2013; Ansari and Mahmood 2019a, b), ranging from the extremely dry desert habitats (Al-Yahya'ei et al. 2011) to the extremely cold arctic regions, and while species are known to occur only within natural communities, many others are of true cosmopolitans (Varga et al. 2015). This phenomenon has been attributed to either innate dissemination or a sequel to human action. The existence of most cosmopolitan AM fungi species as genetically differentiated species complexes suggests that they are extremely adaptable to a wide range of hosts as well as different environmental conditions (Fig. 6.1). AM fungi are acknowledged to be central in protection against adverse abiotic conditions like drought (Chitarra et al. 2016), heat stress (Bunn et al. 2009) and nutrient shortage (Wu 2017). These attributes have made them to be exceptionally useful in the wild in conjunction with conservative agriculture (Van der Heijden et al. 2015; Wu 2017). Hence, AM fungi have established themselves as having an excellent influence within both natural and controlled environmental conditions (Van der Heijden et al. 2015), as well as in agriculture, forestry and horticulture. This chapter looks at the inherent potentials of AM fungi and how these could be harnessed in different areas of agriculture soil and plant health, crop improvement, crop protection and commercial agriculture to ensure food security through sustainable farming practices.

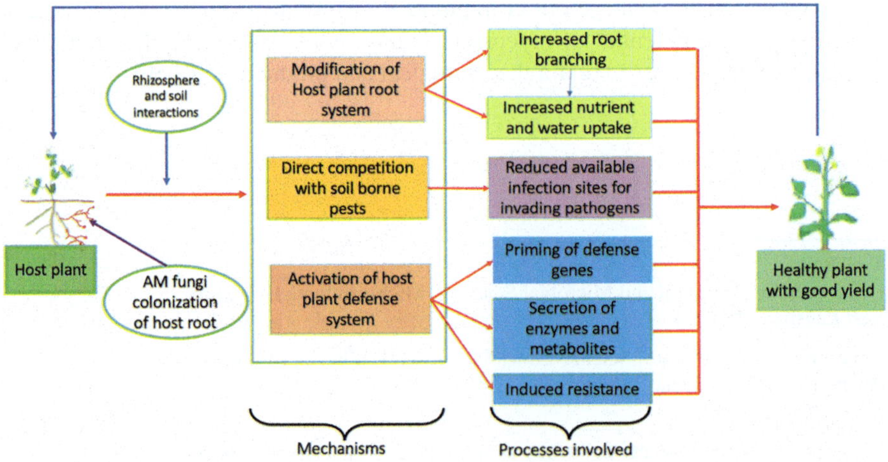

Fig. 6.1 A schematic representation of the mechanisms implicated behind improved plant biomass and yield, and the different roles of AM fungi in the management of various economically important plant diseases

6.2 Effects of AM Fungi in Improving Soil Properties

AM fungi possess a compact network of hyphae mycelium which creates a three-dimensional matrix that entangles soil particles without inducing soil compaction. They are able to achieve this due to the possession of glomalin-related soil proteins (GRSPs) which serve as an agent for stabilising soil aggregates (Gao et al. 2019). GRSPs serve as a crucial determinant of soil quality and carbon sink which confers the soil stabilising potential on AM fungi. Furthermore, the hyphal network initiated by AM fungi effectively promotes root development and plant growth, which protects from wind and soil erosion as well as leaching (Kumar et al. 2016). AM fungi also influence soil water-holding capacity which in turn promotes improved plant growth through increased nutrient supply. These qualities are particularly critical for plants within marginal soils of dry and arid regions, which are characterised by low fertility and high susceptibility to erosion (Yadav et al. 2015; Kumar et al. 2016; Ansari et al. 2017a, b). Thus, AM fungi have proved themselves as excellent eco-friendly biological medium for augmenting water and nutrient usability of crops. Hence, the utilisation of AM fungi may be a sustainable method for the penurious farmers of developing regions to counteract these adverse conditions at no additional cost (Cavagnaro et al. 2015; Chen et al. 2018). Nutrient leaching from soil is a devastating phenomenon within several agroecological zones and has led to the loss of soil fertility through the downward movement of soil nutrients deep into the horizon causing pollution of surface water, especially springs and lakes (Zhang et al. 2020a; Ansari et al. 2019a, b). Good agricultural soil requires a decent nutrient retention ability which could be largely enhanced by the collective activities of AM fungi and other soil microorganisms. However, in practice, agricultural soils are largely exposed to disturbance as a result of tillage activities which break up the soil and make it liable to nutrient leaching in soils with low nutrient retention ability (Awe et al. 2020). The valuable intervention of AM fungi in preventing leaching in soils is known to occur at different stages. The first stage involves fortifying the soil structure to permit enhanced nutrient division to both macro- and micro-composite induced in mycorrhizal soil (Querejeta 2017). Second, AM fungi, through their complex mycelium, take up nutrients from one end of the soil and recycle them to another end, and in the process, they make nutrients available to all parts of the plant roots with their reach and network while also improving soil water retention in the process (Bitterlich et al. 2018; Sileshi et al. 2020). This has been particularly accentuated to enhance tomato yield under dry spell pressure situations (Bitterlich et al. 2018). To sum all these up, AM fungi have shown great potential for maintaining and improving soil fertility by blending all nutrient fluxes within the soil by initiating a closed nutrient cycle within the plant rhizosphere which ultimately results in long-term and sustainable soil fertility maintenance (Cavagnaro et al. 2015; Bitterlich et al. 2018; Chen et al. 2018). AM fungi have a more efficient nutrient uptake ability than most plants since their hyphal networks shorten the distance needed for nutrient diffusion. Hence, the speed by which available phosphorus moves into the mycorrhizae could increase by as much as five times that of the plant roots. Some AM fungi have been found to completely take over the role of phosphorus uptake as a result of their hyphal structure ramifications. A remarkable advantage of this to host plants is the initiation of increased phosphorus uptake (MacLean et al. 2017; Zhang et al. 2020b). This is attributed to an enlargement in the expanse and contact with the soil causing a steady and continuous flow of nutrients into the mycorrhizal structures which provides a modified environment for increased phosphorus storage. Furthermore, some AM fungi isolates possess the ability to alter the phytohormone balance in some host plants which in turn influences the activities of some bioregulators responsible for plant tolerance to stress conditions through improved root production (Rouphael et al. 2015; Lehmann et al. 2020).

6.3 AM Fungi and Plant Yield

The impact of AM fungi on plant/soil and water relations has been exposed where host plants were reported to exhibit increased stomatal conductance and transpiration. AM fungi were also able to alter some physiological processes in colonised plants. These include inducing an increase in the photosynthetic rates and leaf hydration efficiency of host plants. With these, AM fungi have displayed a remarkable ability to influence several growth and plant reproduction characteristics by differentially affecting the reproductive processes of affected plant species (Bennett and Meek 2020). The most commonly reported reproductive responses induced by AM fungi in associated plant species include but are not limited to early flowering, increase in flower bud numbers and inflorescences as well as fruits and seed production. These have been largely attributed to changes in host plant phenology, shoot architecture, onset of flowering, fruit production and number of seeds per fruit (Gao et al. 2020). The degree of benefits a plant can derive from AM fungi association largely depends on the prevailing environmental conditions. Since most natural environments experience nutrient deficiency coupled with the attendant biotic stress conditions, AM fungi colonised host plants tend to have an advantage over the non-host ones of equal species. Thus, AM fungi can invariably increase intraspecies competition in favour of AM fungi host plants (Chen et al. 2018). Due to different plants being connected through a common web of mycorrhizal networks, this bond tends to promote stability of the weaker plant individuals who benefit from the increased nutrient supply with less competition from the stronger ones (Selosse et al. 2017; Chen et al. 2018).

6.4 AM Fungi in Crop Protection

On a global scale, plant diseases stand as a major limiting factor in crop production (Ansari and Khan 2012a, b; Atolani and Fabiyi 2020; Fabiyi 2020). This menace is further aggravated by an upsurge in the resistance of plant pathogens to conventional pesticides which is largely due to excessive continuous use over time with their devastating consequences on human health and the environment (Fabiyi and Olatunji 2021a). In order to ensure food sufficiency and eradicate poverty through increased crop production, especially in the developing worlds, efforts must be put in place to explore all available alternatives to chemical pesticide use in agricultural disease management (Atolani et al. 2014a, b; Fabiyi et al. 2020a, 2022a, b; Fabiyi 2021a). Several biocontrol agents have been studied for their potential as suitable alternatives to chemical pesticides in economically important plant disease management (Fabiyi et al. 2019, 2020b; Fabiyi and Olatunji 2021b; Fabiyi 2021b, c, d, 2022a, b, c, d). AM fungi are known to have provided some levels of protection for agricultural crops due to the fact that most colonised plants have experienced a reduction in damage caused by soil-borne pathogens (Chahal et al. 2021; Himaya et al. 2021; Rizvi et al. 2015; Solanki et al. 2020; Sumbul et al. 2017). Potentials of AM fungi as biocontrol agents of major pathogenic species like *Fusarium, Phytophthora, Rhizoctonia, Guaemannomyces, Verticilium, Sclerotium* and *Phytuim* as well as plant parasitic nematodes such as the root-knot (*Meloidogyne* spp), cyst nematode (*Heterodera* spp.), root lesion (*Pratylenchus* spp.) and the reniform nematode (*Rotylenchulus* spp.), etc., as well as some viruses have been adequately documented (Karagiannidis et al. 2002; Al-Askar and Rashad 2010; Affokpon et al. 2011; Yuan et al. 2016; Aseel et al. 2019; El-Sharkawy et al. 2021). A list of applications of AM fungi in managing some economically important fungal plant pathogens has been enlisted (Table 6.1). However, the level of protection from individual pathogen species varies in respect of host plant species and AM fungi strains involved. Generally, disease reduction experienced due to AM fungi colonisation of host plants results from a complex chemical and biological relationship linking the AM fungi, pathogen and the host plant (Pandey et al. 2018; Gupta 2020). Although some authors concluded that plant defence response activated by AM fungi is usually uncoordinated and transient in nature when compared to the usual response exhibited by naturally resistant plant species, it is widely agreed that AM fungi colonised plants respond more rapidly to most pathogen attacks expressing the phenomenon known as 'induced systemic resistance' (Vos et al. 2012; Jacott et al. 2017; Gupta 2020). There is emerging evidence which proves that AM fungi also confer certain protection from some insects on host plants basically by causing chemical changes within the roots which in turn affects the growth and or reproduction of such insect species (Selvaraj and Thangavel 2021). Furthermore, some chemicals like terpenoids and phenolics which are found to be active pest killers have also been isolated from mycorrhizal roots (Zeng et al. 2013; Pandey et al. 2018).

6.4.1 Mycorrhiza Mediated Fungal Disease Management

The utilisation of mycorrhiza in managing several important fungal diseases of plants in diverse agroecosystem has been extensively studied (Whipps 2004; Ronsheim 2016; Singh and Giri 2017). The activities of mycorrhiza fungi alone or in combination with other microorganisms, especially in most common soil-borne fungal pathogens like *Aphanomyces, Alternaria, Botrytis, Colletotrichum, Fusarium, Macrophomina, Pythium, Phytophthora, Rhizoctonia, Sclerotium* and

Table 6.1 Applications of AM fungi in managing some economically important plant pathogens

Plant pathogen	Crop affected	AM fungi	Disease	Reference
Aphanomyces euteiches	Pea (*Pisium sativum*)	*Glomus intraradices*	Root rot	Bødker et al. (2002)
Alternaria triticina	Wheat	*Glomus mosseae*	Leaf blight	Siddiqui and Singh (2005)
Botrytis cinerea	Tomato	*Glomus mosseae*	Bunch rot	Fiorilli et al. (2011)
Colletotrichum gloeosporioides	Strawberry	*G. mosseae*	Bitter rot	Li et al. (2010)
Fusarium oxysporum	Garden asparagus (*Asparagus officinalis* L.)	*G. mosseae*	Root rot	Matsubara et al. (2001, 2002)
Fusarium oxysporum	Common bean (*Phaseolus vulgaris* L.)	*G. mosseae, G.intraradices*	Root rot	Al-Askar and Rashad (2021)
Fusarium oxysporum f. sp. *pisi*	Pea (*Pisum sativum*)	*G. mosseae*	Root rot	El-Sharkawy et al. (2021)
Macrophomina phaseolina	Chickpea	*Glomus fasciculatum*	Damping off	Shakoor et al. (2015)
Pythium ultimum	Cucumber	*Glomus etunicatum*	Root rot	Rosendahl and Rosendahl (1990)
Phytophthora capsici	Pepper	*G. mosseae*	Blight and fruit rot	Pereira et al. (2016)
Rhizoctonia solani	Mung bean (*Vigna radiata*)	*G. intraradices*	Root and stem rot	Kjøller and Rosendahl (1996)
Rhizoctonia solani	Cowpea (*Vigna unguiculata*)	*G. intraradices*	Root rot	Abdel-Fattah and Shabana (2002)
Sclerotinia sclerotiorum	Bean (*Phaseolus vulgaris* L.)	*G. mosseae*	White mould	Aysan and Demir (2009)
Sclerotium cepivorum	Onions (*Allium cepa*)	*Glomus* spp.	White rot	Torres-Barragán et al. (1996)
Verticillium dahliae	Tomato (*Solanum lycopersicum* (L) H. karst)	*G. mosseae*	Verticillium wilt	Karagiannidis et al. (2002)
Verticillium sp.	Garden egg (*Solanum melongena*)	*G. mosseae*	Wilt	Matsubara et al. (2000)

(continued)

Table 6.1 (continued)

Plant pathogen	Crop affected	AM fungi	Disease	Reference
Helicobasidium mompa	Garden asparagus (*Asparagus officinalis* L.)	*Glomus coronatum*	Violet root rot	Kasiamdari et al. (2002)
Pseudomonas syringae	Soybean	*Entrospora infrequens*	Canker	Malik et al. (2016)
Ralstonia solanacearum	Tobacco (*Nicotiana tabacum*)	*G. mosseae*	Bacterial wilt	Yuan et al. (2016)
Ralstonia solanacearum	Tomato	*G. versiforme*	Wilt	Tahat et al. (2012)
Xanthomonas campestris pv. *alfalfae*	Medicago	*G. intraradices*	Leaf and stem spot	Liu et al. (2007)
Tomato mosaic virus	Tomato (*Solanum lycopersicum* (L) H. karst)	*G. mosseae, G. clarum*	Tomato mosaic disease	Aseel et al. (2019)
Meloidogyne spp.	Vegetables	*Glomus mosseae, Glomus etunicatum, Kuklospora kentinensis, Acaulospora scrobiculata*	Root-knot disease	Affokpon et al. (2011)
Meloidogyne spp.	Tomato	*G. margarita*	Root-knot	Labeena et al. (2002)
Meloidogyne spp.	Okra	*G. fasciculatum*	Root-knot	Vos et al. (2013)
Pratylenchus coffeae	Banana	*G. intraradices*	Root lesion	Elsen et al. (2008)
Radopholus similis	Banana	*G. intraradices*	Toppling disease	Elsen et al. (2008)
Xiphinema index	Grapevine	*G. intraradices*	Secondary infection	Hao et al. (2012)
Nacobbus abberans	Tomato	*G. intraradices*	Swollen root	Marro et al. (2014)

Verticillium has been recorded (Table 6.1). Most of the success stories recorded so far have been largely ascribed to the ability of the AM fungi to alter the hormonal balance of the host plant, effectively induce resistance and or enhance its tolerance to external aggression of invading pathogens (Singh and Giri 2017).

Glomus mosseae was combined with *Pseudomonas fluorescens* to manage *Rhizoctonia solani* (Berta et al. 2005). Also, a combination of two AM fungi: *G. mosseae* and *G. deserticola,* respectively, produced a reduction in bacterial infection in maize plants (Vázquez et al. 2000). Furthermore, combined inoculation of *G.mosseae* and *Aspergillus fumigatus* suppressed diseases caused by *Phytophthora* sp. and *Rhizoctonia solani* which led to an increase in shoot and root length of cultivated wheat plants (Berta et al. 2005; Bhale et al. 2018).

6.4.2 Mycorrhiza Mediated Bacterial Disease Management

When compared to fungi, very few reports are available on the activities of AM fungi in managing bacterial pathogens in plants. However, substantial successes have been reported in the application of AM fungi in the control of common plant bacterial pathogens like *Pseudomonas syringae* on soybean, *Ralstonia solanacearum* on tobacco and tomato plants (Tahat et al. 2012; Yuan et al. 2016), and *Xanthomona campestris* pv. *alfalfae* on Alfafa (Liu et al. 2007; Avis et al. 2008) (Table 6.1).

6.4.3 AM Fungi Mediated Nematode Management

Plant-feeding nematodes are known to pose a significant threat to crop production worldwide by causing an estimated 10–15% yield loss per annum which translates to approximately 170–180 billion USD (Elling 2013). In addition to the direct damage they inflict on crops, researches have shown that they predispose host crops to secondary infection by other pathogens like fungi, viruses and bacteria (Singh et al. 2013, Jones et al. 2013). In view of the increasing health and environmental concerns about the use of synthetic nematicides in nematode pest management, there is a need for other eco-friendly alternatives; the application of AM fungi stands as one of the suggested environment-friendly approaches to nematode pest management. The suppressive influence of AM fungi on some economically important plant nematodes has been reported in several field and greenhouse studies involving crops like tomato, banana, okra and grapevine (Alban et al. 2013; Marro et al. 2014). The activities of AM fungi in managing some plant nematodes classified to the genera *Meloidogyne, Xiphinema, Radopholus, Heterodera, Nacobbus,* etc., on some agricultural crops (Table 6.1).

6.4.4 AM Fungi Mediated Insect Pest Management

The application of AM fungi in managing agricultural insect pests is well substantiated. The majority of the available articles proved that AM fungi colonisation induces an increased nutrient assimilation potential in host plants which has implications on the plant physiology, metabolism and hormonal balance resulting in changes in host plant nutrient composition (Cardoso Filho et al. 2017; Frew 2019), or by altering gene expression independently of plant nutrition (Vannette and Hunter 2009), thereby naturally affecting the behaviour of natural enemies, mostly insect pest. Some reports further suggest that AM fungi influence the host plant's tolerance to insect pest damage by nourishing the host crop with the required nutrients especially nitrogen and phosphorus needed to synthesise defense-related enzymes and or metabolites as well as modified physical structures (Andrade et al. 2010; Seguel et al. 2015). Furthermore, an increase in host plant's nutrient acquisition induced by AM fungi most of the time means more investment in plant defences. Recent findings also reported that AM fungi colonised hosts accumulate more lignins, tannins, cellulose, silicates and some phenolic contents which help to reduce host plant palatability to insect pests (MacLean et al. 2017; da Trindade et al. 2019). A practical example was reported by Formenti and Rasmann (2019) where *Rhizophagus irregularis* symbiosis with tomato plants resulted in an increase in trichomes density of the host plant as a counter response to attack by the oriental leaf worm (*Spodoptera litura*) This further emphasises the fact that AM fungi association is largely responsible for causing a significant improvement in plant defence mechanisms towards invading insect pests by modulating some plant traits which confer tolerance or resistance against such insect pest attack (Koricheva et al. 2009). The application of AM fungi in managing agricultural insect pests has been studied in a diversity of crops worldwide, together with maize (*Zea mays*) (De Lange et al. 2020), potato (*Solanum tuberosum*) (Schoenherr et al. 2019), fava bean (*Vicia faba*) (Cabral et al. 2018), strawberry (*Fragaria* species) (Gange 2000), narrowleaf plantain (*Plantago lanceolata*) (Wang et al. 2015), urad bean (*Vigna mungo*) (Selvaraj et al. 2020), etc. However, available information has shown that the abilities of AM fungi to improve tolerance or resistance in host plants differ from one AM fungi to another and also the efficacy of the protection conferred varies among pathogens. Furthermore, several soil and environmental conditions also play their roles either by promoting or hindering the effectiveness of the protection.

6.4.5 Mechanisms Involved in AM Fungi Plant Disease Management

The mechanisms involved in the process of AM fungi-mediated control of plant diseases and pests may be either directly which usually takes the form of competition for nutrients or space which gives rise to increased nutrient intake by the host plant, or indirectly which involved plant system manipulation processes in areas like alteration in rhizosphere interactions, inducing plant defence and improving tolerance to adverse effects of pathogens invasion (Cameron et al. 2013; Pieterse et al. 2014). These mechanisms, most of the time, are not considered as being independent of one another. However, the comparative contributions of a specified technique will depend largely on the individual AM fungi–host plant-pathogen interactions involved. The previously highlighted mechanisms will be presented in detail within the next four paragraphs.

6.4.6 Mechanisms Implicated behind Mycorrhiza Mediated Disease Resistance

AM fungi have the ability to increase the uptake of the host plant through their extensive extraradical hyphae which functions as an add-on of the host plant's root structure with which they are able to explore more soil area and depth to take up more mineral nutrients such as nitrogen and phosphate in addition with water (Baum et al. 2015). Multiple studies have analysed the nutritional aspect of AM fungi-host plant interactions and concluded that AM fungi substantially improve the uptake of both macro and micronutrients by host plants especially under pressure circumstances (Smith and Smith 2011). Findings have also shown that mycorrhizal-associated host plants displayed better growth, survival and improved resistance to pathogens compared to non-AM fungi-associated plants (Declerck et al. 2002). Several authors have identified increased nutrient uptake as one important AM fungi-mediated plant disease control mechanism (Pettigrew et al. 2005; Wehner et al. 2010; Smith and Smith 2011; Schouteden et al. 2015), some others hold contradictory views in this regard (Linderman 1994; Coyne et al. 2004). These contradictions suggest that host plant pathogen suppression by associated AM fungi should not be considered as being solely due to improved nutrient uptake but rather to a combination of mechanisms that confer reduced pathogen incidences.

6.5 Modification of Host Plant Root System

Modification of the host plant root system by associated AM fungi is mostly described in root number and length which vary according to the AM fungi symbiosis. The formation of a highly branched lateral root system by host plants due to AM fungi colonisation coupled with an increasing number of sites available for root colonisation which is triggered by the production of some metabolites together with auxins that have been recognised to play a prominent part in the modification of root design and this change has been largely linked to altering the dynamics of some soil-borne pathogens (Jung et al. 2012). The intensification observed in AM fungi colonised host plant has also been attributed directly to the activities of some exudates turned out by the AM fungi (Fusconi et al. 1999), or contingently due to the surge in nutrient intake and or transformation induced in the hormonal balance of the host plant which all have implications on infection by pathogens (Vos et al. 2014). A typical example is the infection of the root tips of a hugely branched root structure of strawberry plant by the *Phytophthora* fungus which showed lower infection in the regions of the root which had AM fungi association when compared to the non-mycorrhizal roots (Bhale et al. 2018). On the other hand, depending on the type of pathogen and host plant involved, increased root branching might also work to increase potential infection sites. In some pathogens, especially the migratory endoparasitic nematodes like the *Radopholus* spp., the key roots are the major spot of infection while the sedentary endoparasitic nematodes such as *Meloidogyne* and *Heterodera* species would target the root tips and sites of lateral root formation for infection. Hence, an increase in these root regions due to AM fungi interaction might tend to increase damage potential to the host plant since it provides more avenues for pathogen invasion (Curtis et al. 2009). These opposing scenarios make it almost impossible to have a lack of clear and generalisable conclusion on the mechanisms involving host plant root modification in managing soil-borne pathogens of plants.

6.5.1 Mechanism of Competition

Competition for nutrients and space by AM fungi has been proposed as having a direct effect on the infection of host plants by pathogenic agents. This was proposed that both the AM fungi and the invading pathogenic agent would probably seek and depend on the host plant nutrition, photosynthate, root rhizosphere and or infection site for their survival (Hammer et al. 2011; Jung et al. 2012); hence, if the availability of any of these growth factors becomes limited, competition would result in reduced multiplication of pathogens within the AM fungi colonised root system (Vos et al. 2014). Some studies suggested a positive effect of AM fungi competition in reducing population levels of some plant-feeding nematodes (Vierheilig et al. 2008; Schouteden et al. 2015); other reports in respect of AM fungi competition with plant parasitic nematodes went contrary (Elsen et al. 2003; Dos Anjos et al. 2010; Koffi et al. 2013). Apart from the study by Cordier et al. (1998), who demonstrated a positive effect of the competition between the AM fungi, *Glomus mosseae* the plant morbific fungus, *Phytophthora parasitica* on tomato plant where the competition resulted in the complete exclusion of the pathogenic fungus from the tomato roots, most other available studies were of the views that depending on the AM fungi species, contention for nutrients and space may additionally impact negatively on the AM fungi in favour of the invading pathogen (Elsen et al. 2003; Vos 2012). Hence, authenticity of assigning competition for nutrient and space as a mechanism in AM fungi-mediated biocontrol process is still in doubt due to the many contradictory and inconsistent reports in this regard (Singh and Giri 2017).

6.5.2 Activation of Host Plant Defence System

It is quite typical of most soil microbes, especially AM fungi to exhibit a sustained induction of tolerance termed: induced systemic resistance in response to any other invading pathogenic organism. This form of resistance, which is mostly initiated by the priming of defence genes of the host plant, affords it some degree of protection from infection by different pathogenic agents such as bacteria, fungi, viruses and nematodes as well as different species of herbivorous insects (Pineda et al. 2010). Furthermore, the induced aversion often works by increasing the sensitivity of the host crop towards some plant growth regulators like ethylene and jasmonic acid (Pineda et al. 2010). Successes of AM fungi-induced resistance (AIR) has been demonstrated in many plants involving several plant pathogens; AIR was reported as being the major mechanism involved in managing the pathogenic fungus, *Gaeumannomyces graminis* causing take-all disease (Khaosaad et al. 2007). Also, the activities of AIR were linked with the suppression of different strains of the fungi species such as *Fusarium oxysporum* in transformed carrot root (Benhamou et al. 1994); and in date palm trees; *Alternaria solani* (causing early blight disease in tomatoes) (Fritz et al. 2006); *Colletotrichum* in cucumber plants (Jaiti et al. 2008).

AIR is associated with the production of several chemicals such as chitinases, phytoalexins, glucanases, glycoproteins, phenolic compounds, calloses and enzymes of phenylpropanoid with a diversified range of biochemical properties which are mostly synthesised as secondary metabolites (Singh 2017).

6.6 AM Fungi Application in Commercial Agriculture

In view of the numerous benefits that AM fungi confer on their hosts, it is clear that AM fungi bear substantial potential to be applied in improving agricultural yield. Diverse studies have underscored the great potential of commercial AM fungi inoculum in agricultural production (Berruti et al. 2016; Igiehon and Babalola 2017; Wang et al. 2017) either as a catalyst to improve plant health (Hijri 2016) or in mineral nutrients absorption and translocation (Huang et al. 2020) as well as nutrient mobilisation beyond depletion zones of the crop rhizosphere (Bender et al. 2015). AM fungi have the ability to cause moderate to significant alterations in the secondary metabolism of host plants leading to what is described as enhanced nutraceutical potentials. It is a well-established fact that AM fungi confer great benefits to their host which shows great promise for application in agriculture in different ecological zones of the world. Many studies have revealed several applications of commercial mycorrhizal inoculum in agriculture (Rouphael et al. 2015; Hijri 2016). Most modern agricultural practices are characterised by excessive tillage operations, fertiliser application, use of pesticides and poor rotation sequence which limit the power of plants to effectively establish mutualisms with AM fungi. Although most modern agronomic practices tend to be disruptive to AM fungi, there might still be a great opportunity for cost-effective agriculture to oversee the cropping structure in a course of action that advances AM fungi production. Although commercial assembly and utilisation of AM fungi inoculum are considered exhaustive, AM fungi implementation is worth the stress, especially in the cultivation of high-value horticultural crops and also during adaptation of micropropagated plantlets in nurseries (Tchabi et al. 2016; Chen et al. 2018). Inoculation of micropropagated plants like peach and apple at transplant will help to induce rapid nutrient uptake at the critical weaning stage so as to achieve plants with desirable commercial traits thereby rendering crop production more profitable. AM fungi, thus, invariably cause a decrease in the amount of artificial fertiliser application with less fear of reduced yield thereby increasing farmers' profit. In spite of the great potentials inherent in the application of AM fungi in agriculture, its effectiveness depends largely on the degree of some external factors especially tillage operations and the application of some specific biocides which interfere with AM fungi proliferation in soils and cropping systems involving non-host plants (Hartmann et al. 2015). Furthermore, each crop has its best corresponding AM fungi which would provide diverse benefits and not in every single amalgamation of crop and mycorrhiza, the benefit of appeal is incontrovertibly and emphatically impacted (Berruti et al. 2016; Chen et al. 2018).

6.7 Conclusions and Future Outlooks

Sustainable agriculture is one that relies on biological procedures to accomplish optimal positions of food output in terms of grade and quantity with negligible disturbance to the ecosystem. The several benefits of AM fungi have also positioned them with great potential for significant ecological services within the environment. Application of AM fungi in plant diseases management caused by plant pathogenic fungi, bacteria, nematodes, etc. is a very viable option. However, some setbacks, especially biotrophic nature of AM fungi creates a significant challenge in their application. The world's current mycorrhizal products market is yet to achieve its full potential due to problems arising from regulatory constraints, product efficacy, quality assurance issues, customer awareness and other technical issues. In order to harness and maximise the inherent potential of AM fungi to be used in plant disease management, efforts should be geared up at providing an environment for AM fungi through manipulation of agricultural systems to favour AM fungi colonisation and proliferation. Therefore, understanding and managing the mycorrhizal symbiosis dynamics is crucial for maintaining the growth and health of crops.

References

Abdel-Fattah GM, Shabana YM (2002) Efficacy of the arbuscular mycorrhizal fungus Glomus clarum in protection of cowpea plants against root rot pathogen Rhizoctonia solani/Wirksamkeit des arbusculären Mykorrhizapilz *Glomus darum* als Schutz von Kundebohnenpflanzen gegen den Wurzelfäuleerreger *Rhizoctonia solani*. Zeitschrift für Pflanzenkrankheiten und Pflanzenschutz/J Plant Dis Protect 109:207–215

Affokpon A, Coyne DL, Lawouin L, Tossou C, Agbèdè RD, Coosemans J (2011) Effectiveness of native West African arbuscular mycorrhizal fungi in protecting vegetable crops against root-knot nematodes. Biol Fertil Soils 47(2):207–217

Al-Askar A, Rashad Y (2010) Arbuscular mycorrhizal fungi: a biocontrol agent against common. Plant Pathol J 9(1):31–38

Al-Askar AA, Rashad EM, Ghoneem KM, Mostafa AA, Al-Otibi FO, Saber WI (2021) Discovering Penicillium polonicum with high-lytic capacity on Helianthus tuberosus tubers: oil-based preservation for mold management. Plan Theory 10(2):413

Alban R, Guerrero R, Toro M (2013) Interactions between a root-knot nematode (*Meloidogyne exigua*) and arbuscular mycorrhizae in coffee plant development (*Coffea arabica*). Am J Plant Sci 4:19–23. https://doi.org/10.4236/ajps.2013.47A2003

Al-Yahya'ei M, Oehl F, Vallino M, Lumini E, Redecker D, Wiemken A et al (2011) Unique arbuscular mycorrhizal fungal communities uncovered in date palm plantations and surround-

ing desert habitats of Southern Arabia. Mycorrhiza 21:195–209. https://doi.org/10.1007/s00572-010-0323-5
Andrade LBDS, Oliveira AS, Ribeiro JKC, Kiyota S, Vasconcelos IM, De Oliveira JTA et al (2010) Effects of a novel pathogenesis-related class 10 (PR-10) protein from *Crotalaria pallida* roots with papain inhibitory activity against root-knot nematode *Meloidogyne incognita*. J Agric Food Chem 58:4145–4152. https://doi.org/10.1021/jf9044556
Ansari RA, Khan TA (2012a) Parasitic association of root-knot nematode, *Meloidogyne incognita* on guava. e J Sci Technol 5(12):65–67
Ansari RA, Khan TA (2012b) Diversity and community structure of Phytonematodes associated with guava in and around Aligarh, Uttar Pradesh. Trends Biosci 5(3):202–204
Ansari RA, Mahmood I (2019a) Plant health under biotic stress. In: Ansari RA, Mahmood I (eds) Microbial interactions, vol II. Springer Nature, Singapore. https://doi.org/10.1007/978-981-13-6040-4
Ansari RA, Mahmood I (2019b) Plant health under biotic stress. In: Ansari RA, Mahmood I (eds) Organic strategies, vol I. Springer Nature, Singapore. https://doi.org/10.1007/978-981-13-6043-5
Ansari RA, Mahmood I, Rizvi R, Sumbul A (2017a) PGPR: current vogue in sustainable crop production. In: Kumar V (ed) Probiotics and plant health. Springer Nature, Singapore, pp 455–472
Ansari RA, Mahmood I, Rizvi R, Sumbul A, Safiuddin (2017b) Siderophores: augmentation of soil health and crop productivity. In: Kumar V (ed) Probiotics in agroecosystem. Springer Nature, Singapore, pp 291–312
Ansari RA, Sumbul A, Rizvi R, Mahmood I (2019a) Potential role of plant growth promoting Rhizobacteria in alleviation of biotic stress. In: Ansari RA, Mahmood I (eds) Plant health under biotic stress, Microbial interactions, vol II. Springer Nature, Singapore, pp 177–188
Ansari RA, Sumbul A, Rizvi R, Mahmood I (2019b) Organic soil amendments: potential tool for soil and plant health management. In: Ansari RA, Mahmood I (eds) Plant health under biotic stress, Organic strategies, vol I. Springer Nature, Singapore, pp 1–35
Aseel DG, Rashad YM, Hammad SM (2019) Arbuscular mycorrhizal fungi trigger transcriptional expression of flavonoid and chlorogenic acid biosynthetic pathways genes in tomato against tomato mosaic virus. Sci Rep 9(1):1–10
Atolani O, Fabiyi OA (2020) Plant parasitic nematodes management through natural products: current progress and challenges. In: Ansari RA, Rizvi R, Mahmood I (eds) Management of phytonematodes: recent advances and future challenges. Springer, Singapore, pp 297–315
Atolani O, Fabiyi OA, Olatunji GA (2014a) Nematicidal isochromane glycoside from *Kigelia pinnata* leaves. Acta Agric Slovenica 104(1):25–31
Atolani O, Fabiyi OA, Olatunji GA (2014b) Isovitexin from *Kigelia pinnata*, a potential eco-friendly nematicidal agent. Trop Agric 91(2):67–74
Avis TJ, Gravel V, Antoun H, Tweddell RJ (2008) Multifaceted beneficial effects of rhizosphere microorganisms on plant health and productivity. Soil Biol Biochem 40(7):1733–1740
Awe GO, Reichert JM, Fontanela E (2020) Sugarcane production in the subtropics: seasonal changes in soil properties and crop yield in no-tillage, inverting and minimum tillage. Soil Tillage Res 196:104447
Aysan E, Demir S (2009) Using arbuscular mycorrhizal fungi and Rhizobium leguminosarum biovar phaseoli against Sclerotinia sclerotiorum (Lib.) de Bary in the common bean (Phaseolus vulgaris L.). Plant Pathol J (Faisalabad) 8(2):74–78
Baum C, El-Tohamy W, Gruda N (2015) Increasing the productivity and product quality of vegetable crops using arbuscular mycorrhizal fungi: a review. Sci Hortic (Amsterdam) 187:131–141. https://doi.org/10.1016/j.scienta.2015.03.002
Bender SF, Conen F, Van der Heijden MG (2015) Mycorrhizal effects on nutrient cycling, nutrient leaching and N2O production in experimental grassland. Soil Biol Biochem 80:283–292
Benhamou N, Fortin JA, Hamel C, St-Arnaud M, Shatilla A (1994) Resistance responses of mycorrhizal Ri T-DNA-transformed carrot roots to infection by Fusarium oxysporum f. sp. chrysanthemi. Phytopathology 84(9):958–968

Bennett AE, Meek HC (2020) The influence of arbuscular mycorrhizal fungi on plant reproduction. J Chem Ecol 46(8):707–721
Berruti A, Lumini E, Balestrini R, Bianciotto V (2016) Arbuscular mycorrhizal fungi as natural biofertilizers: let's benefit from past successes. Front Microbiol 6:1559
Berta G, Sampo S, Gamalero E, Massa N, Lemanceau P (2005) Suppression of *Rhizoctonia* root-rot of tomato by *Glomus mossae* BEG12 and *Pseudomonas fluorescens* A6RI is associated with their effect on the pathogen growth and on the root morphogenesis. Eur J Plant Pathol 111:279–288
Bhale UN, Bansode SA, Singh S (2018) Multifactorial role of arbuscular mycorrhizae in agroecosystem. In: Fungi and their role in sustainable development: current perspectives. Springer, Singapore, pp 205–220
Bitterlich M, Sandmann M, Graefe J (2018) Arbuscular mycorrhiza alleviates restrictions to substrate water flow and delays transpiration limitation to stronger drought in tomato. Front Plant Sci 9:154. https://doi.org/10.3389/fpls.2018.00154
Bødker L, Kjøller R, Kristensen K, Rosendahl S (2002) Interactions between indigenous arbuscular mycorrhizal fungi and Aphanomyces euteiches in field-grown pea. Mycorrhiza 12:7–12
Bunn R, Lekberg Y, Zabinski C (2009) Arbuscular mycorrhizal fungi ameliorate temperature stress in thermophilic plants. Ecology 90:1378–1388. https://doi.org/10.1890/07-2080.1
Cabral C, Wollenweber B, António C, Rodrigues AM, Ravnskov S (2018) Aphid infestation in the phyllosphere affects primary metabolic profiles in the arbuscular mycorrhizal hyphosphere. Sci Rep 8(1):14442
Cameron D, Neal A, van Wees S, Ton J (2013) Mycorrhiza-induced resistance: more than the sum of its parts? Trends Plant Sci 18:539–545
Cardoso Filho JA, Sobrinho RR, Pascholati SF (2017) Arbuscular mycorrhizal symbiosis and its role in plant nutrition in sustainable agriculture. In: Agriculturally important microbes for sustainable agriculture. Springer, New York, pp 129–164
Cavagnaro TR, Bender SF, Asghari HR, van der Heijden MGA (2015) The role of arbuscular mycorrhizas in reducing soil nutrient loss. Trends Plant Sci 20:283–290. https://doi.org/10.1016/j.tplants.2015.03.004
Chahal K, Gupta V, Verma NK, Chaurasia A, Rana B (2021) Arbuscular mycorrhizal (AM) fungi as a tool for sustainable agricultural system. In: Mycorrhizal fungi: utilization in agriculture and forestry. IntechOpen, pp 19
Chen ECH, Morin E, Beaudet D, Noel J, Yildirir G, Ndikumana S et al (2018) High intraspecific genome diversity in the model arbuscular mycorrhizal symbiont *Rhizophagus irregularis*. New Phytol 220:1161. https://doi.org/10.1111/nph.14989
Chitarra W, Pagliarani C, Maserti B, Lumini E, Siciliano I, Cascone P et al (2016) Insights on the impact of arbuscular mycorrhizal symbiosis on tomato tolerance to water stress. Plant Physiol 171(2):1009–1023
Cordier C, Pozo MJ, Barea JM, Gianinazzi S, Gianinazzi-Pearson V (1998) Cell defense responses associated with localized and systemic resistance to *Phytophthora parasitica* induced in tomato by an arbuscular mycorrhizal fungus. Mol Plant-Microbe Interact 11:1017–1028. https://doi.org/10.1094/MPMI.1998.11.10.1017
Coyne DL, Sahrawat KL, Plowright RA (2004) The influence of mineral fertilizer application and plant nutrition on plant-parasitic nematodes in upland and lowland rice in Côte d'Ivoire and its implications in long term agricultural research trials. Exp Agric 40:245–256. https://doi.org/10.1017/S0014479703001595
Curtis R, Robinson A, Perry R (2009) Hatch and host location. In: Perry RN, Moens M, Starr JL (eds) Root-knot nematodes. CAB International, Wallingford, pp 139–162
da Trindade R, Almeida L, Xavier L, Lins AL, Andrade EH, Maia JG, da Silva JK (2019) Arbuscular mycorrhizal fungi colonization promotes changes in the volatile compounds and enzymatic activity of lipoxygenase and phenylalanine ammonia lyase in *Piper nigrum* L. 'Bragantina'. Plant (Basel) 8:442

De Lange ES, Laplanche D, Guo H, Xu W, Vlimant M, Erb M, Turlings TC (2020) *Spodoptera frugiperda* caterpillars suppress herbivore-induced volatile emissions in maize. J Chem Ecol 46:344–360

Declerck S, Risede JM, Rufyikiri G, Delvaux B (2002) Effects of arbuscular mycorrhizal fungi on the severity of root rot of bananas caused by *Cylindrocladium spathiphylli*. Plant Pathol 51:109–115

Dos Anjos ÉCT, Cavalcante UMT, Gonçalves DMC, Pedrosa EMR, dos Santos VF, Maia LC (2010) Interactions between an arbuscular mycorrhizal fungus (*Scutellospora heterogama*) and the root-knot nematode (*Meloidogyne incognita*) on sweet passion fruit (*Passiflora alata*). Braz Arch Biol Technol 53:801–809. https://doi.org/10.1590/S1516-89132010000400008

Elling AA (2013) Major emerging problems with minor Meloidogyne species. Phytopathology 103(11):1092–1102

Elsen A, Beeterens R, Swennen R, De Waele D (2003) Effects of an arbuscular mycorrhizal fungus and two plant-parasitic nematodes on *Musa* genotypes differing in root morphology. Biol Fertil Soils 38:367–376. https://doi.org/10.1007/s00374-003-0669-3

Elsen A, Gervacio D, Swennen R, De Waele D (2008) AMF-induced biocontrol against plant parasitic nematodes in Musa sp.: a systemic effect. Mycorrhiza 18:251–256

El-Sharkawy HH, Abbas MS, Soliman AS, Ibrahim SA, El-Nady IA (2021) Synergistic effect of growth-promoting microorganisms on bio-control of *Fusarium oxysporum* f. sp. pisi, growth, yield, physiological and anatomical characteristics of pea plants. Pestic Biochem Physiol 178:104939

Fabiyi OA (2020) Growth and yield response of groundnut *Arachis hypogaea* (Linn.) under *Meloidogyne incognita* infection to furfural synthesized from agro-cellulosic materials. J Trop Agric 58(2):241–245

Fabiyi OA (2021a) Evaluation of plant materials as root-knot nematode (*Meloidogyne incognita*) suppressant in okro (*Abelmuscous esculentus*). Agric Conspec Sci 86(1):51–56

Fabiyi OA (2021b) Sustainable management of *Meloidogyne incognita* infecting carrot: green synthesis of silver nanoparticles with *Cnidoscolus aconitifolius*: (*Daucus carota*). Vegetos 34(2):277–285

Fabiyi OA (2021c) Application of furfural in sugarcane nematode pest management. Pak J Nematol 39(2):151–155. https://doi.org/10.17582/journal.PJN/2021.39.2.151.155

Fabiyi OA (2021d) Evaluation of Nematicidal activity of *Terminalia glaucescens fractions* against *Meloidogyne incognita* on *Capsicum chinense*. J Hortic Res 29(1):67–74. https://doi.org/10.2478/johr-2021-0006

Fabiyi OA (2022a) Fractions from *Mangifera indica* as an alternative in *Meloidogyne incognita* management. Pak J Nematol 40(1):65–74

Fabiyi OA (2022b) Cytotoxicity and nematicidal potential of leaf extracts of *Adansonia digitata* and *Khaya senegalensis* on root-knot nematode (*Meloidogyne incognita*) associated with cabbage *(Brassica oleracea)*. J Agric Sci Sri Lanka 17(3):425–436

Fabiyi OA (2022c) Evaluation of weeds against root-knot nematode (*Meloidogyne incognita*) in vegetables. Sarhad J Agric 38(4):1289–1299

Fabiyi OA (2022d) Application of composited municipal refuse dump site soil, orange, potato and pineapple peels in the control of root knot nematode (*Meloidogyne incognita*) infecting carrots (*Daucus carota* L.). J Solid Waste Manag 48(3):474–485

Fabiyi OA, Olatunji GA (2021a) Environmental sustainability: bioactivity of *Leucaena leucocephala* leaves and pesticide residue analysis in tomato fruits. Acta Univ Agric Silvic Mendelianae Brun 69(4):473–480

Fabiyi OA, Olatunji GA (2021b) Toxicity of derivatized citrulline and extracts of water melon rind (*Citrullus lanatus*) on root-knot nematode (*Meloidogyne incognita*). Trop Agric 98(4):347–355

Fabiyi OA, Olatunji GA, Daodu IO (2019) Nematicidal effect of organic extract metal complex on *Meloidogyne incognita* infecting groundnuts (*Arachis hypogea*). Sci Agric Bohem 50(3):191–196

Fabiyi OA, Saliu OD, Claudius-Cole AO, Olaniyi IO, Oguntebi OV, Olatunji GA (2020a) Porous starch citrate biopolymer for controlled release of Carbofuran in the management of root knot nematode *Meloidogyne incognita*. Biotechnol Rep 25(e00428):1–9

Fabiyi OA, Atolani O, Olatunji GA (2020b) Toxicity effect of *Eucalyptus globulus* to *Pratylenchus* spp of *Zea mays*. Sarhad J Agric 36(4):1244–1253

Fabiyi OA, Baker MA, Olatunji GA (2022a) Application of fatty acid esters on *Meloidogyne incognita* infected Jew's mallow. Pak J Nematol 40(2):127–137

Fabiyi OA, Baker MT, Claudius-Cole AO, Alabi RO, Olatunji GA (2022b) Application of cyclic imides in the management of root-knot nematode (*Meloidogyne incognita*) on cabbage. Indian Phytopathol 75:457. https://doi.org/10.1007/s42360-022-00480-1

Fiorilli V, Catoni M, Francia D, Cardinale F, Lanfranco L (2011) The arbuscular mycorrhizal symbiosis reduces disease severity in tomato plants infected by Botrytis cinerea. J Plant Pathol 93:237–242

Formenti L, Rasmann S (2019) Mycorrhizal fungi enhance resistance to herbivores in tomato plants with reduced jasmonic acid production. Agronomy 9(3):131

Frew A (2019) Arbuscular mycorrhizal fungal diversity increases growth and phosphorus uptake in C3 and C4 crop plants. Soil Biol Biochem 135:248–250

Fritz M, Jakobsen I, Lyngkjaer MF, Thordal-Christensen H, Pons-Kühnemann J (2006) Arbuscular mycorrhiza reduces susceptibility of tomato to *Alternaria solani*. Mycorrhiza 16:413–419. https://doi.org/10.1007/s00572-006-0051-z

Fusconi A, Gnavi E, Trotta A, Berta G (1999) Apical meristems of tomato roots and their modifications induced by arbuscular mycorrhizal and soilborne pathogenic fungi. New Phytol 142:505–516. https://doi.org/10.1046/j.1469-8137.1999.00410.x

Gange AC (2000) Species-specific responses colonization of plant a and mycorrhizal insect to arbuscular of its host plant. New Phytol 150:611–618. https://doi.org/10.1046/j.1469-8137.2001.00137.x

Gao WQ, Wang P, Wu QS (2019) Functions and application of glomalin-related soil proteins: a review. Sains Malaysiana 48(1):111–119

Gao C, El-Sawah AM, Ali DFI, Alhaj Hamoud Y, Shaghaleh H, Sheteiwy MS (2020) The integration of bio and organic fertilizers improve plant growth, grain yield, quality and metabolism of hybrid maize (*Zea mays* L.). Agronomy 10(3):319

Gupta MM (2020) Arbuscular mycorrhizal fungi: the potential soil health indicators. In: Soil health. Springer, Cham, pp 183–195

Hammer EC, Pallon J, Wallander H, Olsson PA (2011) Tit for tat? A mycorrhizal fungus accumulates phosphorus under low plant carbon availability. FEMS Microbiol Ecol 76:236–244. https://doi.org/10.1111/j.1574-6941.2011.01043.x

Hao Z, Fayolle L, van Tuinen D, Chatagnier O, Li X, Gianinazzi S, Gianinazzi-Pearson V (2012) Local and systemic mycorrhiza-induced protection against the ectoparasitic nematode Xiphinema index involves priming of defence gene responses in grapevine. J Exp Bot 63(10):3657–3672

Hartmann M, Frey B, Mayer J, Mäder P, Widmer F (2015) Distinct soil microbial diversity under long-term organic and conventional farming. ISME J 9(5):1177–1194

Hijri M (2016) Analysis of a large dataset of mycorrhiza inoculation field trials on potato shows highly significant increases in yield. Mycorrhiza 26:209–214. https://doi.org/10.1007/s00572-015-0661-4

Himaya, S. M. M., Sivasubramaniam, N., & Afreen, S. M. M. (2021). A review on role of mycorrhizal fungi in plant disease management

Huang GM, Zou YN, Wu QS, Xu YJ, Kuča K (2020) Mycorrhizal roles in plant growth, gas exchange, root morphology, and nutrient uptake of walnuts. Plant Soil Environ 66(6):295–302

Igiehon NO, Babalola OO (2017) Biofertilizers and sustainable agriculture: exploring arbuscular mycorrhizal fungi. Appl Microbiol Biotechnol 101(12):4871–4881

Jacott CN, Murray JD, Ridout CJ (2017) Trade-offs in arbuscular mycorrhizal symbiosis: disease resistance, growth responses and perspectives for crop breeding. Agronomy 7(4):75

Jaiti F, Kassami M, Meddich A, El Hadrami I (2008) Effect of arbuscular mycorrhization on the accumulation of hydroxycinnamic acid derivatives in date palm seedlings challenged with *Fusarium oxysporum* f. sp. *albedinis*. J Phytopathol 156:641–646

Jones JT, Haegeman A, Danchin EGJ, Gaur HS, Helder J, Jones MGK et al (2013) Top 10 plant-parasitic nematodes in molecular plant pathology. Mol Plant Pathol 14:946–961. https://doi.org/10.1111/mpp.12057

Jung SC, Martinez-Medina A, Lopez-Raez JA, Pozo MJ (2012) Mycorrhiza-induced resistance and priming of plant defenses. J Chem Ecol 38:651–664. https://doi.org/10.1007/s10886-012-0134-6

Karagiannidis N, Bletsos F, Stavropoulos N (2002) Effect of Verticillium wilt (Verticillium dahliae Kleb.) and mycorrhiza (Glomus mosseae) on root colonization, growth and nutrient uptake in tomato and eggplant seedlings. Sci Hortic 94:145–156

Kasiamdari RS, Smith SE, Smith FA, Scott ES (2002) Influence of the mycorrhizal fungus, Glomus coronatum, and soil phosphorus on infection and disease caused by binucleate Rhizoctonia and Rhizoctonia solani on mung bean (Vigna radiata). Plant Soil 238:235–244

Khaosaad T, García-Garrido JM, Steinkellner S, Vierheilig H (2007) Take-all disease is systemically reduced in roots of mycorrhizal barley plants. Soil Biol Biochem 39:727–734. https://doi.org/10.1016/j.soilbio.2006.09.014

Kjøller R, Rosendahl S (1996) The presence of the arbuscular mycorrhizal fungus Glomus intraradices influences enzymatic activities of the root pathogen Aphanomyces euteiches in pea roots. Mycorrhiza 6:487–491

Koffi MC, Vos C, Draye X, Declerck S (2013) Effects of Rhizophagus irregularis MUCL 41833 on the reproduction of *Radopholus similis* in banana plantlets grown under *in vitro* culture conditions. Mycorrhiza 23:279–288. https://doi.org/10.1007/s00572-012-0467-6

Koricheva J, Gange AC, Jones T (2009) Effects of mycorrhizal fungi on insect herbivores: a meta-analysis. Ecology 90:2088–2097

Kumar A, Choudhary AK, Suri VK (2016) Influence of AM fungi, inorganic phosphorus and irrigation regimes on plant water relations and soil physical properties in okra (*Abelmoschus esculentus* l.)–pea (*Pisum sativum* l.) cropping system in Himalayan acid Alfisol. J Plant Nutr 39:666–682

Labeena P, Sreenivasa MN, Lingaraju S (2002) Interaction effects between arbuscular mycorrhizal fungi and root-knot nematode Meloidogyne incognita on tomato. Indian Journal of Nematology 32(2):118–120

Lehmann A, Leifheit EF, Feng L, Bergmann J, Wulf A, Rillig MC (2020) Microplastic fiber and drought effects on plants and soil are only slightly modified by arbuscular mycorrhizal fungi. Soil Ecol Lett 4:32–44

Li LF, Li T, Zhang Y, Zhao ZW (2010) Molecular diversity of arbuscular mycorrhizal fungi and their distribution patterns related to host-plants and habitats in a hot and arid ecosystem, Southwest China. FEMS Microbiol Ecol 71(3):418–427

Linderman RG (1994) Role of VAM fungi in biocontrol. In: Pfleger FL, Linderman RG (eds) Mycorrhizae and plant health. APS, St Paul, pp 1–26

Liu J, Wu L, Wei S, Xiao X, Su C, Jiang P et al (2007) Effects of arbuscular mycorrhizal fungi on the growth, nutrient uptake and glycyrrhizin production of licorice (Glycyrrhiza uralensis Fisch). Plant Growth Regul 52:29–39

MacLean AM, Bravo A, Harrison MJ (2017) Plant signaling and metabolic pathways enabling arbuscular mycorrhizal symbiosis. Plant Cell 29:2319–2335

Malik RJ, Dixon MH, Bever JD (2016) Mycorrhizal composition can predict foliar pathogen colonization in soybean. Biol Control 103:46–53

Marro N, Lax P, Cabello M, Doucet ME, Becerra AG (2014) Use of the arbuscular mycorrhizal fungus *Glomus intraradices* as biological control agent of the nematode *Nacobbus aberrans* parasitizing tomato. Braz Arch Biol Technol 57:668–675

Matsubara Y, Kayukawa Y, Yano M, Fukui H (2000) Tolerance of asparagus seedlings infected with arbuscular mycorrhizal fungus to violet root rot caused by Helicobasidium mompa. J Jpn Soc Hortic Sci 69:552–556

Matsubara Y, Ohba N, Fukui H (2001) Effect of arbuscular mycorrhizal fungus infection on the incidence of fusarium root rot in asparagus seedlings. J Jpn Soc Hortic Sci 70:202–206

Matsubara Y, Hasegawa N, Fukui H (2002) Incidence of fusarium root rot in asparagus seedlings infected with arbuscular mycorrhizal fungus as affected by several soil amendments. J Jpn Soc Hortic Sci 71:370–374

Öpik M, Moora M, Liira J, Zobel M (2006) Composition of root-colonizing arbuscular mycorrhizal fungal communities in different ecosystems around the globe. J Ecol 94:778–790. https://doi.org/10.1111/j.1365-2745.2006.01136.x

Öpik M, Zobel M, Cantero JJ, Davison J, Facelli JM, Hiiesalu I et al (2013) Global sampling of plant roots expands the described molecular diversity of arbuscular mycorrhizal fungi. Mycorrhiza 23:411–430. https://doi.org/10.1007/s00572-013-0482-2

Pandey DK, Kaur P, Dey A (2018) Arbuscular mycorrhizal fungi: effects on secondary metabolite production in medicinal plants. In: Gehlot P, Singh J (eds) Fungi and their role in sustainable development: current perspectives. Springer, Singapore, pp 507–538

Pettigrew WT, Meredith WR, Young LD (2005) Potassium fertilization effects on cotton lint yield, yield components, and reniform nematode populations. Agron J 97:1245–1251. https://doi.org/10.2134/agronj2004.0321

Pereira JAP, Vieira IJC, Freitas MSM, Prins CL, Martins MA, Rodrigues R (2016) Effects of arbuscular mycorrhizal fungi on capsicum spp. J Agric Sci 154(5):828–849

Pieterse CMJ, Zamioudis C, Does DV, Van Wees SCM (2014) Signalling networks involved in induced resistance. In: Induced resistance for plant defense. Wiley, Hoboken, NJ, pp 58–80

Pineda A, Zheng SJ, van Loon JJA, Pieterse CMJ, Dicke M (2010) Helping plants to deal with insects: the role of beneficial soil-borne microbes. Trends Plant Sci 15:507–514

Querejeta JI (2017) Soil water retention and availability as influenced by mycorrhizal symbiosis: consequences for individual plants, communities, and ecosystems. In: Collins-Johnson N, Gehring C, Jansa J (eds) Mycorrhizal mediation of soil - fertility, structure, and carbon storage. Elsevier, Amsterdam, pp 299–317. https://doi.org/10.1016/B978-0-12-804312-7.00017-6

Rizvi R, Singh G, Safiuddin, Ansari RA, Tiyagi SA, Mahmood I (2015) Sustainable management of root-knot disease of tomato by neem cake and *Glomus fasciculatum*. Cogent Food Agric 1:1008859

Ronsheim ML (2016) Plant genotype influences mycorrhiza benefits and susceptibility to a soil pathogen. Am Midl Nat 175(1):103–112

Rosendahl CN, Rosendahl S (1990) The role of vesicular-arbuscular mycorrhiza in controlling damping-off and growth reduction in cucumber caused by Pythium ultimum. Symbiosis

Rouphael Y, Franken P, Schneider C, Schwarz D, Giovannetti M, Agnolucci M et al (2015) Arbuscular mycorrhizal fungi act as biostimulants in horticultural crops. Sci Hortic 196:91–108. https://doi.org/10.1016/j.scienta.2015.09.002

Schoenherr AP, Rizzo E, Jackson N, Manosalva P, Gomez SK (2019) Mycorrhiza-induced resistance in potato involves priming of defense responses against cabbage Looper (Noctuidae: Lepidoptera). Environ Entomol 48:370–381

Schouteden N, De Waele D, Panis B, Vos CM (2015) Arbuscular mycorrhizal fungi for the biocontrol of plant-parasitic nematodes: a review of the mechanisms involved. Front Microbiol 6:1280

Seguel A, Barea JM, Cornejo P, Borie F (2015) Role of arbuscular mycorrhizal symbiosis in phosphorus-uptake efficiency and aluminium tolerance in barley growing in acid soils. Crop Pasture Sci 66:696–705

Selosse MA, Charpin M, Not F (2017) Mixotrophy everywhere on land and in water: the grand ecart hypothesis. Ecol Lett 20:246–263. https://doi.org/10.1111/ele.12714

Selvaraj A, Thangavel K (2021) Arbuscular mycorrhizal fungi: potential plant protective agent against herbivorous insect and its importance in sustainable agriculture. In: Shrivastava N, Mahajan S, Varma A (eds) Symbiotic soil microorganisms. Springer, Cham, pp 319–337

Selvaraj A, Thangavel K, Uthandi S (2020) Arbuscular mycorrhizal fungi (*Glomus intraradices*) and diazotrophic bacterium (*Rhizobium* BMBS) primed defense in blackgram against herbivorous insect (*Spodoptera litura*) infestation. Microbiol Res 231:126355

Shakoor S, Inam-ul-Haq M, Bibi S, Ahmed R (2015) Influence of root inoculations with vasicular arbuscular mycorrhizae and rhizomyx for the management of root rot of chickpea. Pakistan Journal of Phytopathology 27(2):153–158

Siddiqui Z, Singh LP (2005) Effects of fly ash and soil micro-organisms on plant growth, photosynthetic pigments and leaf blight of wheat. J Plant Dis Protect 112:146–155

Sileshi GW, Mafongoya PL, Nath AJ (2020) Agroforestry systems for improving nutrient recycling and soil fertility on degraded lands. Agroforestry for Degraded Landscapes: Recent Advances and Emerging Challenges-Vol 1:225–253

Singh I (2017) Antimicrobials in higher plants: classification, mode of action and bioactivities. Chem Biol Lett 4:48–62

Singh I, Giri B (2017) Arbuscular mycorrhiza mediated control of plant pathogens. Mycorrhiza-Nutrient uptake, biocontrol, ecorestoration:131–160

Singh R, Soni SK, Kalra A (2013) Synergy between *Glomus fasciculatum* and a beneficial Pseudomonas in reducing root diseases and improving yield and forskolin content in *Coleus forskohlii* Briq. under organic field conditions. Mycorrhiza 23:35–44

Smith SE, Smith FA (2011) Roles of arbuscular mycorrhizas in plant nutrition and growth: new paradigms from cellular to ecosystem scales. Annu Rev Plant Biol 62:227–250. https://doi.org/10.1146/annurev-arplant-042110-103846

Solanki MK, Kashyap PL, Ansari RA, Kumari B (2020) Microbiomes and plant health: panoply and their applications. Academic Press, London. https://www.elsevier.com/books/microbiomes-and-plant-health/solanki/978-0-12-819715-8

Sumbul A, Mahmood I, Rizvi R, Safiuddin, Ansari RA (2017) Mycorrhiza: an alliance for the nutrient management in plants. In: Kumar V (ed) Probiotics in agroecosystem. Springer Nature, Singapore, pp 371–386

Tahat MM, Sijam K, Othman R (2012) The potential of endomycorrhizal fungi in controlling tomato bacterial wilt Ralstonia solanacearum under glasshouse conditions. Afr J Biotechnol 11(67):13085–13094

Tchabi, A., Hountondji, F. C., Ogunsola, B., Lawouin, L., Coyne, D. L., Wiemken, A., & Oehl, F. (2016). Effect of two species of arbuscular mycorrhizal fungi inoculation on development of micro-propagated yam plantlets and suppression of *Scutellonema bradys* (Tylenchideae)

Torres-Barragán A, Zavale-Tamejia E, Gonzalez-Chavez C, Ferrera Cerrato R (1996) The use of arbuscular mycorrhizae to control onion white rot (Sclerotium cepivorum) under field conditions. Mycorrhiza 6:253–257

Vázquez MM, César S, Azcón R, Barea JM (2000) Interactions between arbuscular mycorrhizal fungi and other microbial inoculants (Azospirillum, pseudomonas, Trichoderma) and their effects on microbial population and enzyme activities in the rhizosphere of maize plants. Appl Soil Ecol 15(3):261–272

Van der Heijden MG, Martin FM, Selosse MA, Sanders IR (2015) Mycorrhizal ecology and evolution: the past, the present, and the future. New Phytol 205(4):1406–1423

Vannette RL, Hunter MD (2009) Mycorrhizal fungi as mediators of defencedefense against insect pests in agricultural systems. Agric For Entomol 11:351

Varga S, Finozzi C, Vestberg M, Kytoviita MM (2015) Arctic arbuscular mycorrhizal spore community and viability after storage in cold conditions. Mycorrhiza 25:335–343. https://doi.org/10.1007/s00572-014-0613-4

Vierheilig H, Steinkellner S, Khaosaad T (2008) The biocontrol effect of mycorrhization on soilborne fungal pathogens and the autoregulation of the AM symbiosis: one mechanism, two effects? In: Varma A (ed) Mycorrhiza. Springer-Verlag, Berlin, pp 307–320. https://doi.org/10.1007/978-3-540-78826-3_15

Vos C (2012) Arbusculaire Mycorrhizenschimmels in De Biocontrole Van Plantenparasitaire Nematoden. University of Leuven (KU Leuven), Leuven

Vos CM, Tesfahun AN, Panis B, De Waele D, Elsen A (2012) Arbuscular mycorrhizal fungi induce systemic resistance in tomato against the sedentary nematode Meloidogyne incognita and the migratory nematode Pratylenchus penetrans. Appl Soil Ecol 61:1–6

Vos C, Schouteden N, Van Tuinen D, Chatagnier O, Elsen A, De Waele D et al (2013) Mycorrhiza-induced resistance against the root–knot nematode Meloidogyne incognita involves priming of defense gene responses in tomato. Soil Biol Biochem 60:45–54

Vos CM, Yang Y, De Coninck B, Cammue BPA (2014) Fungal (-like) biocontrol organisms in tomato disease control. Biol Control 74:65–81. https://doi.org/10.1016/j.biocontrol.2014.04.004

Wang M, Bezemer TM, Van der Putten WH, Biere A (2015) Effects of the timing of herbivory on plant defense induction and insect performance in ribwort plantain (*Plantago lanceolata* L.) depend on plant mycorrhizal status. J Chem Ecol 41:1006–1017

Wang C, White PJ, Li C (2017) Colonization and community structure of arbuscular mycorrhizal fungi in maize roots at different depths in the soil profile respond differently to phosphorus inputs on a long-term experimental site. Mycorrhiza 27(4):369–381

Wehner J, Antunes PM, Powell JR, Mazukatow J, Rillig MC (2010) Plant pathogen protection by arbuscular mycorrhizas: a role for fungal diversity? Pedobiologia 53:197–201

Whipps JM (2004) Prospects and limitations for mycorrhizas in biocontrol of root pathogens. Can J Bot 82:1198–1227

Wu QS (2017) Arbuscular mycorrhizas and stress tolerance of plants. Springer, Singapore. https://doi.org/10.1007/978-981-10-4115-0

Yadav A, Suri VK, Kumar A, Choudhary AK, Meena AL (2015) Enhancing plant water relations, quality and productivity of pea (Pisum sativum L.) through AM fungi, inorganic phosphorus and irrigation regimes in a Himalayan acid Alfisol. Commun Soil Sci Plant Anal 46:80–93

Yuan S, Li M, Fang Z, Liu Y, Shi W, Pan B et al (2016) Biological control of tobacco bacterial wilt using Trichoderma harzianum amended bioorganic fertilizer and the arbuscular mycorrhizal fungi Glomus mosseae. Biol Control 92:164–171

Zeng Y, Guo LP, Chen BD, Hao ZP, Wang JY, Huang LQ et al (2013) Arbuscular mycorrhizal symbiosis and active ingredients of medicinal plants: current research status and prospective. Mycorrhiza 23(4):253–265

Zhang Q, Ren F, Li F, Chen G, Yang G, Wang J et al (2020a) Ammonia nitrogen sources and pollution along soil profiles in an in-situ leaching rare earth ore. Environ Pollut 267:115449

Zhang S, Guo X, Yun W, Xia Y, You Z, Rillig MC (2020b) Arbuscular mycorrhiza contributes to the control of phosphorus loss in paddy fields. Plant Soil 447(1):623–636

Arbuscular Mycorrhizal Fungi: A Potential Agent for Phytonematodes Management in Diverse Agro-climatic Zones

7

Nishi Keshari, K. V. V. S. K. Kranti, Narendra Kumar Gunda, and Rizwan Ali Ansari

Abstract

Plants face continuous environmental stress at different stages of their life cycle. These stresses may be biotic or abiotic. Various strategies are to be devised to fulfill the food requirements of the population, which are increasing day by day. The stress tolerance potential of crop plants with regard to environments needs to be improved to maximize crop production and productivity. In this connection, biological means are one of the most promising means to persuade plants of their stress-resistant ability. Among several biological means, arbuscular mycorrhizal fungi (AMF) has enough potential for mitigating plant stress by increasing their vigor. These AMF are the mycorrhizae that form mutual relationships with higher plants by changing their root morphology and physiology. These plant symbionts assist the crop plants in acquiring essential nutritive composition from the soil, such as phosphorus, sulfur, and nitrogen. Generally, higher plants are colonized by plant symbionts and benefit from these microorganisms. AMF are significantly involved in nutrient management and also help plants synthesize appropriate amounts of heterogenous plant hormones. AMF helps plants improve their developmental characters and vigor to sustain even in extreme environmental conditions. It also improves the different aspects of soil, like soil nutrition,

N. Keshari (✉)
Department of Nematology, Faculty of Agriculture, RPCAU, Samastipur, Bihar, India

K. V. V. S. K. Kranti
AICRP (Nematodes on Agriculture), ICAR-IARI, New Delhi, India

N. K. Gunda
APPO, National Quarantine Centre, New Delhi, India

R. A. Ansari
Department of Plant Protection, Faculty of Agricultural Sciences, Aligarh Muslim University, Aligarh, India

R. A. Ansari et al. (eds.), *Mycorrhizal Symbiosis and Agroecosystem Restoration*, https://doi.org/10.1007/978-981-99-5030-0_7

soil aggregation, physical, chemical, and biological properties. Eventually, these qualities will help reduce soil erosion caused by wind and water. In this way, the nutrients' leaching reduces and groundwater contamination due to leachates lessens to a minimum. However, there is a need for much researches to be done in identifying different AMF suitable for different crops.

Keywords

Arbuscular mycorrhizal fungi · Bio-fertilizer · Plant parasitic nematodes · Interaction · Rhizosphere · Mycorrhizae

7.1 Introduction

Agricultural soil comprises a diverse range of microbes such as fungi, bacteria, viruses, protozoans, and nematodes. The rhizospheric soil, which is under the direct influence of plants, is the most important area where the biotic and abiotic components interact. During evolution, plants form a wide array of positive interactions with several microorganisms harboring in the rhizospheric zone that help in the uptake of nutrients and the protection of plants against pathogenic biotic factors (Smith and Read 2008). Henceforth, the feeder roots, which are infected, got transformed into a different structure called mycorrhizae ("fungus roots"). The interactions between mycorrhizal fungi or nodulating rhizobial bacteria and plants are mutual and have a significant role in plant health improvement (Smith and Read 2008; Hayat et al. 2010). The information with respect to fossil and atomic phylogenies shows the presence of AMF in roots since the very beginning, i.e., 460 MYA (Remy et al. 1994; Redecker et al. 2000). Unfortunately, due to contamination with organic compounds at the time of the green revolution, the soil conditions, like soil moisture, soil pH, and soil mineral constituents, have changed, and because of these affects, there is an imbalance of these interactions, such as water, pH, and mineral nutrient contents (Dudal et al. 2002; Ansari and Mahmood 2017a). So, the most effective way to nullify the harmful impact of chemicals is to search for alternative methods, and bioagents may be a potential source of exploitation in the future. Researchers showed paramount interest in mycorrhizae for use in sustainable plant protection systems.

In the year 1885, A.P. Frank coined the term mycorrhizae, which is derived from a Greek word. This serves as a very perfect example of mutualistic symbiosis, where the plants provide energy in the form of carbohydrates to the fungus, and in return, the fungus provides important nutrients, specifically phosphorus, to the plant, which are usually inaccessible (Smith and Read 2008). There are two kinds of mycorrhiza, ectomycorrhiza (ECM) and endomycorrhiza (EDM); other distinguished mycorrhizae also exist, such as ectendomycorrhizae, orchids, arbutoids, and monotropoids. Hence, AM symbiosis is of greater importance and interest for agriculture and horticulture as it delivers mineral nutrients, particularly phosphate, to plants. In addition to this, a bio-protective effect has also been reported in the last thirty years

(Dehne and Schönbeck, 1975; Rosendahl 1985), and its potential role in the management of plant diseases has also been confirmed (Whipps 2004). However, mechanisms implicated in controlled/regulated disease are still poorly understood. Ectomycorrhizal fungi are key players in forestry and agronomy (Smith and Read 2008).

Plant growth is increased by mycorrhizae apparently because of the enhanced root surface area, which enables it to absorb more water and nutrients because the water stress becomes lessened on plants through selective nutrient absorption and accumulation (Ansari and Mahmood 2019a, b). This makes the root more impervious to fungal infections such as *Phytophthora, Pythium,* and *Fusarium* (Smith and Read 2008) and also to some important plant parasitic nematodes. AMF need hosts for their multiplication and growth, but in the case of non-availability of hosts, they can also survive through spores or resistant hyphae.

7.1.1 Ectomycorrhiza (ECM)

ECM belong to the phylum, Basidiomycota, with a few Ascomycota and very few Zygomycota (Tedersoo et al. 2010). ECMs are shaped principally on timberland trees, generally by mushroom and puffball-delivering basidiomycetes and by certain ascomycetes. The spores of these organisms are created on the ground and are scattered by the wind. These symbionts enter the roots, yet they just multiply around the cortical cells, supplanting the piece of the center lamella between the cells and framing "Hartig net" (Domínguez-Núñez and Albanesi 2019). Interwoven "fungus mantles" are generally produced outside the feeder roots, which have one or two hyphal diameters thickness to as many as thirty to forty.

7.1.2 Endomycorrhiza (EDM)

This is the normally found mycorrhizae, which seem like non-mycorrhizal (NM) establishes in shape and shading remotely, however it develops into the cortical cells of the feeder roots inside either by framing a specific taking care of hyphae (haustoria) called arbuscules by setting up the nearby contact with plant cells (Berruti et al. 2016) or by forming vesicles, which are large swollen food-storing hyphal swellings with varying shape from globose to irregular and contain lipids. It belongs to the phylum glomeromycota. On the root surface, they are encircled by a free mycelial development from which hyphae and huge pearl-covered zygospores, or chlamydospores, are delivered underground. EDMs are developed on most developed plants and on a few woody trees for the most part by zygomycetes, essentially of the genus *Glomus* and other genera such as *Acaulospora*.

The fungus that is capable of establishing AM symbiosis is glomeromycota (Schubler et al. 2001). The strategy of colonizing the root differs in different families. Glomerales primarily colonize root systems starting with hyphal fragments. Diversisporales start from the spores; hence, they are slower colonizers than

Glomerales. Hence, the glomerales possess the greater ability to increase phosphorus uptake, promote biomass production and protect the plant against pathogens as well as drought stress (Hart and Reader 2002). AM has the ability to develop extensive hyphal growth in soil, which extends the nutrient to all zones of plants. AM association with the root enhances soil nutrient uptake (Islam et al. 1980), improves plant growth and yields (Daft and Nicolson 1972; Chhabra et al. 1990), improves stress tolerance, and helps suppress infection caused by plant parasitic nematodes (Rich and Bird 1974; Jain and Sethi 1988a; Sharma et al. 1994). This plant symbiont can also reduce nematode populations and disease production rates (Bagyaraj et al. 1979; Cooper and Grandison 1986; Jain and Hasan 1988). There are some reports where the population of nematodes did not change (O'Bannon et al. 1979; Caron et al. 1983) under the influence of AMF. Plant symbiont *Glomus* spp. is one of the most predominant genera in India, and *Glomus fasciculatum* has been phenomenal in managing plant pathogenic nematodes (Jalali and Chand 1990; Sharma and Trivedi 2001; Nehra et al. 2003; Trivedi 1995, 2003). *G. intraradices* is a very competent species that manages the nutrient uptake by individual hyphae, which also depends on the nutrient status of the surrounding soil (Cavagnaro et al. 2005). *G. intraradices*-inoculated plants exhibited increased levels of phosphorus in a wide array of crops and also improved soil health (Cardoso and Kuyper 2006). Improved plant health, including significant crop production and productivity, is obtained if microbial consortia are introduced into the agroecosystem (Ansari et al. 2017a, b, 2019a, b).

Losses inflicted by nematodes continue to increase and are becoming a limited factor in maximizing crop yields (Ansari and Khan 2012a, b; Ansari et al. 2020a, b). The AMF is not restricted to use as biofertilizer but recently crop protection aspects of these plant symbionts have also been realized, which has drawn the attention of researchers due to a wide range of environmental adaptation capabilities.

Plant parasitic nematodes and plant symbionts share the common ecological niche of agroecosystems. It acts as an obligate symbiont that enhances the biomass of plants, and the nematodes are an obligate parasite that depletes important nutrients from the root, resulting in reduced plant growth. Even if the plant parasitic nematodes are present at high damaging levels, the heavily colonized mycorrhizal plants have the ability to grow well. Specialized structures (mycelia, arbuscules, vesicles, auxiliary cells and spores) are also formed:

- *Hyphae*: The filamentous organization framed are called hyphae. They structure various shapes like "H" or "Y" shapes that develop along the roots, colonize the soil and structure new spores, which are known as fertile hyphae (Akiyama et al. 2005).
- *Arbuscules*: Formed by glomalean fungus within the plant root cells, here the plant exchanges food and nutrients with each other. It also has many small projections that get extended inside the plant cells and are formed by repeated branching of hyphae when they enter the root cells (Balestrini et al. 2015).
- *Vesicles*: A vesicle is the swollen end seen between the cells of the root of the wall. They serve as storage organs for fungal foods and appear as oval bags.

Glomus, Acaulospora and *Entrophospora* are vesicles producing plant symbionts and have remarkable role in agroecosystem reconciliation. Arbuscules last for a few days as they are digested by plants, while vesicles may remain viable for longer being an infective propagules (Wang and Qiu 2006).
- *Spores*: They too function as storage structures, propagules, reproductive, and resting structures. Their size is between 10 and 1000 μm, and they form as swellings on hyphae in the roots and soil (Alizadeh 2011).
- *Auxiliary cells*: Thin-walled cells are the key composition of auxiliary cells, and their function is to store carbon, which is found only in the suborder Gigasporinae (Alizadeh 2011).

7.1.3 Characteristics of AMF

- The arbuscules present within the cortical cells of the roots of the colonized plants supply the inorganic water and nutrients from the soil; in exchange, they derive the organic compounds and the CHO supplied by the plants.
- It helps in soil health ameliorations through glomalin exudation and the binding of soil particles through hyphal bindings.
- Hyphae have the capacity to absorb nutrients because of active transporters, so they are able to colonize other zones of soil where the roots of plants cannot reach.
- Besides the supply of phosphorus to plants from soil, it also helps in the acquisition of immobile nutrients such as ammonia, copper, potassium, iron, sulfur, zinc, and molybdenum; hence, (AMF) can be termed a natural bio fertilizer.
- It acts as bioprotectant against bacterial, nematode, and fungal pathogens.
- AMF colonization builds ferulic and gallic acids in higher plants.
- Ferulic acid was found to be toxic to burrowing nematodes, *Radopholus similis* (Wuyts et al. 2006), and gallic acid showed nematicidal potential against *Meloidogyne incognita* (Seo et al. 2013).
- Plant functional groups such as non-nitrogen-fixing woody plants and forbs, as well as C_4 grass respond positively to mycorrhizae compared to plants with N-fixing bacterial symbionts, and C_3 grasses. Nitrogen-fixing forbs had a greater mycorrhizal growth response than woody ones.

AMF make certain root architectural changes that contribute to acquiring more nutrients that are otherwise not available to plants. This characteristic helps the plants mitigate their stress tolerance under varied extreme conditions (Abass et al. 2014).

7.2 Systemic Acquired Resistance (SAR)

AMFs induce systemic resistance and also possess a bio protectional effect against different plant nematodes. Elsen et al. (2008) conducted studies to determine if *G. intraradices* could induce SAR against *R. similis* and *Pratylenchus coffeae* infesting banana using a split root compartment setup. When both the AMF and nematodes were spatially separated, AMF reduced both nematodes by more than 50%, suggesting that AMF has the ability to SR against the tested nematodes. For AM-induced immunity, systemic mechanisms are involved. AMF-induced resistance generates SAR such as SA priming-dependent genes (Pozo and Azcón-Aguilar 2007; Glazebrook 2005) and preconditioning of Jasmonates (JA)-dependent defenses (Van der Heijden and Horton 2009; Song et al. 2011; Jung et al. 2012), as well as cell wall defenses (Cordier et al. 1998; Pozo et al. 2002). AMF colonized plants have shown higher production of defense compounds like phenolics and β-1 and 3-glucanase are upregulated in pathogen-infected AM plants (Pozo et al. 2002).

Pieterase et al. (2012) reported that initially the host assumes AMF as potential invaders, thus activating the defensive molecules, but later on they are downregulated to allow colonization. The analogous structures of AMF, called arbuscules, are bound by the host plasma membrane. Thus, the host immune system is activated, and the plant responds to pathogen infection through a two-part innate system (Jones and Dangl 2006). The plant defense instrument includes acknowledgment of microorganism/organism-related molecules (PAMPs/MAMPs) at the first site that trigger the safeguard reactions of plants, which incorporate quick ion fluxes across the plasma layer, MAP kinase activation, gene expression changes, support of the cell wall, and the creation of reactive oxygen species (Zipfel et al. 2006). This response is called MTI, i.e., molecular pattern (MAMP)-triggered immunity. The biotrophic pathogens have the capacity to conquer the MTI component by releasing effector proteins into the plants. The recognition of these effectors by so called "R genes" (resistance genes) by plants is the second line of plant immunity, hence generating effector triggered immunity, which is a stronger resistance response. The R-genes and their products control the disease resistance responses at the site of infection. The plant immune element response is used to differentiate between pathogens and symbionts before the establishment of symbiosis. Thus, there is an accumulation of defensive molecules like phytoalexins, callose, lignin, suberin, terpenoids, PR proteins, phenolic compounds, ribonucleases, PALs, peroxidases, and chalcone synthases (Hohmann and Messmer 2017; Jacott et al. 2017; Hill et al. 2018).

AMF do not directly interact with plant parasitic nematodes (PPNs); even though they are close to each other, they change the host physiologically and physically and indirectly affect the host–nematode relationship. De la Pena et al. (2006) observed the locally operating mechanisms involved in plant nematode and AM fungi interaction on Ammophila arenaria (dune grass). Lopez Raez et al. (2010) stated that functional diversity exists among AMF species. Plant growth-promoting molecules are instrumental biomolecules that help in morphological, physiological, and biochemical modifications. The gene LePT4 of tomato plants which encodes the phosphate transporter was encountered in arbusculated cells and used in functional symbiosis as a marker (Balestrini et al. 2007). JA (Jasmonates) plays an important role in AM symbiosis, as JA levels were found to be more in the mycorrhizal roots of *Medicago truncatula* compared with plants receiving no mycorrhizal spores (Hause et al. 2002; Meixner et al. 2005). Riedel et al. (2008), however, reported that the said hormone remains unaltered in *Nicotiana attenuata*. However, a few earlier workers reported that JA biosynthesis or JA flagging showed positive or negative administrative roles of the JA pathway in symbiotic interaction utilizing reverse genetic methodologies with plant mutants (Isayenkov et al. 2005; Herrera-Medina et al. 2008; Jejeda-Sartorius et al. 2008). These contentions were raised because of covering of distinct signaling exercises of its pioneer called oxo-phytodienoic acid (OPDA), and their derivatives of jasmonate, for example, isoleucine form (JA-Ile) (Wang et al. 2008). The UPLC-tied mass spectrometry (LC-MS/NS) analysis studies quantified free JA, which is the JA precursor OPDA, JA-Ile, ABA and salicylic acid (SA) and the results obtained by Lopez et al. (2015) cleared that levels of OPDA, a bioactive derivative of JA-Ile, and SA content were higher on *G. mossae* and *G. intaradices* colonized roots compared to NM roots, while JA levels and ABA levels remained unaltered and ET levels were significantly impaired in AMF inoculated plants (Barazani et al. 2007; Reidel et al. 2009).

The researchers have pointed out markers encoding chitinase, glutathione S-transferase, β-D-xylosidase, PR10-like protein, and a key compound, DXS-2, of the mevalonate autonomous pathway of carotenoid biosynthesis (Hohnjee et al. 2007; Walter et al. 2007; Floss et al. 2008). There is also induction of some genes in the biosynthesis of oxylipins. Two fundamental parts of the oxylipin pathway exist in plants and are consequently governed by two unique Lipoxygenases (LOXs), i.e., 9-LOX and 13 LOXs. The same is true for OPDA content in mycorrhizal barley and medicago roots. Subsequently, one might say that stronger induction of 9 LOX pathways and induction of JA-Ile, many JA-based markers, and enhanced degrees of SA prompt lower levels of mycorrhizal colonization.

The early MTI reaction, including jasmonate-linked 9-LOX pathway, impacts the population of root knot nematodes. Gao et al. (2008) found in maize that the expression of 9 LOX genes (ZMLOX3) is mandatory for resistance against root-knot nematode (*M. incognita*). The successful symbiosis with the plants becomes possible due to the weak and transient initiation of the MTI response during the early stages of plants. AMF circumvent their recognition by plants and also vigorously conquer MTI responses through effector proteins. Hence, to understand this, it is pertinent to know the pre symbiosis part. In asymbiotic stage, AMF with no requirement of plant factors after spore germination and enter the presymbiotic phase, where the fungus shows its activity in the presence of host plant roots by forming many branches. This suggests that AMF perceives the signals released by host plants. As far as nematode and AMF interaction is considered, research was mainly focused on root knot, root lesion, and burrowing nematodes (Hol and Cook 2005). The split root experiment by Hao et al. (2012) provided a demonstration that *R. intraradices* induced bioprotection through both local and systemic mechanisms against *Xiphinema index* activity. They studied the 14 genes to understand the transcriptional modification in *X. index* and myco interactions and observed that CHI, GST, STS, PR-10, ESPS, and HIP (seven genes) were upregulated.

PR-10 proteins express their genes in both the early and late phases of root–AMF interactions (Brechenmacher et al. 2004; Siciliano et al. 2007a, b). In mycorrhizal and *X. index* interactions, the EST91, PR 10 isoform was produced early and prior to the identified bioprotective benefits, while it is missing in roots colonized exclusively by *G. intraradices* or nematodes. PR-proteins have antimicrobial activity and RNAse activity to combat biotic and abiotic stresses (Van Loon and Van strain 1999; Liu and Ekramoddoullah 2006). The PR-10 protein extracted from *Crotolaria pallida* shows nematostatic effects against *M. incognita* via inhibiting digesting proteinase (Andrade et al. 2010). The ESPS enzyme is engaged in the shikimate pathway, which creates aromatic secondary metabolites and amino acids in plants, as well as phenylamine, tyrosine, and tryptophan precursors for aromatic secondary metabolites and plant hormones (Tzin and Galili 2010). According to Ferrari et al. (2007), enhanced expression of ESPS throughout defensive responses generated by oligosaccharide administration in *Arabidopsis thaliana* resulted in *Botrytis cinerea* resistance. As a result, HsP70 protects proteins against degeneration in stressful situations. Rivera-Becerril et al. (2005) exposed mycorrhizal roots to Cd (heavy metal) stress, resulting in the activation of the Hsp70 gene, which preserved protein membrane integrity in arbuscule-containing cells, allowing VAM plants to tolerate contaminated soils and contributing to symbiotic functioning. When the pathogen is present, the conservation of protein integrity may be done when the pathogen is subjected to biotic stress (*X. index*) (Smith 2000; Marshall and Keegstra 1992; Neumann et al. 1994; Hottiger et al. 1992; Rivera-Becerril et al. 2005).

The AMF not only acts systematically in controlling the nematode infection but also enhances the resistance of plants against pathogens that are assumed to be infected by plant nematodes. Hao et al. (2018) reported that *R. intraradices* induces systemic protection against Grape Fan Leaf transmitted by *X. index*. Li et al. (2006) conducted the studies to identify which grapevine chitinase genes among class I chitinase gene (VCHIT 1b), class III chitinase gene (VCH3) & class IV chitinase genes (VvChi4A, VvChi4B, and VvChi4D) were induced following inoculation with *G. versiforme* and to confirm if the increase in gene activity provides resistance to mycorrhizal grapevines against *M. incognita*. The relative quantitative Polymerase chain reaction (PCR) analysis revealed upregulation of class III chitinase gene (VCH3) transcripts, and after infection with root knot nematode, the increase was much higher.

7.3 Interaction of AM Fungi with Plant Parasitic Nematodes

Different plant responses, such as physical and physiological responses to the activity of fungus, are the result of AMF action, which is meant for plant protection (Smith and Read 2008). When both nematodes and AMF compete for root space and feeding sites, AMF have direct effect on root feeding nematodes. The AMF directly compete with roots and create local changes in the root chemistry and root exudates. This makes inhibition of nematodes multiplication (Graham 2001). The majority of the PPNs live in the soil, except *Ditylenchus* spp., which resides in aboveground parts. Based on feeding strategy, they are classified as (a) ectoparasitic nematodes: they feed the plant roots from epidermal cells or cortical cells, keeping their whole body outside the root by inserting only stylets (example). (b) Endoparasitic nematodes: they enter the roots with their whole body and feed inside (example). (c) Sedentary endoparasites: feed inside the roots, displaying the most complex feeding strategy by converting the cells of the vascular cylinder into the feeding sites and thus becoming sedentary parasites with the onset of feeding (cysts and root knot nematodes, for example). (d) Migratory endoparasites: They migrate inter or intracellularly and feed on the cells and thus cause damage along the migration path, e.g., *R. similis* and *Pratylenchus* spp. (Jones et al. 2013; Gheysen and Mitchum 2011).

The nature of interaction between them varied from antagonistic to neutral to synergistic. The primary effect of AMF has generally been to increase host resistance during the infection of plant nematodes. This variation in the nematode population or disease incidence in AM-colonized plants was found to be due to the varying nematode host symbiont combinations. Variables that influence the interaction include the order in which crops are attacked with nematodes compared to the time of AM infection, the infection levels of both worms and fungus, and soil fertility.

7.3.1 AM Fungi and Their Impact on Root-Knot Nematodes

Endoparasitic nematodes and AMF both present in soil and also colonize the same area in host plant roots; hence, They interact with each other (Hussey and Roncandori 1982; Elsen et al. 2003; de la Pena et al. 2006; Rizvi et al. 2015). The effects of nematodes on mycorrhizal development are different (Ansari and Mahmood 2017b). Atilano et al. (1981) reported poor mycorrhizal colonization and the formation of vesicles and extra-metrical mycelia in *G. fasciculatum* in the presence of root-knot nematodes (De-souza 1979; Germani et al. 1980). Prior inoculation of AMF resulted in greater suppression of *M. incognita* multiplication in soil, especially when applied 20 days before nematode inoculation. The higher initial densities of nematodes affected the colonization of *G. fasciculatum*; hence, the initial inoculum of fungus exerted an effect on nematodes that penetrated the seedlings, as nematodes require only a few hours to penetrate, whereas fungus requires a minimum of 10 days to get them established (Sankaranarayanan and Sundarababu 2009). Kellam and Schenk (1980) reported that chlamydospores of *G. macrocarpus* impaired the gallings of *M. incognita* on Pickett soybean plants. AM plants have a larger root system than non-AM plants, and the total number of galls produced on AM plants was less than that on non-AM plants.

Mycorrhizal roots reduced the penetration ability of nematodes, and if penetration was done, interference in giant cell formation was observed. The compatible results were obtained in the findings of Sikora and Schoenbeck (1975), where they reported a significant reduction in the number of *M. incognita* larvae due to

Endogene mossae that were able to reach adulthood in plants containing mycorrhizal roots. Roncadori and Hussey (1977) observed the *Gigaspora margarita* and *M. incognita* interaction and reported that combined inoculation nullified the stunting caused by nematodes. The nematode reproduction, i.e. egg production, was higher on cotton roots with mycorrhiza compared to NM roots because the root systems of mycorrhizal cotton were considerably larger than those of controls. The suppressive effects of *G. intraradices* on *M. incognita* in cucumber have been reported by many workers. Castillo et al. (2006a, b) found that this AMF markedly reduced gall formation (6.3–36.8%) in the root system of the plants and the multiplication of both nematodes, *M. incognita* and *M. javanica*, by 11.8–35.7% in olive trees. The cucumber plants developed tolerance to *M. incognita* by increasing plant growth and yield when pre-inoculated with *G. intraradices* and P fertilizers during the early stages of plant growth (Zhang et al. 2009). It was studied that AMF inoculation reduced infestation and reproduction of root knot nematodes infesting papaya (Jaizme-Vega et al. 2006), tomato infested with *M. incognita* (Siddiqe and Akhtar 2007), and pyrethrum infested with *M. hapla* (Waceke et al. 2001). The olive plants infected with *M. incognita/javanica* were managed by AMF, *G. intraradices*, *G. mossae,* and *G. visocum* (Castillo et al. 2006a, b). Three AMFs, *G. intraradices*, *G. mossae,* and *G. versiforme,* decreased root galling index, egg masses per root system, mature females, and eggs or egg masses. Among the three, *G. mossae* produced the largest shoot biomass, on the other hand, *G. versiforme* plants registered maximum nutrient conditions. Likewise, similar results were obtained in tobacco plants when inoculated with *G. fasciculatum* that increased growth, biomass uptake ability of phosphorus, and total yield (Subhashini and Ramakrishnan 2011).

Researchers have published conflicting data (reduction, increase, or no effect of AMF inoculations on nutrient uptake), and nematode effects on plant development, such as AM inoculations, have frequently decreased the negative effects of root knot nematodes on plant growth and nutrient uptake. *G. fasciculatum* inoculations reduced nitrogen, phosphorus, potassium, calcium, zinc, and manganese absorption in tomato plants infested with root-knot nematodes. Altered zinc, copper, and boron accumulation in coffee plants (Bhagyaraj et al. 1979) infested with nematodes; poor nitrogen, phosphorus, and potassium in grape leaves (de Souze 1979) were observed, and this depends on nematode species (Kesba and Al-Sayed 2005). There also exists a synergistic interaction between the biocontrol agents. Researchers have also found that AMF sporulation and ramification have been significantly improvised by the spore activator, which is generally used during inoculum preparation of other fungal bioagents.

The combination of two or more species of AMF provides good results, such as *G. etunicatum and G. deserticola,* along with the dual introduction of bionematicide on tomato against *M. incognita* race 1 in green house experiments, which resulted in greater inhibition of galls, egg masses, and subsequently growth and reproduction parameters of plants. AMF is a phosphate-solubilizing microorganism and has the potential for the biomanagement of plant pathogens too. They help change the phosphatic compounds, which are insoluble, into soluble forms. *Glomus intraradices* combined application of *Aspergillus awemori* and *Pseudomonas aeruginosa* increased pod numbers, chlorophyll, nitrogen, phosphorus, and potassium contents, and reduced plant nematode-related parameters of *M. incognita* in chickpea (Siddiqui and Akhtar 2009). Some compounds, such as phenylamine and serine, have an inhibitory role against nematodes. These compounds were significantly increased in the plants of tomato inoculated with plant symbionts. It was reported that *G. mossae*, *G. fasciculatum,* and *G. mykovam* increased the biomass of two onions (Yellow Granex and Red Creole) in the existence/absence of P-deficient soils. Yellow Granex exhibited good tolerance to *M. graminicola* with improved bulb weight (262%) and bulb diameter (96.4%). Owing to alterations in biochemistry and physiology, this additional supply of phytoalexins, phenols, lignin, phenylalanine, serine chitinase, and decreased leakage of CHO takes place. (Graham 2001). The compatibility of AMF with some fungi parasitizing the eggs of nematodes differs among the various species of AMF. Root knot nematode causes the root galls, and the root galls hinder the uptake of water and nutrients, and thus photosynthesis and other related mechanisms are impaired. Hence, the growth and yield of plants are increased due to the lowering down of the nematode population, and systemic resistance is induced by improving the host nutrition through modification of the mycorrhizosphere by AM.

To maximize the potential of biocontrol agents, the combined application of two or more beneficial microbes in the preparation of biocontrol agents has been recommended. Oyekanmi et al. (2007) applied the microbes in full factorial combinations that suppressed nematode reproduction. After 7 days, the *M. incognita* penetration rate in cotton plants inoculated with *G. intraradices* and non-AM roots was the same; however, after 28 days, the population of root knots in AMF-colonized roots decreased. The development rate of second-stage juveniles to adult females was not impacted by AMF when *M. incognita* was added at planting, but it was slowed when the juveniles were introduced 28 days after planting in AM-mediated soil (Smith et al. 1986). The species of AMF determines whether or not nematode reproduction is suppressed. For example, AMF differently reduced *M. javanica* reproduction on chickpea, with AMF being more prominent with *G. manihotis* compared to *G. margarita*, very minimally with *Gigaspora gigantean*, and zero with *Entrophospora Columbiana* (Diederichs 1987).

7.3.2 AM Fungi and Their Impact on Cyst Nematodes

Plant symbionts affect potato cyst nematode (PCN) hatching in soil by influencing the production of hatching factors like potato root leachates (PRLs; Deliopoulos et al. 2008). Both PCN and AM occupy the same locations of root systems, which interferes with the PCN life cycle. A range of *in vitro* or in soil studies on various aspects of PCN-AMF interactions and the different mechanisms implicated have been unraveled (Deliopoulos 2004). Jones and Ryan (2004) compared the PRLs collected from the roots of AMF-mediated and non-mediated potatoes and gave the confirmation that potato microplots inoculated with Vaminoc (mixture of three *Glomus* species) promoted the early hatching of *G. pallida,* but there was no effect on the hatching of *G. rostochiensis*. Also, the hatching factors were more active towards both species of cyst nematodes.

The juveniles hatch in response to the chemicals that leach out of the roots. The PRLs isolated from the AMF-inoculated plants contain a large number of HFs and HS, which are more active against *G. pallida* as compared with the PRLs of control plants (Ryan and Jones 2004b). AM colonization stimulated potato root growth and enhanced the tolerance level of plants against *G. rostochiensis* (Ryan et al. 2003). Francil and Dropkin (1985) reported the occurrence of chlamydospores of *G. fasciculatum* in the cysts of the soybean cyst nematode, *H. glycines,* i.e., 1–24% of cysts recovered from field soil samples. Gf1 isolate decreased the population of adult females of nematodes in the first generation by 26% over control. The soybean plants inoculated with AMF and *H. glycines* registered more biomass than NM plants inoculated with nematodes. Similar AMF exhibited negative effects on the production and multiplication of *H. cajani* infecting cowpea (Jain and Sethi 1988a). Earlier, the workers reported the abundance of AMF in soybean production soils, maximizing their roles as nematode antagonism and strategy for management of *H. glycines* on soybean. *H. glycines* in the roots of soybeans and soil was decreased by AM up to 73% at high soybean cyst inoculum levels after 45 days of planting. Also, *H. glycines* was suppressed by AMF and increased soybean tolerance to nematodes under greenhouse conditions, but no such suppression was observed in the field (Tylka et al. 1991; Winkler et al. 1994). The effects of *H. glycines* and *G. mossae* on soybean growth were additive, and the growth stimulation by *G. mossae* was reduced in the presence of *H. glycines* (Todd et al. 2002). The potential for suppression of *H. glycines* populations by AMF is less than that of root knot nematode on the same host, as improved P nutrition is suppressive to *M. incognita* but not to *H. glycines*. The antagonistic nature of *H. glycines* was found in many AMF colonizations as Todd detected the consistent suppression of root colonization by *G. mossae* at high levels of *H. glycines* infection.

7.3.3 AM Fungi and Their Impact on Root Lesion Nematodes

Another migratory endoparasites, root lesion nematodes, move through the cortex and, with the help of stylets, penetrate the parenchymatous cells. With the release of cell-degrading enzymes, they ingest the cellular contents, thus destroying the cortical tissues and resulting in necrotic lesions. After root knot nematodes, the *Pratylenchus* species are of great economic importance to agricultural crops. All the stages of root lesion nematodes, i.e., eggs, juveniles, and adults, and the AM structures such as hyphae, arbuscules, and vesicles, co-occur in the root cortex and tissues. Compared to *Rhizophagus* and *Claroideoglomus*, *Glomus* or *Funneliformis* belonging to the order glomarales decreases or has a non-significant effect on populations of root lesion nematodes. This might be due to a difference in the production of secondary metabolites by AM. A high genetic variability exists within the populations of AMF single species and thus affects the plant mycorrhizal relationships. It was reported that the population of *Pratylenchus* species was significantly impaired in the presence of AMF isolated from Wales and Belgium (Rodríguez-Echeverría et al. 2009). Elsen et al. (2003) observed greater variations in the effect of solo species when compared with the findings of Jaizme-Vega and Pinochet (1997). In spite of using the same banana cultivars and *F. mossae*, the results were quite different. To this end, they interpreted that the AMF strain and environmental conditions could have produced different results between the two experiments.

Commercial AMF protected *Leymus arenarius* against migratory endoparasites (Greipsson and El-Mayas 2002). The results of interaction studies between AMF and *Pratylenchus* are quite inconsistent, as some showed higher plant tolerances/resistance to root lesion nematodes as a result of AMF inoculations to the plants, while others reported no protective effects (Forge et al. 2001; Elsen et al. 2003). The migratory endoparasites did not get affected by AMF root colonization; hence, it can be inferred that there is no mutual inhibition between them. Few researchers have reported that AM is only effective in plants that are colonized with fungi before the nematode attack, as the plants already got improved in respect of nutrition and health. This ability of the plants allows them to survive even in the presence of higher densities of pathogenic nematodes. But De la Pena et al. (2006) reported no maximum. No higher folds of carbon and nitrogen content were observed when they were reinoculated with fungus 2 and 5 weeks before nematode inoculation, whereas the plant biomass was more when simultaneous inoculation of AMF and nematodes was done (De la Pena et al. 2006).

7.3.4 Effect of AM Fungi on Reniform Nematode

Plant parasitic nematodes stimulate the sporulation of AMF by inducing physiological changes (Roncadori and Hussey 1977; Smith 1987).

When co-infecting grape roots, AMF, *Glomus macrocarpus* and plant nematodes, *Meloidogyne incognita* and *Tylenchulus semipentrans*, a concrete finding came out. The plant nematodes badly affected the spore formation, while AMF reciprocally affected the nematodes development and reproduction ability. *G. macrocarpus* increased root protein, total soluble sugars, total carbs, and reduced lipids and dry matter when grown alone or in combination with *M. incognita* or *T. semipenetrans*. *G. macrocarpus* and *R. reniformis* combined reduced protein, total soluble sugars, total carbs, and dry matter while significantly increasing lipids (Kesba and Al-Sayed 2005).

The delay in the development of gelatinous matrix, female eggs, and egg sac production was observed in tomato plant roots of *G. fasciculatus* inoculated plants, leading to enhanced resistance of tomatoes against *R. reniformis* (Sitaramaiah and Sikora 1996). This AMF induced tolerance in cowpea to reniform when the nematodes were present at a damaging level under P-deficient conditions (Lingaraju and Goswami 1993). In the early 1990s, many workers made attempts to select efficient species of AMF for the management of reniform nematodes in cultivated cotton species because of the assumption of intraspecific and interspecific host preferences for EDM. Sreenevasa et al. (2003) screened some AMF isolates that are commonly encountered in cultivated soil: *G. mossae, G. fasciculatum,* and *G. intraradices,* and found that the first two AMF nicely performed in terms of improved plant growth and yield character enhancement.

7.4 Commercialization of AM Fungi

AMF is getting increased attention for their role as biofertilizers, bioregulators, and bioprotectants as they are highly capable of nutrient uptake and P storage ability (Rouphael et al. 2015). AMF are found mostly in higher plants except in some families and crop species, and their major biological character is their obligate biotrophic nature. Understanding AMFs and their symbiosis and species identification at early stages of development is very challenging. Also, the important factor in the inoculum production is isolation and selection of AMF species, and obtaining pure cultures cannot be overlooked. Hence, appropriate host plants is needed to maintain the continuous supply of the spores of AMFs. For large production of AMF, *in vitro* cultivation methods like hydroponic system and root organ culture are mostly used. Conventional methods like mass production in soil-based media and living hosts are popular and used for rapid production. AMF are propagated on substrate-based production systems, substrate-free production systems, and *in vitro* production systems. For commercialization, techniques like hydroponics, aeroponics, and soil less culture produce high-quality inoculums with more number of propagules (Lee and George 2005a, b).

7.4.1 (I) Conventional Methods

Due to the obligate nature of AMF, there are big complications to develop economic production for large-scale supply and getting high-quality AMF inoculum. This is a major bottleneck for commercialization. Several techniques given by workers across the world, such as farm production, pot culture using traps, nutrient film technique (NFT), and aeroponics, have been attempted (Fortin et al. 2002; Lee and George 2005a, b). AMF are cultured on the roots of plants on many substrates like vermiculites, soilarite, sand peat, rockwool, glass beads, and clay. In aeroponic systems, the pre-colonized AM fungal seedlings with roots are used for production and modified Hoagland's nutrition with lower phosphorus levels (Mohammad et al. 2000). For large-scale AMF, the NFT and modified NFT were developed by improving aeration with the supply of nutrients, optimum phosphorus, and the use of glass beads (Lee and George 2005a, b). Pot culture is the most commonly used technique for AMF propagation on the right host in disinfected soils.

7.4.2 (II) In Vitro Method/Root Organ Culture Method

It is an amazing bulk reproduction approach that uses less area to produce clean, functional, and contamination-free inoculum. Through transformation by *A. rhizogenes*, a continuous culture of robust ROCs (Ri T-DNA-converted) (soil bacterium) has been obtained.

7.4.3 (III) On Farm Production

After the AMF roots are gathered and utilized as inoculum in the field, the AMF inoculum is replicated on raised beds in the farmers' nursery kitchen garden. This approach saves money over the other ways since it eliminates the need for numerous processes such as AMF inoculation, substrate/potting mixture usage, maintenance, and shipping. As a result, it has an impact on commercialization. The aforementioned processes are eliminated by on-farm reproduction of indigenous species that are already acclimated to the optimum site and habitat, lowering the cost.

7.5 Summary and Future Outlooks

AMF are the most common and important plant symbionts that are usually encountered in a diverse range of rhizosphere. They develop different structures for their survival, growth, and development. It primarily colonizes the root system starting from hyphal fragments, possesses the greater ability to increase the phosphorus uptake, promotes the plant growth, protects the crop plants from various plant pathogens, and has ability to develop extensive mycelium in the soil, thereby extending the nutrient to all the zones of plants. The strategy of root colonization differs in different families. AMF may induce systemic resistance and hold bioprotectional properties against plant parasitic nematodes. AMF have a direct effect on root-feeding nematodes. It competes directly with roots and creates changes in root physiology. This inhibits nematode multiplication. The population of plant nematodes, including root knot nematodes, is reduced when they multiply with AMF, while the plant growth characters significantly enhanced. AMF products should be registered under different categories, like bioprotectants, biofertilizers, or biostimulants, according to their use, so that the eco-friendly management of plant nematodes could be promoted. For appropriate marketing, a proper dosage and the density of the formulation are still poorly understood and required to be taken into consideration. Field-based demonstrations are also required to be performed to prove the beneficial effect of AMF in agriculture with regard to plant nematode management.

References

Abass M, Ahanger A, Elsayed H, Abd-Allah F, Ahmad P (2014) Arbuscular mycorrhiza in crop improvement under environmental stress, 69-95. In: Emerging technologies and management of crop stress tolerance, A sustainable approach, vol 2. Academic Press

Akiyama K, Matsuzaki K, Hayashi H (2005) Plant sesquiterpenes induce hyphal branching in arbuscular mycorrhizal fungi. Nature 435(7043):824–827. https://doi.org/10.1038/nature03608

Alizadeh O (2011) Mycorrhizal symbiosis. Adv Stud Biol 6(3):273–281

Andrade LB, Oliveira AS, Ribeiro JK, Kiyota S, Vasconcelos IM, de Oliveira JT, de Sales MP (2010) Effects of a novel pathogenesis-related class 10 (PR-10) protein from *Crotalaria pallida* roots with papain inhibitory activity against root-knot nematode *Meloidogyne incognita*. J Agric Food Chem 58:4145–4152

Ansari RA, Khan TA (2012a) Parasitic association of root-knot nematode, *Meloidogyne incognita* on guava. e J Sci Technol 5(12):65–67

Ansari RA, Khan TA (2012b) Diversity and community structure of Phytonematodes associated with guava in and around Aligarh, Uttar Pradesh. Trends Biosci 5(3):202–204

Ansari RA, Mahmood I (2017a) Optimization of organic and bio-organic fertilizers on soil properties and growth of pigeonpea. Sci Hortic 226:1–9

Ansari RA, Mahmood I (2017b) Determination of disease incidence caused by *Meloidogyne* spp. and or *Fusarium udum* on pigeonpea in Aligarh district: a survey. Trends Biosci 10(24):5239–5243

Ansari RA, Mahmood I (2019a) Plant health under biotic stress. In: Ansari RA, Mahmood I (eds) Microbial interactions, vol II. Springer Nature, Singapore. https://doi.org/10.1007/978-981-13-6040-4

Ansari RA, Mahmood I (2019b) Plant health under biotic stress. In: Ansari RA, Mahmood I (eds) Organic strategies, vol I. Springer Nature, Singapore. https://doi.org/10.1007/978-981-13-6043-5

Ansari RA, Mahmood I, Rizvi R, Sumbul A (2017a) PGPR: current vogue in sustainable crop production. In: Kumar V (ed) Probiotics and plant health. Springer Nature, Singapore, pp 455–472

Ansari RA, Mahmood I, Rizvi R, Sumbul A, Safiuddin (2017b) Siderophores: augmentation of soil health and crop productivity. In: Kumar V (ed) Probiotics in agroecosystem. Springer Nature, Singapore, pp 291–312

Ansari RA, Sumbul A, Rizvi R, Mahmood I (2019a) Potential role of plant growth promoting Rhizobacteria in alleviation of biotic stress. In: Ansari RA, Mahmood I (eds) Plant health under biotic stress, Microbial interactions, vol II. Springer Nature, Singapore, pp 177–188

Ansari RA, Sumbul A, Rizvi R, Mahmood I (2019b) Organic soil amendments: potential tool for soil and plant health management. In: Ansari RA, Mahmood I (eds) Plant health under biotic stress, Organic strategies, vol I. Springer Nature, Singapore, pp 1–35

Ansari RA, Rizvi R, Mahmood I (2020a) Management of phytonematodes: management of phytonematodes: recent advances and future challenges. Springer Nature, Singapore. https://doi.org/10.1007/978-981-15-4087-5

Ansari RA, Rizvi R, Sumbul A, Mahmood I (2020b) Plant-growth-promoting Rhizobacteria (PGPR)-based sustainable management of phytoparasitic nematodes: current understandings and future challenges. In: Ansari RA et al (eds) Management of phytonematodes: recent advances and future challenges. Springer Nature, Singapore. https://doi.org/10.1007/978-981-15-4087-5_3

Atilano RA, Menge JA, Vangundy SD (1981) Interaction of *Meloidogyne arenaria* and Glomus *fasciculatum* in grape. J Nematol 13:52–57

Bagyaraj DJ, Manjunath A, Reddy DDR (1979) Interaction of vesicular arbuscular mycorrhizae with root-knot nematode in tomato. Plant Soil 51:397–403

Balestrini R, Gómez-Ariza J, Lanfranco L, Bonfante P (2007) Laser microdissection reveals that transcripts for five plant and one fungal phosphate transporter genes are contemporaneously present in arbusculated cells. Mol Plant-Microbe Interact 20:1055–1062. https://doi.org/10.1094/MPMI-20-9-1055

Balestrini R, Lumini E, Borriello R, Bianciotto V (2015) Plant-soil biota interactions. In: Paul EA (ed) Soil microbiology, ecology and biochemistry. Academic Press, London, pp 311–338. https://doi.org/10.1016/b978-0-12-415955-6.00011-6

Barazani O, Von Dahl CC, Baldwin IT (2007) *Sebacina vermifera* promotes the growth and fitness of *Nicotiana attenuata* by inhibiting ethylene signaling. Plant Physiol 144:1223–1232

Begum N, Qin C, Ahanger MA, Raza S, Khan MI, Asharaf M, Ahmad N, Jhang L (2019) Role of arbuscular mycorrhizal fungi in plant growth regulation: implications in abiotic stress tolerance. Front Sci 10:1068. https://doi.org/10.3389/fpls.2019.01068

Berruti A et al (2016) Arbuscular Mycorrhizal fungi as natural biofertilizers: let's benefit from past successes. Front Microbiol 6:1559. https://doi.org/10.3389/fmicb.2015.01559

Bhagyaraj DJ, Manjunath A, Reddy DDR (1979) Interaction of vesicular arbuscular mycorrhizae with root-knot nematodes on tomato. Plant Soil 51:379–403

Blanvillain S, Meyer D, Déjean G, Castaing A, Lautier M, Guynet C, Denancé N, Vasse J, Lauber E, Arlat M (2007) Biology of plant-microbe interaction. In: Lorito M, Woo SL, Scala F (eds) Proceedings of the 13th International Congress on Molecular Plant-Microbe Interactions, vol 6. Sorrento (Naples), Italy, July 21–27, 2007

Borowicz V (2001) Do arbuscular mycorrhizal fungi alter plant–pathogen relations? Ecology 82:3057–3068. https://doi.org/10.1890/0012-9658(2001)082[3057:DAMFAP]2.0.CO;2

Brechenmacher L, Weidman S, Van Tuinen D, Chatagnier O, Gianinazzi S, Franken P, Gianinazzi-Pearson V (2004) Expression profiling of up-regulated plant and fungal genes in early and late stages of *Medicago truncatula–Glomus mosseae* interactions. Mycorrhiza 14:253–262

Cardoso IM, Kuyper TW (2006) Mycorrhizas and tropical soil fertility. Agris 116(1–2):72–84

Castillo P, Nico AI, Azcon-Aguilar C, Del Rio Rincon C, Calvet C, Jimenez-Diaz RM (2006a) Protection of olive planting stocks against parasitism of root-knot nematodes by arbuscular mycorrhizal fungi. Plant Pathol 55:705–713

Castillo C, Rubio R, Rouanet J, Borie F (2006b) Early effects of tillage and crop rotation on arbuscular mycorrhizal fungal propagules in an Ultisol. Biol Fertil Soils 43:83–92. https://doi.org/10.1007/s00374-005-0067-0

Cavagnaro TR, Smith FA, Smith SE, Jakobsen I (2005) Functional diversity in arbuscular mycorrhizas: exploitation of soil patches with different phosphate enrichment differs among fungal species. Plant Cell Environ 28:642–650

Chhabra ML, Singh RP, Jalali, BL (1990) Studies on vesicular arbuscular (VA) mycorrhiza impact on growth and development of cowpea (*Vigna unguiculata* (L) Walp.). In: Proceedings of a national conference on mycorrhiza. Haryana Agricultural University, Hisar, India, Feb 14–16, pp 239

Cooper KM, Grandison GS (1986) Interaction of vesicular-arbuscular mycorrhizal fungi and root-knot nematode on cultivars of tomato and white clover susceptible to *Meloidogyne hapla*. Ann Appl Biol 108:555–565

Cordier C, Pozo MJ, Barea JM, Gianinazzi S, Gianinazzi-Pearson V (1998) Cell defense responses associated with localized and systemic resistance to *Phytophthora* induced in tomato by an arbuscular mycorrhizal fungus. Mol Plant-Microbe Interact 11:1017–1028

Coutos-Thévenot P, Poinssot B, Bonomelli A, Yean H, Breda C, Buffard D, Esnault R, Hain R, Boulay M (2001) *In vitro* tolerance to *Botrytis cinerea* of grapevine 41B rootstock in transgenic plants expressing the stilbene synthase Vst 1 gene under the control of a pathogen-inducible PR 10 promoter. J Exp Bot 52:901–910

Daft MJ, Nicolson TH (1972) Effect of endogone mycorrhiza on plant growth. IV. Quantitative relationships between the growth of the host and the development of the endophyte in tomato and maize. New Phytol 71(2):287–295. www.jstor.org/stable/2430645. Accessed 3 July 2021

Davis RM, Menge JA (1980) Influence of *Glomus fasciculatus* and soil phosphorus on *Phytophthora* root rot of citrus. Phytopathology 70:447–452

De la Pena E, Echeverria SR, van der Putten WH, Freitas H, Moens M (2006) Mechanism of control of root-feeding nematodes by mycorrhizal fungi in the dune grass *Ammophila arenaria*. New Phytol 169:829–840

Dehne HW (1982) Interactions between vesicular-arbuscular mycorrhizal fungi and plant pathogens. Phytopathology 72:1115–1119

Dehne HW, Schonbeck F, Baltruschat H (1978) Investigation on the influence of endotrophic mycorrhiza on crop diseases. III chitinase activity and ornithine cycle. J Phytopathol 95:666–678

Deliopoulos T (2004) Interactions between potato plants, arbuscular mycorrhizal fungi, potato cyst nematodes and a nematicide. Open University (United Kingdom)

Deliopoulos T, Haydock PP, Jones PW (2008) Interaction between arbuscular mycorrhizal fungi and the nematicide aldicarb on hatch and development of the potato cyst nematode, Globodera pallida, and yield of potatoes. Nematology 10(6):783–799

Diederichs C (1987) Interaction between five endomycorrhizal fungi and the root knot nematode *Meloidogyne javanica* on chickpea under tropical conditions. Trop Agric 64:353–355

Domínguez-Núñez JA, Albanesi A (2019) Ectomycorrhizal fungi as biofertilizers in forestry. https://doi.org/10.5772/intechopen.88585; Corpus ID: 202020901

Dudal R, Nachtergaele F, Purnell MF (2002) The human factor of soil formation. Ecosyst Environ 116:72–84

Elsen A, Stephane D, De Waele D (2003) Use of root organ cultures to investigate the interaction between *Glomus intradices* and *Pratylenchus coffeae*. Appl Environ Microbiol 69:4308–4311

Ferlian O, Biere A, Bonfante P, Buscot F, Eisenhauer N, Fernandez I (2018) Growing research networks on mycorrhizae for mutual benefits. Trends Plant Sci 23:975–984. https://doi.org/10.1016/j.tplants.2018.08.008

Ferrari S, Galletti R, Denoux C, Lorenzo G, Ausubel F, Dewdney J (2007) Resistance to *Botrytis cinerea* induced in *Arabidopsis* by elicitors is independent of salicylic acid, ethylene, or jasmonate signaling but requires phytoalexin deficient 3. Plant Physiol 144:367–379

Floss DS, Hause B, Lange PR, Küster H, Strack D, Walter MH (2008) Knock-down of the MEP pathway isogene 1-deoxy-D-xylulose 5-phosphate synthase 2 inhibits formation of arbuscular mycorrhiza-induced apocarotenoids and abolishes normal expression of mycorrhiza-specific plant marker genes. Plant J 56:86–100

Forge T, Muehlchen A, Hackenberg C, Neilsen G, Vrain T (2001) Effects of preplant inoculation of apple (*Malus domestica* Borkh.) with arbuscular mycorrhizal fungi on population growth of the root-lesion nematode, *Pratylenchus penetrans*. Plant Soil 236:185–196

Fortin JA, Bécard G, Declerck S, Dalpé Y, St-Arnaud M, Coughlan AP, Piché Y (2002) Arbuscular mycorrhiza on root-organ cultures. Can J Bot 80:1. https://doi.org/10.1139/b01-139

Francl LJ, Dropkin VH (1985) *Glomus fasciculatum*, a week pathogen of *Heterodera glycines*. J Nematol 17:470–475

Gao X, Starr J, Göbel C, Engelberth J, Feussner I, Tumlinson J (2008) Maize 9-lipoxygenase ZmLOX3 controls development, root-specific expression of defense genes, and resistance to root-knot nematodes. Mol Plant-Microbe Interact 21:98–109. https://doi.org/10.1094/MPMI-21-1-0098

Gao X, Guo H, Zhang Q, Guo H, Zhang L, Zhang C, Gou Z, Liu Y, Wei J, Chen A, Chu Z, Zeng F (2020) Arbuscular mycorrhizal fungi (AMF) enhanced the growth, yield, fiber quality and phosphorus regulation in upland cotton (*Gossypium hirsutum* L.). Sci Rep 10:2084. https://doi.org/10.1038/s41598-020-59180-3

Germani G et al (1980) Influence of 1,2 dibromo-3-chloropropane fumigation on nematode population, mycorrhizal infection, N2-fixation and yield of field-grown groundnut. Rev Nematol 3:75–79

Gheysen G, Mitchum MG (2011) How nematodes manipulate plant development pathways for infection. Curr Opin Plant Biol 14:415–421. https://doi.org/10.1016/j.pbi.2011.03.012

Gianinazzi-Pearson V, Denarié J (1997) Red carpet programmes for root endosymbioses. Trends Plant Sci 2:371–372

Giovannini L, Palla M, Agnolucci M, Avio L, Sbrana C, Turrini A, Giovannetti M (2020) Arbuscular mycorrhizal fungi and associated microbiota as plant biostimulants: research strategies for the selection of the best performing Inocula. Agronomy 10(1):106. https://doi.org/10.3390/agronomy10010106

Glazebrook J (2005) Contrasting mechanisms of defense against biotrophic and necrotrophic pathogens. Annu Rev Phytopathol 43:205–227

Graham JH (2001) What do root pathogens see in mycorrhizas? New Phytol 149:357–359

Greipsson S, El-Mayas H (2002) Synergetic effects of soil pathogenic fungi and nematodes reducing bioprotection of arbuscular mycorrhizal fungi on the grass *Leymus arenarius*. BioControl 47:715

Hao Z, Fayolle L, van Tuinen D, Chatagnier O, Li X, Gianinazzi S, Gianinazzi-Pearson V (2012) Local and systemic mycorrhiza-induced protection against the ectoparasitic nematode *Xiphinema index* involves priming of defence gene responses in grapevine. J Exp Bot 63(10):3657–3672. https://doi.org/10.1093/jxb/ers046

Hao Z, Tuinen DV, Fayolle L, Chatagnier O, Li X, Chen B, Gianinazzi S, Gianinazzi-Pearson V (2018) Arbuscular mycorrhiza affects grapevine fanleaf virus transmission by the nematode vector Xiphinema index. Appl Soil Ecol 129:107–111. https://doi.org/10.1016/j.apsoil.2018.05.007

Hao Z, Xie W, Chen B (2019) Arbuscular mycorrhizal symbiosis affects plant immunity to viral infection and accumulation. Viruses 11(6):534. https://doi.org/10.3390/v11060534

Hart MM, Reader RJ (2002) Taxonomic basis for variation in the colonization strategy of arbuscular mycorrhizal fungi. New Phytol 153:335–334

Hause B, Maier W, Miersch O, Kramell R, Strack D (2002) Induction of jasmonate biosynthesis in arbuscular mycorrhizal barley roots. Plant Physiol 130:1213–1220

Hayat R, Ali S, Amara U, Khalid R (2000) Soil beneficial bacteria and their role in plant growth promotion: a review. Ann Microbiol 60:579–598. https://doi.org/10.1007/s13213-010-0117-1

Hill EM, Robinson LA, Abdul-Sada A, Vanbergen AJ, Hodge A, Hartley SE (2018) Arbuscular mycorrhizal fungi and plant chemical defence: effects of colonisation on aboveground and belowground metabolomes. J Chem Ecol 44:198–208. https://doi.org/10.1007/s10886-017-0921-1

Hohmann P, Messmer MM (2017) Breeding for mycorrhizal symbiosis: focus on disease resistance. Euphytica 213:113. https://doi.org/10.1007/s10681-017-1900-x

Hol WHG, Cook R (2005) An overview of arbuscular mycorrhizal fungi-nematode interactions. Basic Appl Ecol 6:489–503. https://doi.org/10.1016/j.baae.2005.04.001

Hottiger T, Devirgilio C, Bell W, Boller T, Wiemken A (1992) The 70-kilodalton heat-shock proteins of the SSA subfamily negatively modulate heat-shock-induced accumulation of trehalose and promote recovery from heat stress in the yeast, *Saccharomyces cerevisiae*. Eur J Biochem 210:125–132

Hussey RS, Roncandori RW (1982) Vesicular arbuscular mycorrhiza may limit nematode activity and improve plant growth. Plant Dis 66:9–14

Ijdo M, Cranenbrouck S, Declerck S (2011) Methods for large-scale production of AM fungi: past, present, and future. Mycorrhiza 21:1–16. https://doi.org/10.1007/s00572-010-0337-zs

Isayenkov S, Mrosk C, Stenzel I, Strack D, Hause B (2005) Suppression of allene oxide cyclase in hairy roots of *Medicago truncatula* reduces jasmonate levels and the degree of mycorrhization with *Glomus intraradices*. Plant Physiol 139:1401–1410

Islam R, Ayanaba A, Sanders FE (1980) Response of cowpea (*Vigna unguiculata*) to inoculation with VA-mycorrhizal fungi and to rock phosphate fertilization in some unsterilized Nigerian soils. Plant Soil 54:107–117

Jacott C, Murray J, Ridout C (2017) Trade-offs in arbuscular mycorrhizal symbiosis: disease resistance, growth responses and perspectives for crop breeding. Agronomy 7:75. https://doi.org/10.3390/agronomy7040075

Jain RK, Hasan N (1988) Role of vesicular-arbuscular mycorrhizal fungi and nematode activities in forage production. Acta Bot Indica 16:84–88

Jain RK, Sethi CL (1988a) Interaction between vesicular-arbuscular mycorrhiza and Meloidogyne incognita, *Heterodera cajani* on cowpea as influenced by time of inoculation. Indian J Nematol 18:89–93

Jain RK, Sethi CLK (1988b) Influence of endomycorrhizal fungi, *Glomus fasciculatum* and *Glomus epigaeus* on penetration and development of *Heterodera cajani* on cow pea as influenced by time of inoculation. Indian J Nematol 18:263–268

Jaizme-Vega M, Pinochet J (1997) Growth response of banana to three mycorrhizal fungi in *Pratylenchus goodeyi* infested soil. Nematropica 27:69–76. https://doi.org/10.1023/A:1004236310644

Jaizme-Vega MC, Tenoury P, Pinochet J, Jaumot M (1997) Interactions between the root knot nematode *Meloidogyne incognita* and *Glomus mosseae* in banana. Plant Soil 196:27–35

Jaizme-Vega M, Rodríguez-Romero A, Barroso Núñez L (2006) Effect of the combined inoculation of arbuscular mycorrhizal fungi and plant growth-promoting rhizobacteria on papaya (*Carica papaya* L.) infected with the root-knot nematode *Meloidogyne incognita*. Fruits 61(3):151–162. https://doi.org/10.1051/fruits:2006013

Jalali BL, Chand H (1990) Current trends in mycorrhizal research. Proceedings of the national conference on mycorrhiza, HAU, Hisar, pp 210

Jones J, Dangl J (2006) The plant immune system. Nature 444:323–329. https://doi.org/10.1038/nature05286

Jones JT, Haegeman A, Danchin EG, Gaur HS, Helder J, Jones MG (2013) Top 10 plant-parasitic nematodes in molecular plant pathology. Mol Plant Pathol 14:946–961. https://doi.org/10.1111/mpp.12057

Jung SC, Martinez-Medina A, Lopez-Raez JA, Pozo MJ (2012) Mycorrhiza-induced resistance and priming of plant defenses. J Chem Ecol 38:651–664. https://doi.org/10.1007/s10886-012-0134-6

Kasab AS, Taha AHY (1990) Interaction between plant parasitic nematodes, VAM, rhizobia and nematicide on Egyptian clover. Ann Agric Sci 35:509–520

Kellam MK, Schenk NC (1980) Interactions between a vesicular-arbuscular mycorrhizal fungus and root-knot nematode on soybean. Phytopathology 70:293–296

Kesba HH, Al-Sayed ASA (2005) Interactions of three species of plant-parasitic nematodes with arbuscular mycorrhizal fungus, Glomus macrocarpus, and their effect on grape biochemistry. Nematology 7(6):945–952

Lee YJ, George E (2005a) Contribution of mycorrhizal hyphae to the uptake of metal cations by cucumber plants at two levels of phosphorus supply. Plant Soil 278:361–370. https://doi.org/10.1007/s11104-005-0373-1

Lee YJ, George E (2005b) Development of a nutrient film technique culture system for arbuscular mycorrhizal plants. HortScience 40(2):378–380

Li HY, Liu RJ, Shu HR (2002) Interaction between AM fungi and *Meloidogyne incognita* and their effects on the host plant grape. Acta Hortic Sin 6:510–514

Lingaraju S, Goswami BK (1993) Influence of mycorrhizal fungus *Glomus fasciculatum* on the host parasite relationship of *Rotylenchulus reniformis* on cowpea. Indian J Nematol 23(2):137–141

Liu JJ, Ekramoddoullah AK (2006) The family of plant pathogenesis-related proteins: their structure, regulation and function in response to biotic and abiotic stress. Physiol Mol Plant Pathol 68:419–423

López-Bucio J, Pelagio-Flores R, Herrera-Estrella A (2015) *Trichoderma* as biostimulant: exploiting the multilevel properties of a plant beneficial fungus. Sci Hortic 196:109–123. https://doi.org/10.1016/j.scienta.2015.08.043

Marshall JS, Keegstra K (1992) Isolation and characterization of a cDNA clone encoding the major hsp70 of the pea chloroplastic stroma. Plant Physiol 100:1048–1054

Meixner C, Ludwig-Müller J, Miersch O, Gresshoff P, Staehelin C, Vierheilig H (2005) Lack of mycorrhizal autoregulation and phytohormonal changes in the supernodulating soybean mutant Nts1007. Planta 222:709–715. https://doi.org/10.1007/s00425-005-0003-4

Millet YA, Danna CH, Clay NK, Songnuan W, Simon MD, Werck Reichhart D et al (2010) Innate immune responses activated in *Arabidopsis* roots by microbe-associated molecular patterns. Plant Cell 22:973–990. https://doi.org/10.1105/tpc.109.069658

Mohammad A, Khan AG, Kuek C (2000) Improved aeroponic culture of inocula of arbuscular mycorrhizal fungi. Mycorrhiza 9:337–339

Nehra S, Pandey S, Trivedi PC (2003) Interaction of arbuscular mycorrhizal fungi and different levels of root-knot nematode on ginger. Indian Phytopathol 56:297–299

Neumann D, Lichtenberger O, Günther D, Tschiersch K, Nover L (1994) Heat-shock proteins induce heavy-metal tolerance in higher plants. Planta 194:360–367

O'Bannon JH, Inserra RN, Nemec S, Volvas N (1979) The influence of *Glomus mosseae* on *Tylenchulus semipenetrans*- infected and uninfected Citrus limon seedlings. J Nematol 11:247–250

Oyekanmi EO, Coyne DL, Fagade OE, Osonubi OE (2007) Improving root-knot nematode management on two soybean genotypes through the application of *Bradyrhizobium japonicum*, *Trichoderma pseudokoningii* and *Glomus mosseae* in full factorial combinations. Crop Prot 26:1006–1012

Pieterse CMJ, Koornneef A, Leonreyes HA, Ritsema T, Verhage A, Joosten RG, Vos MD, Oosten VRV, Dicke M (2008) Cross-talk between signaling pathways leading to defense against pathogens and insects. Biol Plant Microbe Interact 6:1–9

Pozo MJ, Azcón-Aguilar C (2007) Unraveling mycorrhiza-induced resistance. Curr Opin Plant Biol 10:393–398. https://doi.org/10.1016/j.pbi.2007.05.004

Pozo MJ, Azcón-Aguilar C, Dumas-Gaudot E, Barea JM (1999) β-1,3-glucanase activities in tomato roots inoculated with arbuscular mycorrhizal fungi and/or *Phytophthora parasitica* and their possible involvement in bioprotection. Plant Sci 141:149–157

Pozo MJ, Cordier C, Dumas-Gaudot E, Gianinazzi S, Barea JM, Azcón-Aguilar C (2002) Localized versus systemic effect of arbuscular mycorrhizal fungi on defence responses to *Phytophthora* infection in tomato plants. J Exp Bot 53(368):525–534. https://doi.org/10.1093/jexbot/53.368.525

Redecker D, Kodner R, Graham L (2000) Glomalean fungi from the ordovician. Science 289:1920–1921. https://doi.org/10.1126/science.289.5486.1920

Remy W, Taylor TN, Hass H, Kerp H (1994) Four hundred million year old vesicular arbuscular mycorrhizae. Proc Natl Acad Sci U S A 91:11841–11843

Rich JR, Bird GW (1974) Association of early-season VA mycorrhizae and increased growth and development of cotton. Phytopathology 64:1421–1425

Riedel T, Groten K, Baldwin IT (2008) Symbiosis between *Nicotiana attenuata* and *Glomus intraradices:* ethylene plays a role, jasmonic acid does not. Plant Cell Environ 31:1203–1213

Rivera RA, Fernaández F (2005) Inoculation and management of mycorrhizal fungi within tropical agroecosystems. In: Uphof N (ed) Biological approaches to sustainable soil systems. Cornell University, New York, NY, pp 479–488

Rivera-Becerril F, van Tuinen D, Martin-Laurent F, Metwally A, Dietz K-J, Gianinazzi S, Gianinazzi-Pearson V (2005) Molecular changes in *Pisum sativum* L. roots during arbuscular mycorrhiza buffering of cadmium stress. Mycorrhiza 16:51–60

Rizvi R, Singh G, Safiuddin ARA, Tiyagi SA, Mahmood I (2015) Sustainable management of root-knot disease of tomato by neem cake and *Glomus fasciculatum*. Cogent Food Agric 1:1008859

Rodríguez-Echeverría S, de La Peña E, Moens M, Freitas H, Van Der Putten WH (2009) Can root-feeders alter the composition of AMF communities? Experimental evidence from the dune grass *Ammophila arenaria*. Basic Appl Ecol 10:131–140. https://doi.org/10.1016/j.baae.2008.01.004

Roncadori RW, Hussey RS (1977) Interaction of the endomycorrhizal fungus *Giga- spora margarita* and root-knot nematode on cotton. Phytopathology 67:1507–1511

Rosendahl S (1985) Interactions between the vesicular-arbuscularmycorrhizal fungus *Glomus fasciculatum* and Aphanomyceseuteiches root rot of peas. Phytopathol Z 114:31–40

Rouphael Y, Franken P, Schneider C, Schwarz D, Giovannetti M, Agnolucci M et al (2015) Arbuscular mycorrhizal fungi act as biostimulants in horticultural crops. Sci Hortic 196:91–108

Ryan NA, Jones PW (2004a) The effect of mycorrhiza-tion of potato roots on the hatching chemicals active towardsthe potato cyst nematodes, *Globodera pallida* and *G. rostochiensis*. Nematology 6:335–342

Ryan NA, Jones PW (2004b) The ability of rhizosphere bacteria isolated from nematode host and non-host plants to influence the hatch in vitro of the two potato cyst nematode species, *Globodera rostochiensis* and *G. pallida*. Nematology 6:375–387

Ryan NA, Deliopoulos T, Jones PW, PPJ H (2003) Effects of a mixed-isolate mycorrhizal inoculum on the potato – potato cyst nematode interaction. Ann Appl Biol 143:111–119

Schubler A, Schwarzott D, Walker C (2001) A new fungal phylum, the glomeromycota: phylogeny and evolution. Mycol Res 105:1413–1142

Sharma A, Trivedi PC (2001) Concomitant effect of VAM and neem products on *Heterodera avenae* infected wheat. J Indian Bot Soc 80:9–13

Sharma AK, Srivastava PC, Joshi BN (1994) Contribution of VA mycorrhizal to zinc uptake in plants. In: Manthey JA, Crowley DE, Luster DG (eds) Biochemistry of metal micronutrients in the rhizosphere. Lewis. Publishers, Boca Raton, FL, pp 111–124

Siciliano V, Genre A, Balestrini R, Cappellazzo G, de Wit PJGM, Bonfante P (2007a) Transcriptome analysis of arbuscular mycorrhizal roots during development of the prepenetration apparatus. Plant Physiol 144:1455–1466

Siciliano V, Genre A, Balestrini R, Dewit PJ, Bonfante P (2007b) Pre-penetration apparatus formation during AM infection is associated with a specific transcriptome response in epidermal cells. Plant Signal Behav 2(6):533–535. https://doi.org/10.4161/psb.2.6.4745

Siddiqui ZA, Akhtar MS (2009) Effects of antagonistic fungi, plant growth-promoting rhizobacteria, and arbuscular mycorrhizal fungi alone and in combination on the reproduction of Meloidogyne incognita and growth of tomato. J Gen Plant Pathol 75:144–153

Sikora RA, Schoenbeck F (1975) Effect of vesicular arbuscular mycorrhiza (*Endogone mosseae*) on the population dynamics of the (*Meloidogyne incognita* and *Meloidogyne hapla*). In: VIII international plant protection congress, Reports and information, Section V, Moscow, USSR. pp 158–166

Sitaramaiah K, Sikora RA (1996) Influence of mycorrhizal fungus *Glomus fasciculatum* spore concentration of *Rotylenchulus reniformis* population dynamics and cotton growth. Indian J Nematol 26:1–6

Smith DF (2000) Chaperones in progesterone receptor complexes. Semin Cell Dev Biol 11:45–52

Smith SE, Read DJ (2008) Mycorrhizal symbiosis, 3rd edn. Academic Press, New York

Smith SE, Smith FA (2010) The role of arbuscular mycorrhizal fungi (AMF) in crop phosphorus nutrition: a need for changed ideas, Eyre Peninsula Farming Systems Summary 2010. pp 125–127

Smith GS, Hussey RS, Roncadori RW (1986) Penetration and post-infection development of *Meloidogyne incognita* on cotton as affected by *Glomus intraradices* and phosphorus. J Nematol 18:429–435

Solanki MK, Kashyap PL, Ansari RA, Kumari B (2020) Microbiomes and plant health: panoply and their applications. Academic Press, London. https://www.elsevier.com/books/microbiomes-and-plant-health/solanki/978-0-12-819715-8

Song YY, Ye M, Li CY, Wang RL, Wei XC, Luo SM et al (2013) Priming of anti-herbivore defense in tomato by arbuscular mycorrhizal fungus and involvement of the jasmonate pathway. J Chem Ecol 39:1036–1044. https://doi.org/10.1007/s10886-013-0312-1

Song Y, Chen D, Lu K, Sun Z, Zeng R (2015) Enhanced tomato disease resistance primed by arbuscular mycorrhizal fungus. Front Plant Sci 6:786. https://doi.org/10.3389/fpls.2015.00786

Subhashini DV, Ramakrishnan S (2011) Interaction of VAMycorrhiza and root-knot nematode on tobacco plants: effects of nematode inoculum density, soil texture and soil sterilization. Indian J Nematol 41:56–61

Sumbul A, Mahmood I, Rizvi R, Safiuddin, Ansari RA (2017) Mycorrhiza: an alliance for the nutrient management in plants. In: Kumar V (ed) Probiotics in agroecosystem. Springer Nature, Singapore, pp 371–386

Tedersoo L, May TW, Smith ME (2010) Ectomycorrhizal lifestyle in fungi: global diversity, distribution, and evolution of phylogenetic lineages. Mycorrhiza 20(4):217–263

Tejeda-Sartorius M, de la Vega OM, Délano-Frier JP (2008) Jasmonic acid influences mycorrhizal colonization in tomato plants by modifying the expression of genes involved in carbohydrate partitioning. Physiol Plant 133:339–353. https://doi.org/10.1111/j.1399-3054.2008.01081.x

Trivedi PC (1995) Biocontrol potential of VAM fungi in the management of plant parasitic nematodes. Platinum jublee volume. J Indian Bot Soc 74:165–171

Trivedi PC (2003) Microbes in the management of plant parasitic nematodes. In: Trivedi PC (ed) Advances in microbiology. Scientific Publishers, Jodhpur, pp 341–355

Tylka GL, Hussey RS, Roncadori RW (1991) Axenic germinates of vesicular arbuscular mycorrhizal fungi- effects of selected *Streptomyces* species. Phytopathology 81:754-75H.E

Tzin V, Galili G (2010) New insights into the shikimate and aromatic amino acids biosynthesis pathways in plants. Mol Plant 3:956–972

Van der Heijden MGA, Horton TR (2009) Socialism in soil? The importance of mycorrhizal fungal networks for facilitation in natural ecosystems. J Ecol 97:1139–1150

Van Loon LC, Van Strien EA (1999) The families of pathogenesis-related proteins, their activities, and comparative analysis of PR-1 type proteins. Physiol Mol Plant Pathol 55:85–97

Walter MH, Floss DS, Hans J, Fester T, Strack D (2007) Apocarotenoid biosynthesis in arbuscular mycorrhizal roots: contributions from methylerythritol phosphate pathway isogenes and tools for its manipulation. Phytochemistry 68:130–138

Wang B, Qiu YL (2006) Phylogenetic distribution and evolution of mycorrhizas in land plants. Mycorrhiza 16(5):299–363

Wang L, Allmann S, Wu J, Baldwin IT (2008) Comparisons of LIPOXYGENASE3- and JASMONATE-RESISTANT4/6-silenced plants reveal that jasmonic acid and jasmonic acid-amino acid conjugates play different roles in herbivore resistance. Plant Physiol 146:904–915. https://doi.org/10.1104/pp.107.109264

Wanjohi WJ, Waudo SW, Sikora R (2009) Response of meloidogyne hapla to mycorrhiza fungi inoculation on pyrethrum. Afr J Sci Technol 2:63–70. https://doi.org/10.4314/ajst.v2i2.44673

Whipps JM (2004) Prospects and limitations for mycorrhizas in biocontrol of root pathogens. Can J Bot 1227:1198–1227. https://doi.org/10.1139/B04-082

Winkler HE, Hetrick BA, Todd TC (1994) Interactions of *Heterodera glycines*, *Macrophomina phaseolina*, and mycorrhizal fungi on soybean in Kansas. J Nematol 26:675–682

Wuyts N, Swennen R, De Waele D (2006) Effects of plant phenylpropanoid pathway products and selected terpenoids and alkaloids on the behaviour of the plant-parasitic nematodes *Radopholus similis, Pratylenchus penetrans* and *Meloidogyne incognita*. Nematology 8:89–101. https://doi.org/10.1163/156854106776179953

Zhang LD, Zhang JL, Christie P, Li XL (2009) Effect of inoculation with the arbuscular mycorrhizal fungus *Glomus intraradices* on the root-knot nematode *Meloidogyne incognita* in cucumber. J Plant Nutr 32:967–979

Zipfel C, Kunze G, Chinchilla D, Caniard A, Jones JD, Boller T, Felix G (2006) Perception of the bacterial PAMP EF-Tu by the receptor EFR restricts Agrobacterium-mediated transformation. Cell 125:749–776

Part III

Case Studies/Reports

8 Arbuscular Mycorrhizal Technology in Sustainable Agriculture: Current Knowledge and Challenges in Agroforestry

Gökhan Boyno, Rizwan Ali Ansari, Emre Demirer Durak, Hasret Güneş, Rojbin Çevik, and Semra Demir

Abstract

In agroecosystems, arbuscular mycorrhizal fungi (AMF) are the most common and ubiquitous. Because of their productive and comprehensive symbiotic connections with plants, AM technology looks to be a viable option for sustainable agriculture and agroforestry. The commercialization of this technology may be utilized in agriculture, horticulture, and agroforestry to improve land use management and reduce the need for synthetic chemicals for plant growth and disease control. Furthermore, while mycorrhiza inoculation of plants is a well-known procedure, developing an inoculum consistently under field circumstances remains a bottleneck for their wide range of applications. Mycorrhizal inoculum generation, on the other hand, is a complicated process that necessitates commercial enterprises having the requisite biotechnological skills and capacity to react to ethical, educational, legal, and commercial needs. The aim of this chapter is to compile the available data on the theme of commercialization of AM technology as a tool and its use in increasing plant growth and yield characters.

Keywords

Sustainable agriculture · Agroforestry · Arbuscular mycorrhizal technology · AM technology · Plant symbionts

---

G. Boyno · E. D. Durak · H. Güneş · R. Çevik · S. Demir (✉)
Department of Plant Protection, Faculty of Agriculture, Van Yuzuncu Yil University, Van, Turkey
e-mail: semrademir@yyu.edu.tr

R. A. Ansari
Department of Plant Protection, Faculty of Agricultural Sciences, Aligarh Muslim University, Aligarh, India

R. A. Ansari et al. (eds.), *Mycorrhizal Symbiosis and Agroecosystem Restoration*,
https://doi.org/10.1007/978-981-99-5030-0_8

8.1 Introduction

Nondestructive methods for achieving low costs and high output can be mutually reinforcing in creating a viable system with low external inputs and long-term farming. This is primarily accomplished through a societal intervention that comprises an increase in crop yield, a reduction in pesticide inputs, and a social assessment of welfare and bioethical elements. The efficient use of soil microorganisms contributes to the long-term viability of agricultural ecosystems (Jeffries et al. 2003; Selosse et al. 2004; Bünemann et al. 2006; Barrios 2007; Vosátka and Albrechtová 2009; Gianinazzi et al. 2010). Growing demand for high-quality food production utilizing these eco-friendly farming techniques has led to the introduction of beneficial microorganism-based fertilizers that do not deplete the natural resource base (Ansari and Mahmood 2017a; Ansari et al. 2017a, b, 2020b). In this case, farmers will be able to utilize bio-fertilizers to boost productivity per unit area. Arbuscular mycorrhizal fungi (AMF) stand out in this group due to several mycorrhizal species colonizing at the same time. AMF species are found in 80–90 percent of all plant species known to science (Rakshit et al. 2002; Rakshit 2015). By replenishing reduced carbon (C) from plant photosynthesis and mineral nutrients like nitrogen (N) and phosphorus (P), this relationship includes a bidirectional movement of matter between symbiotic partners (Ferrol et al. 2002; Demir et al. 2015). AMF has a number of "nonnutritive" impacts on plant physiology, including lowering biotic/abiotic stress, functioning as a biocontrol agent, preventing erosion, stabilizing soil aggregates, and altering plant compatibility and the long-term survival of the entire plant–soil system (Smith and Read 2010; Ansari and Mahmood 2017b, 2019a; Ansari et al. 2020a). Therefore, AMF play a very important role not only as bio-fertilizers but also as bio-protectors and bio-regulators either in solo or in mixture with other potential beneficial microorganisms (Pal et al. 2013, 2015; Parewa et al. 2014; Boyno et al. 2022; Ansari et al. 2019a, b), which are caused by pathogens and pests (Ansari and Khan 2012a, b). This chapter entails AMF distribution, methods of multiplication and application, and commercialization at a large scale. Major prevailing challenges and possible answers have also been put forth to get the readers acquainted.

8.2 AM Technology in Sustainable Agriculture and Agroforestry

Research, commercialization, manufacture, marketing, distribution, and the application of AM inoculum are all activities that fall under the umbrella of AM technology (Benami et al. 2020). In applied mycorrhizal research for sustainable agriculture, the application of combinations of minimal effective propagation to crops, the identification of species, the development of AM technology to produce more effectively, and the assessment of mycorrhizal viability are all priorities (Vosátka et al. 2012; Guo 2019). Aside from these divisions, because of the complexity of these operations, the development and application of AM inoculums have been the primary focus in the mycorrhizal sector. Producing cost-effective mycorrhizal inoculants has been a difficult challenge throughout the company's existence. Mycorrhizal inoculation in agricultural areas, on the other hand, has proven considerable yield advantages in various crop kinds, as recorded in several field studies (Pellegrino et al. 2015; Hijri 2016; Benami et al. 2020). However, it may be argued that the development of next-generation mycorrhizal technology should not be limited to issues about production and inoculation (Rillig et al. 2016). Given various situations in which they assist the plants with which they interact in obtaining nutrients, mycorrhizal fungi have great promise in agriculture. Despite this, their potential impacts on products are almost imperceptible, and mycorrhizae are used in a few sectors (Adholeya 2012). Regardless of potential production gains, the use of mycorrhizae for monetization is not currently on the rise. Forestry, on the other hand, is among the few sectors that fully recognize the importance of mycorrhizae in plant growth. Although mycorrhizal symbiosis is required in exotic woods, AMF are critical in agroforestry (Muleta et al. 2008; Araújo et al. 2019). Mycorrhizal infection is commonly used in a variety of different small businesses. Without mycorrhizal inoculation, orchid seedlings will not germinate in the growth media, making mycorrhizae vital for farmers and small-scale firms. Because it can handle higher levels of heavy metals including aluminium, zinc, nikel, iron, lead, and cadmium, land recovery is one of the most recent areas of commercial expansion for mycorrhiza (Pal et al. 2016).

Many sectors are assumed to be affected by the quickly changing AM technology environment, which is influenced by globalization, resistance, economic burdens, and the progress of new innovations. As the market for organic food grows, especially in developed countries, so does interest in technology (Benami et al. 2020). Instead of utilizing inorganic fertilizers, pesticides, and fungicides, inoculation of soil with mycorrhizae can increase growth and disease resistance. Inoculation of soil with an appropriate fungal isolate can also reduce the need for farmers in impoverished nations to repeat expensive fertilizer treatments that they cannot afford. However, the process of converting this concept into a viable firm is impeded by a lack of knowledge dissemination, prospective consulting services, and a lack of hope (Pal et al. 2016). Sustainable agriculture and agroforestry rely heavily on AM technology (Siddiqui and Mahmood 1996; Akhtar and Siddiqui 2008; Futai et al. 2008; Akhtar et al. 2011). Commercialization of AM based on this technology has accelerated in recent years for the following reasons:

1. Plant development and health benefits, as well as land reclamation, plant breeding, and nutrition and disease control,
2. Growing concern over soil microbes and the adoption of mycorrhizal inoculants as a viable agrochemical substitute, and
3. Giving more importance to sustainable agriculture and forestry by the society.

In essence, the commercialization of AM technology is a lengthy process that necessitates the acquisition of technical competence and compliance with legal, ethical, educational, and business criteria. However, according to Gianinazzi and Vosátka (2004), future AM technology should address the following requirements:

1. Development of genetic or sensor technology to track AM inoculum in the field;
2. Increasing data gathering on mycorrhizae ecophysiology in stressed environments;
3. Developing a better knowledge of how mycorrhizae interact with the other soil microbes; and
4. Identifying suitable or innovative plant species with improved mycorrhizal characteristics, as well as supplementing mycorrhizae with new symbiotic properties.

8.3 Use of AM Technology in Sustainable Agriculture and Agroforestry

8.3.1 An Overview of the Market and Products

The economic potential of AM technology for agro-plant production in horticulture, agroforestry, bioremediation in degraded regions (Neill et al. 1991; Vural et al. 2018), and other parts of the plant sector has recently grown due to improved scientific knowledge of mycorrhizal symbioses (Tawaraya 2003). Because many important global food crops are highly mycorrhizal-dependent plant species, they can profit from the addition of appropriate AMF inoculums, improving global food output. Successful firms must establish crucial technical competence as well as the ability to conform to legal, ethical, educational, and marketing standards in order to construct these inoculums. Variable volumes of different fungal species, varied percentages of viable spores, and inputs like fertilizers and hydrogels, among other things, are all possibilities. Some inoculums contain just spores from a single species, whereas others have a diverse mix. When selecting commercially manufactured inoculums, it is also necessary to consider the plant's unique requirements and the current soil conditions.

During the recent decade, AMF inoculum manufacturing, related services, and marketing for the wholesaling markets have increased considerably (Singh et al. 2016; Basiru et al. 2021). Commercial producers, as well as governmental and private entities, are among the clients (Tiwari et al. 2002). While exact sales numbers have yet to be gathered, based on the worldwide biofertilizer industry, it can be determined that there is significant development potential. The worldwide biofertilizer market was valued at 787.8 million dollars in 2016 and is expected to grow to 1.65–2.31 billion dollars by 2022 (Market Analysis Report 2018). During the projected year from 2017–2025, global market demand is estimated to grow by 12.9 percent (Transparency Market Research 2018). Increased usage of biofertilizers in soil management operations, expansion of the organic food sector, and rising financial and environmental expenses connected with biofertilizers are all contributing to this tremendous surge in demand (e.g., nutrient inhibitors). Scientific proof of this plant symbionts' beneficial impacts on plant health, compatibility, and production has fuelled the industry's growth. In addition, when suitable inoculums are created, the economic viability of AM technology becomes increasingly essential. In the present climate-sensitive agrotechnology framework, there has been market awareness that mycorrhizal crops offer a sustainable method for crop production.

8.3.2 Inoculation Strategies and Application Technology

AMF inoculation to a wide range of crop plants is critical especially in nonirrigated locations or in degraded soils where plants have much turmoil in developing root systems. New and more productive AMF isolates may now be utilized to replace the less successful native AMF isolates that are already present in the soil. When inoculated AMF are left in the soil for a long period, their impact is considered to diminish, although they can still be sporulated (Jansa et al. 2006; Rouphael et al. 2015). In the context of sustainable agriculture, it is also proposed that, while perennial plants in agroforestry areas only require one inoculation, it may also be useful to introduce newly chosen AMF isolates at optimal levels. A single propagule can colonize a root in theory, but it may take a longer period. As a result, starting many infections is the greatest way to speed up the inoculum colonization phase (Sharma et al. 1996; Adholeya et al. 2005). Furthermore, fungal propagules must be adjacent to plant roots for efficient mycorrhizal colonization. The faster the root colonization, the more AM fungal propagules are released into the root zone. The effectiveness of this in practice will, of course, be determined by the product, the setting, the distribution mechanism, and various other edaphic factors. The estimation of AMF propagules per zone or per plant is influenced by various factors: (a) the weight or volume of the packet; (b) the quantity of AMF propagules present; (c) the rate at which the inoculum is applied to seeds or soil; (d) how well the product adheres to the seed; and (e) the planting density per hectare (Adholeya et al. 2005).

Various marketed inoculums that function as natural stimulants of plant growth and development have been launched in recent years (Gousterova et al. 2008; Khan et al. 2009). These inoculums are made up of plant growth-promoting microorganisms (PGPM). A marketed inoculum may contain one or more AMF species, as well as other organisms that help the target plant acquire the required parameters, such as beneficial fungi or bacteria. In addition to AMF, two other PGPMs, plant growth-promoting rhizobacteria (PGPR) and *Trichoderma*, play a role in minimizing plant diseases and increasing plant development (Murphy et al. 2003; Harman 2006; Woo et al. 2006; Grover et al. 2011; Calvo-Polanco et al. 2016; Ilangumaran and Smith 2017). Single and mixed-production PGPMs as marketed inoculums might be a sustainable strategy to boost plant growth while reducing external inputs and increasing biotic/abiotic stress tolerance (Daranas et al. 2018).

Simultaneous inoculation with diverse strains of PGPR, *Trichoderma*, and/or AMF typically resulted in improved yield and growth due to increased nutrient absorption when compared to single inoculation (Belimov et al. 1995; Bashan et al. 2004; Kabdwal et al. 2019). In the case of PGPR (Kloepper 1996; Vassilev et al. 2001a, b; Barea et al. 2002; Akköprü et al. 2005) and N_2-fixing bacteria (Biró et al.

2000; Akköprü and Demir 2005), interactions between bacteria and AMF have positive activities in terms of nutrient absorption.

AMF and several PGPR species, including *Azotobacter, Azospirillum, Pseudomonas*, and *Bacillus* species, have been shown to have a synergistic relationship that benefits plant growth (Barea et al. 2005a). Furthermore, when mycorrhizal fungi were co-inoculated with PGPR, AMF root colonization was improved (Gamalero et al. 2004; Toro et al. 1997). Plants infected with a combination of *G. deserticola* and *Rhizobium trifoli* had four times greater nodule counts than single *R. trifoli*, resulting in grafting and increased mycorrization and nodulation with *R. trifoli* and *Yarrowia lipolytica* coencapsulated (Vassilev et al. 2001a, b). Inoculation with AMF and nodule-inducing rhizobia increased the efficiency of P and N uptake (Xavier and Germida 2003). Mycorrhizal and nodule symbiosis have been shown to have synergistic effects on plant development, mineral nutrition, and infection rate (Barea et al. 2005b). Furthermore, the consortia of AMF + *T. harzianum* (Th43) (Kabdwal et al. 2019), and AMF (*Rhizophagus fasciculatus*) + *T. viride* (talc based) (Doley and Jite 2014) boosted the growth and crop productivity. Co-inoculation of both kind of microorganisms enhanced the absorption of mineral nutrients and growth (Gryndler et al. 2002; Medina et al. 2003). PGPM inoculation with commercial biofertilizers comprising consortia of various microorganisms registered significant improvement in the plant growth and yield characters (Malusà et al. 2001; Malusà et al. 2007; Sas-Paszt et al. 2008).

All of this research shows the usefulness and increased efficiency of biofertilizers including a greater number of species with varying growth-boosting mechanisms. The availability of diverse AMF (Ijdo et al. 2011), PGPR (Lucy et al. 2004), and *Trichoderma* (Kabdwal et al. 2019) strains studied in different crop kinds and field circumstances should enable the development of commercially viable consortia. Indeed, it should not be overlooked that as a result of some consortia created, PGPMs may have a detrimental impact on each other (Boyno et al. 2022).

There are just a few techniques for delivering AMF to crops in the field. Farmers are hesitant to invest in specialist equipment for microbial-based goods. As a result, marketed inoculums should be straightforward to apply using normal agricultural gear and procedures. Therefore, the application of these commercialized inoculums can be divided into five main methods: broadcasting method, in-furrow application method, seed dressing method, root dipping method, and seedling/sapling inoculation method (Muresu et al. 2003; Adholeya et al. 2005; Malusá et al. 2012; Basiru et al. 2021).

8.3.2.1 Mycorrhizal fungi in transplanted crops

Seedlings are cultivated in either sterilized or unsterilized soil containing specific mycorrhizal fungi in a slight nursery beds or containers. They are then transplanted when the mycorrhizal colonization is well established. This approach has proven successful in generating significant and economically viable growth responses in crucial crops like tobacco, tomato, finger millet and chili (Rao et al. 1983; Sreeramulu and Bagyaraj 1986). Additionally, it has demonstrated positive outcomes in horticultural crops like citrus, mango, asters, and marigold (Viyanak and Bagyaraj 1990), as well as in forest tree species including *Leucaena* spp., *Tamarindus indica, Acacia nilotica,* and *Calliandra calothyrsus* (Reena and Bagyaraj 1990). This methodology holds promise for application in various transplanted crops significant to agriculture, horticulture, and forestry. Further exploration is warranted to investigate the potential introduction of efficient mycorrhizal fungi to cereals through forest tree species in alley cropping system.

8.3.2.2 In-Furrow Application Method

Other methods that are actively used and promoted globally include various types of in-furrow applications (Bashan 1998; Benami et al. 2020). This approach involves placing the inoculum under or besides seeds within a furrows (Owusu-Bennoah and Mosse 1979; Hayman et al. 1981). Soil is applied to the seeds after they have been put on the inoculum. The inoculum layer will colonize the new roots when the seeds germinate. In fact, when the seeds germinate, exudates such as strigolactones, cutin monomers, and chitin-related compounds are secreted, drawing AMF to the plant (Akiyama et al. 2010; Bonfante and Genre 2015). This is important as it will encourage the formation of colonization and increase the amount of sporulation. However, it should not be ignored that some products negatively affect AMF as a result of the exudates they secrete. In particular, it has been reported that there is no symbiotic interaction between AMF and many plant species belonging to the *Brassicaceae*, *Urticaceae*, *Caryophyllaceae,* and *Chenopodiaceae* families (Brundrett 2009; Tushar and Satish 2013; Güneş et al. 2019).

As a result, the in-furrow treatment is quite effective and results in significant mycorrhizal colonization (Adholeya et al. 2005). However, it can be time-consuming when applied to wide areas (Bashan 1998).

8.3.2.3 Application of mycorrhizal fungi as a seed coating

The seed dressing method is a distinct type of inoculation technique. In this method, the inoculum contains an additive that has good adhesion qualities, such as gum acacia. This additive enhances propagule retention on the seed surface and makes seed dressing technology possible. The inoculated seeds are then allowed to dry. For long-term viability, the drying process and keeping product humidity below 5% are critical (Rivera and Fernandez 2006. Seedlings will be quickly colonized with this approach since the inoculum is in direct contact with the seed (Adholeya et al. 2005). It is also a promising approach since it takes less inoculum and little study (Sieverding 1991; Adholeya et al. 2005). In *Sorghum vulgare*, Rivera and Fernandez (2006) reported that seed dressing with marketed mycorrhizal inoculum (EcoMic) at a low dose of 10% of the stated dose resulted in greater root colonization (percent) and an increase in fungal mycelium. Furthermore, Saleh and El-Akshar (2020) demonstrated that seed dressing with AMF inoculum improved rice plant morphological development and yield, as well as resistance to *Bipolaris oryzae* disease. The most straightforward way to inoculate plants with mycorrhizal fungi would be to coat seeds with mycorrhizal inoculum, employing techniques similar to those used for *Rhizobium,* provided it consistently yields effective infection (Bagyaraj 1992). This involves applying an adhesive, such as methyl cellulose, to the seeds, to which the inoculum is intended to adhere. Regrettably, due to their substantial size, attaching vesicular-arbuscular mycorrhizal propagules in this manner is more challenging than it is for bacteria. Nevertheless, this method has proven effective for large-seeded crops like citrus in field nurseries (Hattingh 1975).

8.3.2.4 Mycorrhizal pellets

Instead of applying vesicular-arbuscular mycorrhizal inoculum onto seeds, a more practical approach for seed inoculation is to create multiseeded pellets. These pellets, approximately 1 cm in diameter, consist of soil or peat inoculum containing vesicular-arbuscular mycorrhizae, stabilized with clay or other binding agents. The inoculum can be produced in a process that involves mixing the soil or peat with mycorrhizal spores, and forming the mixture into pellets. This method has proven to be effective in producing high infection rates of vesicular-arbuscular mycorrhizae on seeds (Hayman et al. 1981). Furthermore, Hall and Kelson (1981) described a system that can produce approximately 5000 of these infected soil pellets per person per day, with seeds attached using gum arabic as an adhesive (Koziol et al. 2017).

8.3.2.5 Fluid drilling in mycorrhiza inoculations

The seed slurry technique for vesicular-arbuscular mycorrhizal inoculation is not only effective, but also presents several advantages over other methods. Firstly, the use of a viscous fluid helps to maintain a uniform mixture of seeds and inoculum, ensuring even distribution and coverage (Hayman et al. 1981) Secondly, the reduction in the bulkiness of the inoculum makes it easier to handle and apply, which can be especially beneficial when working with large areas. Additionally, the ability to combine this technique with rhizobia inoculation provides a more comprehensive approach to promoting healthy crop growth, particularly in leguminous plants. In terms of practical implementation, this method can be scaled up to cover large areas and can be easily integrated into existing seed sowing and soil management practices. Moreover, the benefits of vesicular-arbuscular mycorrhizal associations, such as improved nutrient uptake and stress tolerance, can translate into increased crop yields and reduced inputs, resulting in more sustainable and profitable farming practices.

8.3.2.6 Pre-cropping

Populations of beneficial mycorrhizal fungi can be significantly upscaled directly within the field condition. Mycorrhizal plants are grown and allow their infected roots and associated spores to remain in the soil and colonize upcoming suitable crops. This method along with the judicious crop rotations that incorporate mycorrhizal plants and organic amendments to encourage native fungal populations, gives a promising tactic to improve the mycorrhizal population and inoculum size within the field (Bagyaraj 1990). This technique is effectively applied to enhance the population of a specific, efficient mycorrhizal fungi.

8.4 Commercialization of AM Technology

The approaches utilized in the commercialization of AM technology were classified into different categories (Siddiqui and Kataoka 2011). Important approaches for obtaining efficient AM fungal propagules have been depicted in Fig. 8.1.

8.4.1 Soil-Based Systems

The isolation of the pure culture strain of AMF using the soil-based approach involves the phases of host plant selection and growth environment optimization (Siddiqui and Kataoka 2011; Fig. 8.1). The host plants and the fungi are cultivated in a solid growth medium such as soil, vermiculite, sand, clay, perlite, or other types of mixed bark in this traditional and extensively used technique (Brundrett et al. 1996; Douds Jr et al. 2010). Traditional sand-based pot culture techniques do not generate enough mycorrhizal inoculum, and it is frequently contaminated by other bacteria. Pesticides such as Captan and Furadan, when used at half the authorized dosage in pot cultures, have been shown to reduce other microbial contaminants leaving no pernicious effect on mycorrhizal fungi (Bagyaraj 1992). It has been proven to be quite useful in creating "clean" mycorrhizal inoculum with great potential in a short amount of time (Bagyaraj 1992; Akhtar and Panwar 2011). Solid

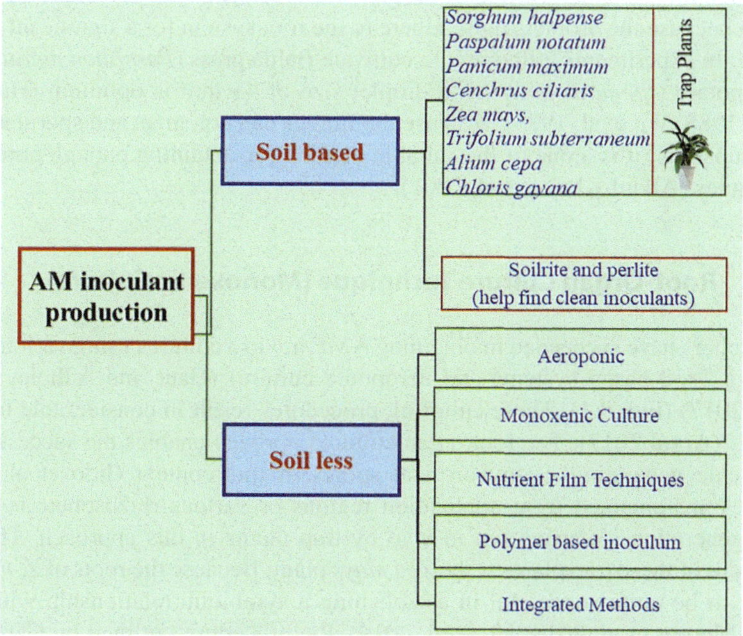

Fig. 8.1 Different methods used for large-scale production of AM fungi. The main logic of the techniques used in commercialization is to obtain a high amount of AM propagules

growth culture inoculum is also heavy, difficult to transport, and too bulky to make it ultra. Inoculum generation is affected by different particle size distributions of substrates. It is also claimed that the best substrate for optimal production has a low nutrient and carbon content (Siddiqui and Kataoka 2011). Phosphorus (P), for example, is rapidly absorbed from soil particles, resulting in Pi-free zones in the plant's rhizosphere soil. Mycorrhizal roots' extraradical hyphae stretch beyond these P-depleted areas, bringing inaccessible Pi to plants and making it available to them (Etesami and Jeong 2021). As a result, in soils low in nutrients, mycorrhizae thrive to reach these nutrients. This aspect is considered an important concept for optimum production.

8.4.2 Aeroponic Culture

It is a soil-free cultivation technique in which plant roots are sprayed with nutritional solutions on a regular or continuous basis (Jarstfer and Sylvia 1995; Mohammad et al. 2000). Several *Glomus* species have been tested through aeroponic cultivation and found promising results (Tiwari et al. 2004, 2020). An inoculum generally takes 12–15 weeks to obtain. The roots are colonized after 9 weeks, and spore production takes 12 weeks (Sylvia and Hubbell 1986; Mohammad et al. 2000). This has several drawbacks, as the system is also susceptible to other undesirable microorganisms. In addition, the nutritional solution and flow must be monitored regularly. Standardization of droplet size is required for successful aeroponic growth because the droplets must adhere to the root system for a significant amount of time. In experiments utilizing it to cultivate Bahia grass (*Paspalum notatum*) and sweet potato (*Ipomoea batatas*), a droplet size of 45 mm is optimum (Hung and Sylvia 1988; Wu et al. 1995). Because the fungus can colonize, and sporulate without a substrate, it is a one of the suitable method for obtaining enough pure AMF-propagules (Abdul-Khaliq et al. 2001).

8.4.3 Root-Organ Culture Technique (Monoxenic Culture)

Researchers have succeeded in obtaining AMF in vitro cultures using various methods (e.g., soil-based systems and aeroponic culture) (Gaur and Adholeya 1994; Aryal 2017) (Fig. 8.1). These culturing procedures result in considerable financial benefits (Aryal 2017). The root organ culture approach enables the successful and large-scale generation of mycorrhizal spores in this context (Ijdo et al. 2011). Samples are obtained from application regions or various rhizosphere soils, and AMF generation is carried out in vivo by trap plants in this approach. The most important of these trap plants is the *Zea mays* plant. Because the roots of *Z. mays* are known to be quite successful in establishing a symbiotic relationship with many AMF (Mathur et al. 2018; Hu et al. 2019), the procedure outlined by Gerdemann and Nicolson (1963) is then used to isolate healthy AMF spores from pot culture using the wet sieving method. These spores are used to inoculate petri dishes with minimal (M) medium (Bécard and Fortin 1988).

Surface sterilization of AMF spores can be done by combining Chloramine-T with Tween-20 (0.1 percent v/v) for 10 min or washing with various antibiotic solutions. Mycorrhizal spores that have been surface-sterilized can be aseptically transplanted onto fine roots of carrots that have been converted with Ri-T-DNA and put on M medium, also known as white medium (Bécard and Fortin 1988; Adholeya et al. 2005) or Strullu-Romand (MSR) medium (Strullu and Romand 1986) (Fig. 8.1). Doner and Bécard (1991) found that the M medium in the two-compartment petri dish is deficient in sucrose, allowing spores to increase in the absence of roots. Every 15 weeks, clonally subculture the spores and root-containing media produced here in a two-compartment petri plate (St-Arnaud et al. 1996). This subcultured media should be injected with *Agrobacterium rhizogenes* bacteria to boost its growth potential (Bécard and Fortin 1988). According to Kumar and Yadav (2018), roots with 10–50 clusters of mycorrhizal spores are cut and transplanted to new receiver operating characteristic (ROC) medium plates with fresh roots in this arrangement. After 3 months of incubation at 26 °C, the spores generated on ROC plates are cut with a sterile knife and transferred to a falcon tube with 15 mL of citrate buffer. After that, the spores are shaken horizontally at 250 U/min for 60 min at 37 °C. To collect the residue at the bottom of the tube, let the spores be at room temperature for 10 min. The supernatant is then discarded, and the spores are washed with autoclaved Milli Q water, filtered through a sieve, and collected in tubes at −20 °C (Kumar and Yadav 2018).

Several species, including *Rhizophagus intraradices,* have been successfully mass-produced using AM technology. After a 4-month growth period in a single-compartment petri dish, Chabot et al. (1992) developed 750 spores in a 30 ml medium using surface-sterilized spores as starting material. After 3 months of incubation, Diop et al. (1994) got around 890 spores utilizing cut roots as the original inoculum. Jolicoeur et al. (1999) used an innovative airlift bioreactor-based manufacturing method. Cultures of the *R. intraradices* in *Daucus carota* roots were produced from spores obtained from soil, as reported by Chabot et al. (1992). Colonized root sections were transferred to a clean solid M medium in petri plates every ~3 months for the cultivation of the root-fungus pair (Bécard and Fortin 1988). At 26 ± 1 °C, all petri plates were incubated in the dark. Mycorrhizal roots were removed, chopped into 1 cm sections using a knife, and placed into a bioreactor without the inoculum gel component. Researchers collected 12,400 spores per litre of media at the end of the operation (Jolicoeur et al. 1999). At 3–4 months, St-Arnaud et al. (1996) collected 15,000 spores in a two-compartment petri plate. Douds (2002) created this two-chamber system by periodically changing the distal, medium chamber with the new medium. This technique yielded 65,000 spores on the distal side of the two chambers over 7 months. The infective propagules of AM fungi were recovered by avoiding severe contaminations (Tiwari et al. 2002; Adholeya et al. 2005).

8.4.4 Technique of Nutrient Film (NFT)

NFT is a specialized commercial agricultural production system that recycles enormous volumes of nutrient fluid on a continuous basis on a film that runs over plant roots. MacDonald (1981) created axenic mycorrhizas between *Glomus caledonium* and *Trifolium parviflorum* and others using a small autoclave hydroponic growth system. However, Mosse and Thompson (1984) modified this method for the generation of AMF inoculum. Furthermore, Lee and George (2005) developed a modified NFT enabling large-scale AMF biomass production combining intermittent nutrient supply, optimized P source, and increased aeration with the utilization of glass beads as support materials. In addition, the average number of spores of total AMF (*G. manihotis, G. etunicatum, Glomus sp, Gigaspora margarita*, and *Acaulospora tuberculata*) was determined to be 1783–2023.30 spores/50 g (Karti et al. 2021).

The nutrient solution in the NFT system must be kept as a thin film (5–10 mm). Mycorrhizal inoculation is also affected by chemical types of nutrients. As a result, it is preferable to employ a well-balanced and appropriate composition. NFT can yield less sporulation than soil-based systems. Contamination issues with undesirable organisms often arise as a result of the nutrient solution utilized. The optimal amounts of various nutritional components vary per mycorrhizal system, based on the plant's size and other characteristics (Sharma et al. 2000). Another factor to consider is the trade-off between growing plants and mycorrhizal colonization, which is impeded by soggy conditions (Tarafdar 1995). The inoculum created by this method, on the other hand, is more concentrated and bulkier than that generated by plants growing in soil or other solid media, and it can be collected more easily (Chellappan et al. 2002; Abdul-Khaliq et al. 2001).

8.4.5 Inoculum Made of Polymers

Polymers are frequently utilized for a variety of applications in biotechnological operations. Gel materials are mostly employed to immobilize live cells, but some are also utilized as components of solid medium for microorganism maintenance.

Hydrogels are the most convenient way to apply polymer materials without having to go through the technical encapsulating process (Vassilev et al. 2005). Many hydrogels were used as transporters of AMF in root-dip and fluid-drill area and greenhouse experiments (Nemec and Ferguson 1985; Johnson and Hummel 1985); however, the pH ranges of the gel substances prevented root colonization and spore germination (Hung et al. 1991; Calvet et al. 1996; Plenchette and Strullu 2003; Jaizme-Vega et al. 2003).

Microbial cells are frequently retained or encapsulated in polymer materials as a strong immobilization technique. The purpose of this method is to keep spores or cells within porous materials created in situ surround biomaterial. Synthetic polymers are not required in mycorrhizal inoculant compositions. The transporter must be reasonably priced and suitable for the materials used in the product's construction. Natural polysaccharides and other hydrophilic hydrogels were utilized as carrier materials. Natural polysaccharides including kappa-carrageenan, agar, and alginates come in a variety of natural, synthetic, and semi-synthetic polymer combinations. Of the roughly 1350 carrier combinations in use, calcium alginates are the most commonly utilized (Vassilev et al. 2005). Alginate beads provide more flexibility in the encapsulation and inoculation of monoxenically generated AMF (Diop 2003). Flavonoids should be included in these capsules as well (Bécard and Piché 1989; Gianinazzi-Pearson et al. 1996; Siddiqui and Kataoka 2011).

8.4.6 Integrated Method

Mycorrhizal symbiosis should be viewed as more than just a bipartite plant–fungus relationship; it should also include the related organisms (Frey-Klett et al. 2007; Tarkka and Frey-Klett 2008). The "mycorrhizosphere" is the result of these mycorrhiza-associated organisms influencing one other (Frey-Klett and Garbaye 2005). The mycorrhizosphere is made up of mycorrhizas, extramatrical mycelium, and related microorganisms. The interaction of bacterial species with AMF increases propagules (AMF structures such as spores, hyphae) and AMF colonization rates, especially in this mycorizosphere (Barea et al. 2002; Akköprü et al. 2005; Pathak et al. 2017). The use of "mycorrhizal helper bacteria (MHB)" in this context enhances AMF symbiosis in a variety of agricultural plants (Tarkka and Frey-Klett 2008).

Several researchers have examined the function of MHB in the genesis and development of various species of AMF (Siddiqui and Mahmood 1998; Vosatka et al. 1999; Frey-Klett et al. 2007; Tarkka and Frey-Klett 2008). The correct establishment of in vitro-generated plantlets in field circumstances can be achieved by combining and carefully applying AMF and PGPR. PGPR improved mycorrhizal colonization, according to Bhowmik and Singh (2004), and might be used to mass-produce AMF cultures. Silva et al. (2007) found that adding Tris–HCl buffer to the substrate improved AMF sporulation. According to these researchers, large-scale inoculum formation may be accomplished by adding Tris–HCl buffer to the nutritional solution and storing it at +4 °C.

One explanation for improved plant growth is the association of nitrogen fixers and P-solubilizers with AMF (Turk et al. 2006), and these connections are useful in increasing micropropagated plant survival rates (Webster et al. 1995). *Bradyrhizobium, Rhizobium,* and *Frankia* are microorganisms that can aid in mass-produce AMF in vitro by improving soil-binding stability, capacity, and qualities that make the soil favourable to the growth of micro-propagated plantlets like mycorrhizae (Varma and Schuepp 1995).

8.5 Challenges to Commercial Use

Even though mycorrhizal research has just achieved a critical mass, it is essential to identify the obstacles in their commercialization. The inability to develop AMF in pure culture in particular is a significant disadvantage (Sharma et al. 2017). It can only be cultivated with plants by adding inoculum under certain conditions, according to the available knowledge, and it cannot be easily mass-produced in laboratory conditions (Sharma et al. 2017; Kadian et al. 2018). Currently, the mycorrhizal inoculum is created as another non-sterile substrate, including a non-sterile medium, soil, and propagule (spores, hyphae, and colonized root fragments) in the majority of the samples. Counterfeit bio-products are another issue in commercializing AM technology. Increased sales of counterfeit bio-products, a dearth of live quality control procedures, and fewer propagule numbers than advertised in many products all hurt AM technology (Nagpal et al. 2021). Counterfeit mycorrhizal products have a major impact on the natural resource driven products. In addition, the composition of the carrier medium and the quantity of active spores per unit weight/volume varies considerably among commercial suppliers. The fact that these fungi grow slower than other microbes, limits their use in large-scale farming. One of the challenges that mycorrhizal inoculum manufacturers confront is finding consumers in the agricultural and agroforestry sectors. In fact, in both established and emerging areas, the "organic" sector is regarded to be one of the most profitable segments in which mycorrhizal technologies may penetrate. Organic agricultural sectors are anticipated to have the largest value and profit margins, at least in industrialized countries, because marketed mycorrhizal inoculums can supplement or even replace conventional and chemical-based fertilizers (Vosátka et al. 2008). However, the market's progress is limited by a lack of awareness in prospective emerging nations, poor infrastructure, money, and a lack of knowledge of critical mycorrhizal characteristics. Plant mycorrhization in agroforestry and sustainable agriculture has drawn a lot of attention in recent years because of its role as a biofertilizer to boost host development. However, further effort is needed to identify acceptable local AM fungal strains for high-quality crop production and educate farmers in developing countries about the function of mycorrhiza in agroforestry and sustainable agricultural systems (Dobo et al. 2018). Also, due to shelf life or unclear storage stability, production constraints and technological challenges, as well as the time and labour needed to cultivate appropriate numbers of propagules, mycorrhizal markets are not very convincing (Benami et al. 2020).

8.6 Formulation of AM Technology

Today, commercialized AM technology is available in several forms. Some businesses sell a single mycorrhiza strain along with a carrier. However, most businesses sell microorganisms in the form of mixtures using different substrates.

Formulation methods account for possible negative environmental impacts as well as ingredients that might render the inoculum ineffective. To create a substance that can be efficiently transported to the intended application, a combination of microbial propagules with a variety of transporters or excipients is utilized. There have been several different mycorrhizal inoculum compositions proposed. Glass beads (Redecker et al. 1995) and expanded clay (Plenchette et al. 1983; Adholeya et al. 2005) have been utilized in research laboratories and the commercial sector, respectively. These formulations benefit from permitting the spontaneous retention of mycorrhizal roots and spores during the growth period in greenhouse settings. Mycorrhizal invaginations can settle in the porous structure of the beads, which has many air gaps. Inoculum can also be mixed with carriers like air-dried sand, vermiculite, and soil (Millner and Kitt 1992). Liquid and powder inoculum, granules or tablets/pellets, granules and gel beads are all examples of mycorrhizal inoculum. *Glomus* spp. intraradical vesicles/spores can likewise be preserved and utilized as such in alginate beads (Redecker et al. 1995). Under controlled settings, intraradical elements isolated in such beads have been found to regenerate and colonize new roots (Strullu and Plenchette 1991). Trapping monoxenically generated spores in alginate particles has also been demonstrated to be successful (Declerck et al. 1996).

8.7 Conclusions and Future Prospects

Mycorrhizal fungi can help restore economic efficiency and environmental safety by increasing natural and managed ecosystems without depleting natural resources. They can also help lower fertilizer prices and energy demands, restoring economic efficiency and environmental protection. Appropriate mycorrhizal inoculums, on the other hand, improve biocontrol potential in a wide range of agricultural and soil characteristics in both academic and commercial settings worldwide. Under traditional agroecology or agroforestry, the main challenges in commercializing AM technology are a lack of large-scale field testing and appropriate finance. Manufacturers and distributors of mycorrhizal inoculum also confront similar problems across the world. To satisfy the needs of a broad client base, these constraints involve the need to modify products, boost market knowledge, and develop more effective distribution tactics. Concerning its commercialization plan, AM technology must be competent, efficient, and enlightening to succeed. Another requirement is to foster an entrepreneurial culture within the company, supported by excellent research infrastructure, networking, and financing. Mycorrhizal bio-fertilizers are expected to become a trustworthy partner with chemical inputs in the upcoming years, benefiting from agricultural, economic, and social perspectives. Carrier cost is a significant factor in commercial process development since the cost of the completed product grows with each stage of the manufacturing process. A suitable formulation carrier should be cheap (preferably from locally available nontoxic waste) and have no negative impacts on mycorrhizal symbiosis. It should also be simple to use and apply so that maximum dispersion is achieved. In potted plants, the formulation should allow for early breakdown or dissemination (for pellets, granules, and tablets). Because the roots and mycorrhizal propagules may not make contact if the transporter is too firmly adherent and does not disintegrate after watering, the impact may be reduced. Growth conditions should be strictly controlled, with specific care devoted to retaining the inoculum's potency. Even a minor error might cause the organism to lose viability, discouraging the end user from using these techniques in agriculture. Growth conditions should be strictly controlled, with specific care devoted to retaining the inoculum's potency. Even a little inaccuracy might result in the organism losing viability, deterring farmers from employing these approaches.

References

Abdul-Khaliq, Gupta ML, Alam A (2001) Biotechnological approaches for mass production of arbuscular mycorrhizal fungi: current scenario and future strategies. In: Mukerji KG, Manoharachary C, Chamola BP (eds) Techniques in mycorrhizal studies. Kluwer Academic Publishers, Dordrecht, pp 299–312

Adholeya A (2012) Development and testing of mycorrhiza in multilocation field trials for improved crop yield under different cultivation systems and soils. In: Book of abstracts, 7th international symbiosis society congress (The earth's vast symbiosphere, July 22–28, 2012), Kraków, Poland, p 254

Adholeya A, Tiwari P, Singh R (2005) Large-scale inoculum production of arbuscular mycorrhizal fungi on root organs and inoculation strategies. In: In vitro culture of mycorrhizas. Springer, Berlin, pp 315–338

Akhtar M, Panwar J (2011) Arbuscular mycorrhizal fungi and opportunistic fungi: efficient root symbionts for the management of plant parasitic nematodes. Adv Sci Eng Med 3(3):165–175

Akhtar MS, Siddiqui ZA (2008) Arbuscular mycorrhizal fungi as potential bioprotectants against plant pathogens. In: Mycorrhizae: sustainable agriculture and forestry. Springer, Dordrecht, pp 61–97

Akhtar MS, Siddiqui ZA, Wiemken A (2011) Arbuscular mycorrhizal fungi and Rhizobium to control plant fungal diseases. In: Alternative farming systems, biotechnology, drought stress and ecological fertilisation. Springer, Dordrecht, pp 263–292

Akiyama K, Ogasawara S, Ito S, Hayashi H (2010) Structural requirements of strigolactones for hyphal branching in AM fungi. Plant Cell Physiol 51(7):1104–1117

Akköprü A, Demir S (2005) Biological control of Fusarium wilt in tomato caused by *Fusarium oxysporum* f. sp. *lycopersici* by AMF *Glomus intraradices* and some rhizobacteria. J Phytopathol 153(9):544–550

Akköprü A, Demir S, Özaktan H (2005) Effect of different fluorescent pseudomonad (FP) isolates and an arbuscular mycorrhizal fungus (AMF) *Glomus intraradices* on some of the morphological parameters of tomato and fusarium wilt (*Fusarium oxysporum* f.sp. *lycopersici* (Sacc) Syd. Et Hans.) in tomato. Yuzuncu Yıl Univ J Agric Sci 15(2):131–138

Ansari RA, Khan TA (2012a) Parasitic association of root-knot nematode, *Meloidogyne incognita* on guava. e J Sci Technol 5(12):65–67

Ansari RA, Khan TA (2012b) Diversity and community structure of Phytonematodes associated with guava in and around Aligarh, Uttar Pradesh. Trends Biosci 5(3):202–204

Ansari RA, Mahmood I (2017a) Optimization of organic and bio-organic fertilizers on soil properties and growth of pigeonpea. Sci Hortic 226:1–9

Ansari RA, Mahmood I (2017b) Determination of disease incidence caused by *Meloidogyne* spp. and or *Fusarium udum* on pigeonpea in Aligarh district: a survey. Trends Biosci 10(24):5239–5243

Ansari RA, Mahmood I (2019a) Plant health under biotic stress. In: Ansari RA, Mahmood I (eds) Microbial interactions, vol II. Springer Nature, Singapore. https://doi.org/10.1007/978-981-13-6040-4

Ansari RA, Mahmood I (2019b) Plant health under biotic stress. In: Ansari RA, Mahmood I (eds) Organic strategies, vol I. Springer Nature, Singapore. https://doi.org/10.1007/978-981-13-6043-5

Ansari RA, Mahmood I, Rizvi R, Sumbul A (2017a) PGPR: current vogue in sustainable crop production. In: Kumar V (ed) Probiotics and plant health. Springer Nature, Singapore, pp 455–472

Ansari RA, Mahmood I, Rizvi R, Sumbul A, Safiuddin (2017b) Siderophores: augmentation of soil health and crop productivity. In: Kumar V (ed) Probiotics in agroecosystem. Springer Nature, Singapore, pp 291–312

Ansari RA, Sumbul A, Rizvi R, Mahmood I (2019a) Potential role of plant growth promoting Rhizobacteria in alleviation of biotic stress. In: Ansari RA, Mahmood I (eds) Plant health under biotic stress, Microbial interactions, vol II. Springer Nature, Singapore, pp 177–188

Ansari RA, Sumbul A, Rizvi R, Mahmood I (2019b) Organic soil amendments: potential tool for soil and plant health management. In: Ansari RA, Mahmood I (eds) Plant health under biotic stress, Organic strategies, vol I. Springer Nature, Singapore, pp 1–35

Ansari RA, Rizvi R, Mahmood I (2020a) Management of phytonematodes: management of phytonematodes: recent advances and future challenges. Springer Nature, Singapore. https://doi.org/10.1007/978-981-15-4087-5

Ansari RA, Rizvi R, Sumbul A, Mahmood I (2020b) Plant-growth-promoting Rhizobacteria (PGPR)-based sustainable management of phytoparasitic nematodes: current understandings and future challenges. In: Ansari RA et al (eds) Management of phytonematodes: recent advances and future challenges. Springer Nature, Singapore. https://doi.org/10.1007/978-981-15-4087-5_3

Araújo TM, Silva KD, Pereira GMD, Curcino A, Stürmer SL, Gomide PHO, Florestas E (2019) Diversity of arbuscular mycorrhizal fungi in agroforestry, conventional plantations and native forests in Roraima state, Northern Brazil. J Agric Sci 11:282

Aryal HP (2017) A protocol on *in vitro* propagation of arbuscular mycorrhizal fungi using root organ culture technique. Tribhuvan Univ J 31(1–2):17–24

Bagyaraj DJ (1990) Soil and plants. In: Arora DK, Rai B, Mukeji KG, Knudsen G (eds) Handbook of applied mycology, vol I. Marcel Dekker, New York, pp 3–34

Bagyaraj DJ (1992) 19 Vesicular-arbuscular mycorrhiza: application in agriculture. Methods Microbiol 24:359–373

Bagyaraj DJ (1992) 19 Vesicular-arbuscular Mycorrhiza: application in agriculture. In: Methods in microbiology, vol 24. Academic Press, pp 359–373

Barea JM, Azcón R, Azcón-Aguilar C (2002) Mycorrhizosphere interactions to improve plant fitness and soil quality. Antonie Van Leeuwenhoek 81(1):343–351

Barea JM, Azcón R, Azcón-Aguilar C (2005a) Interactions between mycorrhizal fungi and bacteria to improve plant nutrient cycling and soil structure. In: Microorganisms in soils: roles in genesis and functions. Springer, Berlin, pp 195–212

Barea JM, Pozo MJ, Azcon R, Azcon-Aguilar C (2005b) Microbial co-operation in the rhizosphere. J Exp Bot 56(417):1761–1778

Barrios E (2007) Soil biota, ecosystem services and land productivity. Ecol Econ 64(2):269–285

Bashan Y (1998) Inoculants of plant growth-promoting bacteria for use in agriculture. Biotechnol Adv 16(4):729–770

Bashan Y, Holguin G, De-Bashan LE (2004) Azospirillum-plant relationships: physiological, molecular, agricultural, and environmental advances (1997-2003). Can J Microbiol 50(8):521–577

Basiru S, Mwanza HP, Hijri M (2021) Analysis of arbuscular mycorrhizal fungal inoculant benchmarks. Microorganisms 9(1):81

Bécard G, Fortin JA (1988) Early events of vesicular–arbuscular mycorrhiza formation on Ri T-DNA transformed roots. New Phytol 108(2):211–218

Bécard G, Piché Y (1989) Fungal growth stimulation by CO2 and root exudates in vesicular-arbuscular mycorrhizal symbiosis. Appl Environ Microbiol 55(9):2320–2325

Belimov AA, Kojemiakov AP, Chuvarliyeva CN (1995) Interaction between barley and mixed cultures of nitrogen fixing and phosphate-solubilizing bacteria. Plant Soil 173(1):29–37

Benami M, Isack Y, Grotsky D, Levy D, Kofman Y (2020) The economic potential of arbuscular mycorrhizal fungi in agriculture. In: Nevalainen H (ed) Grand challenges in fungal biotechnology. Springer, Cham, pp 239–279

Bhowmik SN, Singh CS (2004) Mass multiplication of AM inoculum: effect of plant growth-promoting rhizobacteria and yeast in rapid culturing of Glomus mosseae. Curr Sci:705–709

Biró B, Köves-Péchy K, Vörös I, Takács T, Eggenberger P, Strasser RJ (2000) Interrelations between *Azospirillum* and *Rhizobium* nitrogen-fixers and arbuscular mycorrhizal fungi in the rhizosphere of alfalfa in sterile, AMF-free or normal soil conditions. Appl Soil Ecol 15(2):159–168

Bonfante P, Genre A (2015) Arbuscular mycorrhizal dialogues: do you speak 'plantish'or 'fungish'? Trends Plant Sci 20(3):150–154

Boyno G, Demir S, Danesh YR (2022) Effects of some biological agents on the growth and biochemical parameters of tomato plants infected with *Alternaria solani* (Ellis & Martin) Sorauer. Eur J Plant Pathol 162(1):19–29

Brundrett MC (2009) Mycorrhizal associations and other means of nutrition of vascular plants: understanding the global diversity of host plants by resolving conflicting information and developing reliable means of diagnosis. Plant Soil 320(1):37–77

Brundrett M, Bougher N, Dell B, Grove T, Malajczuk N (1996) Working with mycorrhizas in forestry and agriculture, vol 32. Australian Centre for International Agricultural Research, Canberra, p 374

Bünemann EK, Schwenke GD, Van Zwieten L (2006) Impact of agricultural inputs on soil organisms—a review. Soil Res 44(4):379–406

Calvet C, Camprubí A, Rodríguez-Kábana R (1996) Inclusion of arbuscular mycorrhizal fungi in alginate films for experimental studies and plant inoculation. HortScience 31(2):285–285

Calvo-Polanco M, Sánchez-Romera B, Aroca R, Asins MJ, Declerck S, Dodd IC et al (2016) Exploring the use of recombinant inbred lines in combination with beneficial microbial inoculants (AM fungus and PGPR) to improve drought stress tolerance in tomato. Environ Exp Bot 131:47–57

Chabot S, Bécard G, Piché Y (1992) Life cycle of *Glomus intraradix* in root organ culture. Mycologia 84(3):315–321

Chellappan P, Christy SA, Mahadevan A (2002) Multiplication of arbuscular mycorrhizal fungi on roots. In: Techniques in mycorrhizal studies. Springer, Dordrecht, pp 285–297

Daranas N, Badosa E, Francés J, Montesinos E, Bonaterra A (2018) Enhancing water stress tolerance improves fitness in biological control strains of *Lactobacillus plantarum* in plant environments. PLoS One 13(1):e0190931

Declerck S, Strullu DG, Plenchette C (1996) *In vitro* mass-production of the arbuscular mycorrhizal fungus, *Glomus versiforme*, associated with Ri T-DNA transformed carrot roots. Mycol Res 100(10):1237–1242

Demir S, Şensoy S, Ocak E, Tüfenkci Ş, Durak ED, Erdinc C, Ünsal H (2015) Effects of arbuscular mycorrhizal fungus, humic acid, and whey on wilt diseasecaused by *Verticillium dahliae* Kleb. in three solanaceous crops. Turk J Agric For 39(2):300–309

Diop TA (2003) *In vitro* culture of arbuscular mycorrhizal fungi: advances and future prospects. Afr J Biotechnol 2(12):692–697

Diop SSD, Grizzle JW, Moraal PE, Stefanopoulou A (1994) Interpolation and numerical differentiation for observer design. In: Proceedings of the American control conference, vol 2. American Automatic Control Council, pp 1329–1329

Dobo B, Asefa F, Asfaw Z (2018) Diversity and abundance of arbuscular mycorrhizal fungi under different plant and soil properties in Sidama, Southern Ethiopia. Agrofor Syst 92(1):91–101

Doley K, Jite PK (2014) Interaction effects of *Glomus fasciculatum* and *Trichoderma viride* inoculations on groundnut plants inoculated with pathogen *Macrophomina phaseolina*. Int J Agric Sci 4(9):281–288

Doner LW, Bécard G (1991) Solubilization of gellan gels by chelation of cations. Biotechnol Tech 5(1):25–28

Douds DD (2002) Increased spore production by Glomus intraradices in the split-plate monoxenic culture system by repeated harvest, gel replacement, and resupply of glucose to the mycorrhiza. Mycorrhiza 12(4):163–167

Douds DD Jr, Nagahashi G, Hepperly PR (2010) On-farm production of inoculum of indigenous arbuscular mycorrhizal fungi and assessment of diluents of compost for inoculum production. Bioresour Technol 101(7):2326–2330

Etesami H, Jeong BR (2021) Contribution of Arbuscular mycorrhizal fungi, phosphate–solubilizing bacteria, and silicon to P uptake by plant: a review. Front Plant Sci 12:1355

Ferrol N, Barea J, Azcon-Aguilar C (2002) Mechanisms of nutrient transport across interfaces in arbuscular mycorrhizas. In: Diversity and integration in mycorrhizas. Springer, pp 231–237

Frey-Klett P, Garbaye J (2005) Mycorrhiza helper bacteria: a promising model for the genomic analysis of fungal–bacterial interactions. New Phytol 168(1):4–8

Frey-Klett P, Garbaye J, Tarkka M (2007) The mycorrhiza helper bacteria revisited. New Phytol 176(1):22–36

Futai K, Taniguchi T, Kataoka R (2008) Ectomycorrhizae and their importance in forest ecosystems. In: Mycorrhizae: sustainable agriculture and forestry. Springer, Dordrecht, pp 241–285

Gamalero E, Trotta A, Massa N, Copetta A, Martinotti MG, Berta G (2004) Impact of two fluorescent pseudomonads and an arbuscular mycorrhizal fungus on tomato plant growth, root architecture and P acquisition. Mycorrhiza 14(3):185–192

Gaur A, Adholeya A (1994) Estimation of VAMF spores in soil: a modified method. Mycorrhiza News 6(1):10–11

Gerdemann JW, Nicolson TH (1963) Spores of mycorrhizal Endogone species extracted from soil by wet sieving and decanting. Trans Br Mycol Soc 46(2):235–244

Gianinazzi S, Vosátka M (2004) Inoculum of arbuscular mycorrhizal fungi for production systems: science meets business. Can J Bot 82(8):1264–1271

Gianinazzi S, Gollotte A, Binet MN, van Tuinen D, Redecker D, Wipf D (2010) Agroecology: the key role of arbuscular mycorrhizas in ecosystem services. Mycorrhiza 20(8):519–530

Gianinazzi-Pearson V, Dumas-Gaudot E, Gollotte A, Alaoui AT, Gianinazzi S (1996) Cellular and molecular defence-related root responses to invasion by arbuscular mycorrhizal fungi. New Phytol 133(1):45–57

Gousterova A, Nustorova M, Christov P, Nedkov P, Neshev G, Vasileva-Tonkova E (2008) Development of a biotechnological procedure for treatment of animal wastes to obtain inexpensive biofertilizer. World J Microbiol Biotechnol 24(11):2647–2652

Grover M, Ali SZ, Sandhya V, Rasul A, Venkateswarlu B (2011) Role of microorganisms in adaptation of agriculture crops to abiotic stresses. World J Microbiol Biotechnol 27(5):1231–1240

Gryndler M, Vosátka M, Hrŝelová H, Catská V, Chvátalová I, Jansa J (2002) Effect of dual inoculation with arbuscular mycorrhizal fungi and bacteria on growth and mineral nutrition of strawberry. J Plant Nutr 25(6):1341–1358

Güneş H, Demir S, Durak ED (2019) Relationship between *Brassicaceae, Chenopodiaceae* and *Urticaceae* families with arbuscular mycorrhizal fungi (AMF). KSU J Agric Nat 22:102–108

Guo X (2019) The role of arbuscular mycorrhiza in sustainable environment and agriculture. In: Biofertilizers for sustainable agriculture and environment. Springer, Cham, pp 501–520

Hall IR, Kelson A (1981) An improved technique for the production of endomycorrhizal infested soil pellets. N Z J Agric Res 24(2):221–222

Harman GE (2006) Overview of mechanisms and uses of *Trichoderma* spp. Phytopathology 96(2):190–194

Hattingh MJ (1975) Inoculation of Brazilian sour orange with endomycorrhizal fungus. Phytopathol 65:1013–1016

Hayman DS, Morris EJ, Page RJ (1981) Methods for inoculating field crops with mycorrhizal fungi. Ann Appl Biol 99(3):247–253

Hijri M (2016) Analysis of a large dataset of mycorrhiza inoculation field trials on potato shows highly significant increases in yield. Mycorrhiza 26(3):209–214

Hu ZH, Zhuo F, Jing SH, Li X, Yan TX, Lei LL et al (2019) Combined application of arbuscular mycorrhizal fungi and steel slag improves plant growth and reduces Cd, Pb accumulation in *Zea mays*. Int J Phytoremediation 21(9):857–865

Hung LLL, Sylvia DM (1988) Production of vesicular-arbuscular mycorrhizal fungus inoculum in aeroponic culture. Appl Environ Microbiol 54(2):353–357

Hung LLL, O'Keefe DM, Sylvia DM (1991) Use of hydrogel as a sticking agent and carrier for vesicular-arbuscular mycorrhizal fungi. Mycol Res 95(4):427–429

Ijdo M, Cranenbrouck S, Declerck S (2011) Methods for large-scale production of AM fungi: past, present, and future. Mycorrhiza 21(1):1–16

Ilangumaran G, Smith DL (2017) Plant growth promoting rhizobacteria in amelioration of salinity stress: a systems biology perspective. Front Plant Sci 8:1768

Jaizme-Vega MC, Rodríguez-Romero AS, Hermoso CM, Declerck S (2003) Growth of micropropagated bananas colonized by root-organ culture produced arbuscular mycorrhizal fungi entrapped in ca-alginate beads. Plant Soil 254(2):329–335

Jansa J, Wiemken A, Frossard E (2006) The effects of agricultural practices on arbuscular mycorrhizal fungi. Geol Soc Lond, Spec Publ 266(1):89–115

Jarstfer AG, Sylvia DM (1995) Aeroponic culture of VAM fungi. In: Mycorrhiza. Springer, Berlin, Heidelberg, pp 427–441

Jeffries P, Gianinazzi S, Perotto S, Turnau K, Barea JM (2003) The contribution of arbuscular mycorrhizal fungi in sustainable maintenance of plant health and soil fertility. Biol Fertil Soils 37(1):1–16

Johnson CR, Hummel RL (1985) Influence of mycorrhizae and drought stress on growth of Poncirus× citrus seedlings. HortScience 20(4):754–755

Jolicoeur M, Williams RD, Chavarie C, Fortin JA, Archambault J (1999) Production of Glomus intraradices propagules, an arbuscular mycorrhizal fungus, in an airlift bioreactor. Biotechnol Bioeng 63(2):224–232

Kabdwal BC, Sharma R, Tewari R, Tewari AK, Singh RP, Dandona JK (2019) Field efficacy of different combinations of Trichoderma harzianum, Pseudomonas fluorescens, and arbuscular mycorrhiza fungus against the major diseases of tomato in Uttarakhand (India). Egyptian Journal of Biological Pest Control 29(1):1–10

Kadian N, Yadav K, Aggarwal A (2018) Mass multiplication of arbuscular mycorrhizal fungi associated with some leguminous plants: an ecofriendly approach. Indian J Exp Biol 56:258–266

Karti PDMH, Prihantoro I, Aryanto AT (2021) Evaluation of inoculum arbuscular mycorrhizal fungi in Brachiaria decumbens. In: IOP conference series: earth and environmental science, vol 694, No. 1, p 012048. IOP Publishing

Khan W, Rayirath UP, Subramanian S, Jithesh MN, Rayorath P, Hodges DM et al (2009) Seaweed extracts as biostimulants of plant growth and development. J Plant Growth Regul 28(4):386–399

Kloepper JW (1996) Host specificity in microbe-microbe interactions. Bioscience 46(6):406–409

Koziol L, Schultz PA, Bever JD, House G, Bauer J, Middleton E (2017) USER MANUAL: a practical guide to inoculation with Arbuscular Mycorrhizal fungi in ecological restoration. SERDP Project RC-2330

Kumar S, Yadav S (2018) *In vitro* cultivation of AMF using root organ culture: factory of biofertilizers and secondary metabolites production. Microb Cell Factories:95–108

Lee YJ, George E (2005) Development of a nutrient film technique culture system for arbuscular mycorrhizal plants. HortScience 40(2):378–380

Lucy M, Reed E, Glick BR (2004) Applications of free living plant growth-promoting rhizobacteria. Antonie Van Leeuwenhoek 86(1):1–25

MacDonald RM (1981) Routine production of axenic vesicular-arbuscular mycorrhizas. New Phytol 89(1):87–93

Malusà E, Buffa G, Ciesielska J (2001) Effect of different fertilisation management on photosynthesis, yield and fruit quality of peach. In: Plant Nutrition. Springer, Dordrecht, pp 332–333

Malusà E, Sas-Paszt L, Zurawicz E, Popinska W (2007) The effect of a mycorrhiza-bacteria substrate and foliar fertilization on growth response and rhizosphere pH of three strawberry cultivars. Int J Fruit Sci 6:25–41

Malusá E, Sas-Paszt L, Ciesielska J (2012) Technologies for beneficial microorganisms inocula used as biofertilizers. Sci World J 2012:1–12

Market Analysis Report (2018) Biofertilizers market size, share & trends analysis report by product (nitrogen fixing, phosphate solubilizing), by application (seed treatment, soil treatment), and segment forecasts 2012–2022. Grand View Research, 147 pp

Mathur S, Sharma MP, Jajoo A (2018) Improved photosynthetic efficacy of maize (*Zea mays*) plants with arbuscular mycorrhizal fungi (AMF) under high temperature stress. J Photochem Photobiol B Biol 180:149–154

Medina A, Probanza A, Mañero FG, Azcón R (2003) Interactions of arbuscular-mycorrhizal fungi and *Bacillus* strains and their effects on plant growth, microbial rhizosphere activity (thymidine and leucine incorporation) and fungal biomass (ergosterol and chitin). Appl Soil Ecol 22(1):15–28

Millner PD, Kitt DG (1992) The Beltsville method for soilless production of vesicular-arbuscular mycorrhizal fungi. Mycorrhiza 2(1):9–15

Mohammad A, Khan AG, Kuek C (2000) Improved aeroponic culture of inocula of arbuscular mycorrhizal fungi. Mycorrhiza 9(6):337–339

Mosse B, Thompson JP (1984) Vesicular–arbuscular endomycorrhizal inoculum production. I. Exploratory experiments with beans (*Phaseolus vulgaris*) in nutrient flow culture. Can J Bot 62(7):1523–1530

Muleta D, Assefa F, Nemomissa S, Granhall U (2008) Distribution of arbuscular mycorrhizal fungi spores in soils of smallholder agroforestry and monocultural coffee systems in southwestern Ethiopia. Biol Fertil Soils 44(4):653–659

Muresu R, Sulas L, Caredda S (2003) Legume-Rhizobium symbiosis. Characteristics and prospects of inoculation. Rivista di Agronomia (Italy)

Murphy JF, Reddy MS, Ryu CM, Kloepper JW, Li R (2003) Rhizobacteria-mediated growth promotion of tomato leads to protection against cucumber mosaic virus. Phytopathology 93(10):1301–1307

Nagpal S, Sharma P, Kumawat KC (2021) Microbial bioformulations: revisiting role in sustainable agriculture. In: Biofertilizers. Woodhead Publishing, pp 329–346

Neill EG, Neill RV, Norby RJ (1991) Hierarchy theory as a guide to mycorrhizal research on large-scale problems. Environ Pollut 73(3–4):271–284

Nemec S, Ferguson JJ (1985) A fluid-drilling applicator for applying VAM in the field. In: 6th North American Conference on Mycorrhizae, Bend, Oregon (USA), 25–29 June 1984. Oregon State University, Forest Research Laboratory

Owusu-Bennoah E, Mosse B (1979) Plant growth responses to vesicular-arbuscular mycorrhiza: XI. Field inoculation responses in barley, lucerne and onion. New Phytol 83(3):671–679

Pal S, Singh HB, Rai A, Rakshit A (2013) Evaluation of different medium for producing on farm arbuscular mycorrhizal inoculum. Int J Agric Environ Biotechnol 6(4):557–562

Pal S, Farooqui A, Rakshit A, Rai S, Rai A, Singh HB (2015) Mycorrhiza in a changing environment helps plants to deal stress. Microbial empowerment in agriculture–a key to sustainability and crop productivity. Biotech Books, New Delhi, pp 109–128

Pal S, Singh HB, Farooqui A, Rakshit A (2016) Commercialization of arbuscular mycorrhizal technology in agriculture and forestry. In: Singh HB, Sarma BK, Keswani C (eds) Agriculturally important microorganisms. Springer, Singapore, pp 97–105

Parewa HP, Rakshit A, Ali M, Lal B (2014) Arbuscular mycorrhizal fungi: a way to improve soil quality. Popular Kheti 2:85–92

Pathak D, Lone R, Koul KK (2017) Arbuscular mycorrhizal fungi (AMF) and plant growth-promoting Rhizobacteria (PGPR) Association in Potato (*Solanum tuberosum* L.): a brief review. In: Kumar V, Kumar M, Sharma S, Prasad R (eds) Probiotics and plant health. Springer, Singapore, pp 401–420

Pellegrino E, Öpik M, Bonari E, Ercoli L (2015) Responses of wheat to arbuscular mycorrhizal fungi: a meta-analysis of field studies from 1975 to 2013. Soil Biol Biochem 84:210–217

Plenchette C, Strullu DG (2003) Long-term viability and infectivity of intraradical forms of *Glomus intraradices* vesicles encapsulated in alginate beads. Mycol Res 107(5):614–616

Plenchette C, Fortin JA, Furlan V (1983) Growth responses of several plant species to mycorrhizae in a soil of moderate P-fertility. Plant Soil 70(2):199–209

Rakshit A (2015) Soil biodiversity: stars beneath our feet. SATSA Mukhaptra Annu Tech Issue 19:43–48

Rakshit A, Bhadoria PBS, Das DK (2002) An overview of mycorrhizal symbioses. J Inter Des 6:570–581

Rao YG, Bagyaraj DJ, Rai PV (1983) Selection of an efficient VA mycorrhizal fungus for finger millet: II. Screening under field conditions. Zentralblatt für Mikrobiologie 138(6):415–419

Redecker D, Thierfelder H, Werner D (1995) A new cultivation system for arbuscular mycorrhizal fungi on glass beads. Angew Bot 69(5–6):189–191

Reena J, Bagyaraj DJ (1990) Growth stimulation of Tamarindus indica by selected VA mycorrhizal fungi. World J Microbiol Biotechnol 6:59–63

Rillig MC, Sosa-Hernández MA, Roy J, Aguilar-Trigueros CA, Vályi K, Lehmann A (2016) Towards an integrated mycorrhizal technology: harnessing mycorrhiza for sustainable intensification in agriculture. Front Plant Sci 7:1625

Rivera R, Fernandez F (2006) Inoculation and management of mycorrhizal fungi within tropical agroecosystems. In: Norman Uphoff et al. pp 479–489

Rouphael Y, Franken P, Schneider C, Schwarz D, Giovannetti M, Agnolucci M et al (2015) Arbuscular mycorrhizal fungi act as biostimulants in horticultural crops. Sci Hortic 196:91–108

Saleh MM, El-Akshar YS (2020) Integration between arbuscular mycorrhizal fungi, bacterial and fungal bioagents for controlling rice brown spot disease. Egyptian J Phytopathol 48(1):81–93

Sas-Paszt L, Żurawicz E, Masny A, Filipczak J, Pluta S, Lewandowski M, Basak A (2008) The use of biostimulators in small fruit growing, Biostimulators in modern agriculture. Fruit crops. Editorial House Wieś Jutra, Warszawa, pp 76–90

Selosse MA, Baudoin E, Vandenkoornhuyse P (2004) La diversite des microorganismes symbiotiques: une clef pour la reussite ecologique et la protection des plantes. Comptes rendus-Biologies 7(327):639–648

Sharma MP, Gaur A, Bhatia NP, Adholeya A (1996) Growth responses and dependence of *Acacia nilotica* var. *cupriciformis* on the indigenous arbuscular mycorrhizal consortium of a marginal wasteland soil. Mycorrhiza 6(5):441–446

Sharma AK, Singh C, Akhauri PRERNA (2000) Mass culture of arbuscular mycorrhizal fungi and their role in biotechnology. Proc Indian Natl Sci Acad B 66(4/5):223–238

Sharma S, Sharma S, Aggarwal A, Sharma V, Singh MJ, Kaushik S (2017) Mass multiplication of arbuscular mycorrhizal fungi. In: Aggarwal A, Yadav K (eds) Mycorrhizal Fungi. Astral international (P) Ltd., New Delhi, pp 155–174

Siddiqui ZA, Kataoka R (2011) Mycorrhizal inoculants: progress in inoculant production technology. In: Microbes and microbial technology. Springer, New York, NY, pp 489–506

Siddiqui ZA, Mahmood I (1996) Biological control of *Heterodera cajani* and *fusarium udum* on pigeonpea by *Glomus mosseae, Trichoderma harzianum,* and *Verticillium chlamydosporium.* Israel J Plant Sci 44(1):49–56

Siddiqui ZA, Mahmood I (1998) Effect of a plant growth promoting bacterium, an AM fungus and soil types on the morphometrics and reproduction of *Meloidogyne javanica* on tomato. Appl Soil Ecol 8(1–3):77–84

Sieverding E (1991) Vesicular-arbuscular mycorrhiza management in tropical agroecosystem. Deutshe Gesellschaft Technische Zusammenarbeit (GTZ) GmbH, Eschborn

Silva FSBD, Yano-Melo AM, Maia LC (2007) Production and infectivity of inoculum of arbuscular mycorrhizal fungi multiplied in a substrate supplemented with Tris-HCl buffer. Braz J Microbiol 38(4):752–755

Singh HB, Sarma BK, Keswani C (2016) Agriculturally important microorganisms. Springer, Singapore

Smith SE, Read DJ (2010) Mycorrhizal symbiosis. Academic press

Sreeramulu KR, Bagyaraj DJ (1986) Field response of chilli to VA mycorrhiza on black clayey soil. Plant Soil 93:299–302

St-Arnaud M, Hamel C, Vimard B, Caron M, Fortin JA (1996) Enhanced hyphal growth and spore production of the arbuscular mycorrhizal fungus *Glomus intraradices* in an *in vitro* system in the absence of host roots. Mycol Res 100(3):328–332

Strullu DG, Plenchette C (1991) The entrapment of *Glomus* sp. in alginate beads and their use as root inoculum. Mycol Res 95(10):1194–1196

Strullu DG, Romand C (1986) Méthode d'obtention d'endomycorhizes à vésicules et arbuscules en conditions axéniques. Comptes rendus de l'Académie des sciences. Série 3, Sciences de la vie 303(6):245–250

Sylvia DM, Hubbell DH (1986) Growth and sporulation of vesicular-arbuscular mycorrhizal fungi in aeroponic and membrane systems. Symbiosis 1:259–267

Tarafdar JE (1995) Role of a VA mycorrhizal fungus on growth and water relations in wheat in presence of organic and inorganic phosphates. J Indian Soc Soil Sci 43(2):200–204

Tarkka MT, Frey-Klett P (2008) Mycorrhiza helper bacteria. In: Mycorrhiza. Springer, Berlin, Heidelberg, pp 113–132

Tawaraya K (2003) Arbuscular mycorrhizal dependency of different plant species and cultivars. Soil Sci Plant Nutr 49(5):655–668

Tiwari P, Prakash A, Adholeya A (2002) Commercialization of arbuscular mycorrhizal fungi. Handbook of fungal biotechnology. Marcel Dekker, New York

Tiwari P, Adholeya A, Prakash A, Arora DK (2004) Commercialization of arbuscular mycorrhizal biofertilizers. Fungal biotechnology in agricultural, food, and environmental applications 21:195–203

Tiwari JK, Sapna DEVI, Buckseth T, Nilofer ALI, Singh RK, Zinta R, Chakrabarti SK (2020) Precision phenotyping of contrasting potato (*Solanum tuberosum* L.) varieties in a novel aeroponics system for improving nitrogen use efficiency: in search of key traits and genes. J Integr Agric 19(1):51–61

Toro M, Azcon R, Barea J (1997) Improvement of arbuscular mycorrhiza development by inoculation of soil with phosphate-solubilizing rhizobacteria to improve rock phosphate bioavailability ((sup32) P) and nutrient cycling. Appl Environ Microbiol 63(11):4408–4412

Transparency Market Research (2018) Global biofertilizers market: snapshot. Transparency Market Research, Albany, NY

Turk MA, Assaf TA, Hameed KM, Al-Tawaha AM (2006) Significance of mycorrhizae. World J Agric Sci 2(1):16–20

Tushar K, Satish B (2013) Incidences of arbuscular mycorrhizal fungi (AMF) 'in urban farming of Mumbai and Subursbs, India. Int Res J Environ Sci 2(1):12–18

Vassilev N, Vassileva M, Azcon R, Medina A (2001a) Interactions of an arbuscular mycorrhizal fungus with free or co-encapsulated cells of *Rhizobium trifoli* and *Yarowia lipolytica* inoculated into a soil-plant system. Biotechnol Lett 23(2):149–151

Varma A, Schüepp H (1995) Mycorrhization of the commercially important micropropagated plants. Crit Rev Biotechnol 15(3–4):313–328

Vassilev N, Vassileva M, Azcon R, Medina A (2001b) Preparation of gel-entrapped mycorrhizal inoculum in the presence or absence of *Yarowia lipolytica*. Biotechnol Lett 23(11):907–909

Vassilev N, Nikolaeva I, Vassileva M (2005) Polymer-based preparation of soil inoculants: applications to arbuscular mycorrhizal fungi. Rev Environ Sci Biotechnol 4(4):235–243

Viyanak K, Bagyaraj DJ (1990) Selection of efficient VA mycorrhizal fungi for trifoliate orange. Biol Agric Hortic 6(4):305–311

Vosátka M, Albrechtová J (2009) Benefits of arbuscular mycorrhizal fungi to sustainable crop production. In: Microbial strategies for crop improvement. Springer, Berlin, Heidelberg, pp 205–225

Vosatka M, Jansa J, Regvar M, Sramek F, Malcova R (1999) Inoculation with mycorrhizal fungi-a feasible biotechnology for horticulture. Phyton (Horn) 39(3):219–224

Vosátka M, Albrechtová J, Patten R (2008) The international market development for mycorrhizal technology. In: Mycorrhiza. Springer, Berlin, Heidelberg, pp 419–438

Vosátka M, Látr A, Gianinazzi S, Albrechtová J (2012) Development of arbuscular mycorrhizal biotechnology and industry: current achievements and bottlenecks. Symbiosis 58(1):29–37

Vural A, Demir S, Boyno G (2018) Bioremediation and using of fungi in bioremediation. YYU J Agric Sci 28(4):490–501

Webster G, Poulton PR, Cocking EC, Davey MR (1995) The nodulation of micro-propagated plants of *Parasponia andersonii* by tropical legume rhizobia. J Exp Bot 46(9):1131–1137

Woo SL, Scala F, Ruocco M, Lorito M (2006) The molecular biology of the interactions between *Trichoderma* spp., phytopathogenic fungi, and plants. Phytopathology 96(2):181–185

Wu CG, Liu YS, Hung LL (1995) Spore development of *Entrophospora kentinensis* in an aeroponic system. Mycologia 87(5):582–587

Xavier LJ, Germida JJ (2003) Selective interactions between arbuscular mycorrhizal fungi and *Rhizobium leguminosarum* bv. *viceae* enhance pea yield and nutrition. Biol Fertil Soils 37(5):261–267

Unraveling the Mysteries of Mycorrhiza-Plant Interactions: Mechanisms of Protection and Ecological Factors Influencing Symbioses

9

Dina S. S. Ibrahim, Shaimaa N. Riad, Kamal A. M. Abo-Elyousr, Sallam M. A. Nashwa, Hadeel M. M. Khalil Bagy, Sahar Abdelrazek, and Aya A. Abdellatif

Abstract

Mycorrhizal fungi are root symbionts that embrace many benefits to the associated plant host. Protecting plants from devastating plant pathogens and pests like fungi, bacteria, and nematodes is among the numerous significant attributes besides plant health and yield ameliorations. The protection that mycorrhiza holds for their hosts extended to cover a large number of economic crops worldwide that possess the possibility of using it as a potential bio-protector. However, in order to maximize the efficacy of mycorrhizal application, the mechanism implicated behind protection and factors that affect mycorrhizal symbioses should be well examined. Several mechanisms and factors have been proposed to

D. S. S. Ibrahim (✉) · S. N. Riad
Department of Nematodes Diseases and Central Lab of Biotechnology, Plant Pathology Institute, Agricultural Research Center (ARC), Giza, Egypt

K. A. M. Abo-Elyousr
Plant Pathology Department, Faculty of Agriculture, Assiut University, Assiut, Egypt

Department of Arid Land Agriculture, Faculty of Meteorology, Environment and Arid Land Agriculture, King Abdul-Aziz University, Jeddah, Saudi Arabia

S. M. A. Nashwa · H. M. M. Khalil Bagy
Plant Pathology Department, Faculty of Agriculture, Assiut University, Assiut, Egypt

S. Abdelrazek
Plant Pathology Department, Faculty of Agriculture, Assiut University, Assiut, Egypt

School of Plant and Environmental Science, Virginia Tech, Blacksburg, VA, USA

A. A. Abdellatif
Central Lab of Organic Agriculture, Agricultural Research Center (ARC), Giza, Egypt

R. A. Ansari et al. (eds.), *Mycorrhizal Symbiosis and Agroecosystem Restoration*, https://doi.org/10.1007/978-981-99-5030-0_9

explain this protective relationship. Most of them are either conditional or case specific. Most plants rely on several mechanisms at the same time. In this chapter, we go over the main mechanisms reported to involve in mycorrhizal host protection against fungal, bacterial, and nematode diseases. Spotting the light on the main ecological factors that affect the outcome of mycorrhiza interaction with hosts has also been taken into consideration.

Keywords

Mycorrhiza · Bio-protector · AM fungi · Plant disease resistance · Sustainable agriculture

9.1 Introduction

Mycorrhizal fungi are the oldest symbiotic relationship between fungi and plants. The symbiotic relation was discovered in 1879–1882 by Kamieński (Frank 1885). The word mycorrhiza originated from the word "mukēs", which means fungus in Greek, and the word "rhiza" means root (Deacon). There are two types of arbuscular mycorrhizal (AM) fungi: ectomycorrhizae and endomycorrhizal (EM). Ectomycorrhizae fungi produce hyphae that are externally connected and cover the root of the plant. They develop among the outside surface of root epidermal cells creating a complicated intercellular system that looks like a net of hyphae, called the Hartig net (Smith and Read 2010), whereas EM creates relationships within the cells of the host. The apoplectic space separates plant and fungal cells within the Hartig net and acts as an interchange area for plant carbon and fungus nutrients (Nehls 2008; Smith and Read 2010). AM fungi are obligate symbiosis fungi with wide host ranges and a significant role in curving ecological systems and related productivity. A large number of tracheophyte plants can create a symbiotic relationship with AM fungi (Sumbul et al. 2017; Dowarah et al. 2021). AM fungi mainly depend on their host for photosynthesis, but they also provide many other beneficial properties for their host through many processes. The most important one is to boost host tolerance to biotic stresses caused by harmful nematodes, fungi, bacteria, and many other pathogens. Various studies on the decrease of damage caused by soil-borne necrotrophic pathogens have been conducted through the connection of host with mycorrhiza, including *Fusarium oxysporum,* the pathogenic fungi of cucurbits (Jun-Li et al. 2010), *Pythium aphanidermatum* in pawpaws (Olawuyi et al. 2014), and *Fusarium oxysporum* pathogen of strawberries (Yang et al. 2015). The primary mechanism of mycorrhiza involves competing with the pathogen for food, space, and hosted plant root colonization sites, rhizosphere microbial change, and induction of host defense (Harrier and Watson 2004). Other mechanisms include improvement in nutrient uptake ability in host plants, healing pathogen infection site damage, modification in host root architecture, and durability as well as competition with soil pathogens for the colonization sites and also for the maximum exploitation of host photosynthates (Harrier and Watson 2004). Finally, the performance of a mycorrhizal association in mitigating different biological stresses is totally dependent and crafted by many other environmental, microbial, and compatibility elements. This chapter narrates numerous mechanisms contributed to AM fungi mediating plant stress tolerance, and different factors that affect the ability and effectiveness of mycorrhizal association.

9.2 Interactions Between Mycorrhizal Associations and Root Exudates

The rhizosphere is a complex environment of numerous microorganisms as well as the site of many chemical and physical reactions (Odelade and Babalola 2019; Oliver et al. 2021). Root exudation plays an important role in the organization of plant interactions with the soil environment (Canarini et al. 2019a, b). Polysaccharides, which form a layer of mucilage around the root, are among the exudate components of plants, enzymes including acid phosphatases; volatile compounds like ethylene; and low molecular weight metabolites including sugars, amino acids, organic acids, and phenolic acids (Bais et al. 2006).

The AM fungi pre-symbiosis process is stimulated by the secretion of specific plant root exudates that are required for fungal development and root colonization. This relationship between mycorrhizal fungi and plants provides many benefits to the plants such as increased plant development and tolerance. Thus, root exudates can be illustrated as messenger molecules that initiate and coordinate biocommunication between the host's roots and various soil organisms (Vigo et al. 2000; Tahat et al. 2012). Root exudation and sensing of specific secondary metabolites have to be managed to develop a successful mycorrhizal association (Bais et al. 2006). Once the symbiotic relationship is recognized, plants begin to supply plenty of carbon (recently photo-assimilated sugars in addition to fatty acids) to fungi. The fungus in turn provides nutrients, especially phosphorus, to the plant (van der Heijden et al. 2015). It is unknown whether this significant diversion of belowground carbon flux causes a qualitative or quantitative change in the ratio of the plant root exudates (Canarini et al. 2019a, b). During the colonization phase, the communication between the roots of the plant and the mycorrhizal fungi occurred via the exchange of chemical signals (Gutjahr and Parniske 2013; Smith et al. 2011). Many researchers have exhibited the defensive impact of mycorrhizal symbioses for various root diseases (Harrier and Watson 2004; Goicoechea 2020; Rodriguez-Heredia et al. 2020).

9.2.1 Modulation in Root Architecture

Majority of AM fungi are well known for their ability to facilitate the uptake of water, and nutrients by the plant, causing modification in root architecture and affecting the rhizosphere interactions (Poveda 2020; Poveda et al. 2020). Plants acquire nutrients through the joint activity of roots and associated microbes, which show a major role in nutrient solubilization and absorption as well as the prevention from soil-borne diseases (Ismail et al. 2013). In symbiotic relationship, plants provide organic nutrients to the fungal symbiont in exchange for inorganic nutrients absorbed by hyphae extending from roots into the soil. The AM fungi could augment the surface area of plant roots more than 100 times compared to plants without mycorrhizal association (Smith and Read 2010; Yang et al. 2015). Plant's responses to AM fungi colonization are influenced by the root structure (Smith and Read 2010; Smith and Smith 2011). Plants species characterized by coarse root architecture have large-diameter roots and little root hair (Fitter 2004; Smith and Read 2010). The ability of such plants to absorb nutrients is limited (Bates and Lynch 2001); thus the presence of AM fungal hyphae maximizes the surface area accessible for nutrient absorption, particularly phosphorus in these plants (Raven and Edwards 2001). Newsham et al. (1995) stated that AM fungi function is affected by the morphology of the root. Plants with simple root systems rely on AM fungi for the uptake of nutrients, while plants having roots with more complex structures are less reliant on mycorrhizal fungi for their uptake of nutrients and are much more susceptible to causal agents of root diseases due to the increased number of infection sites. The architecture of the plant root is changed by the mycorrhizal symbiosis that aids in the control of various phytodiseases and supporting plants to absorb valuable mineral elements from the soil. It produces a mechanical barrier against the penetration of phytopathogen and subsequent spread, as well as thickening the cell wall by lignification and the formation of additional polysaccharides, which prevents root pathogen entry (Singh et al. 2000). Additionally, such changes in root morphology, for instance, increasing the total root length, enhance the plant's capacity to absorb and utilize nutrients in the soil (Xia et al. 2020). Moreover, suppressive effect of AM fungi on root pathogens may result from the direct competition for space and different resources (Bødker et al. 2002).

9.2.2 Nutrient Uptake and Accumulation

To combat various plant infections, agricultural techniques including bio-fertilizers, resistant cultivars, plant quarantine, fumigation of the soil, and crop rotation are the techniques that can be used to improve soil quality. Alternative measures such as the addition of microorganisms like AM fungi must be used to strengthen plant disease resistance (Grosch et al. 2005). There is an increased interest in biological fertilizers that comprises of a large number of microorganisms. These types of fertilizers are environmentally friendly, inexpensive, and widely available throughout the country. The importance of using bio-fertilizers in agriculture is reducing mineral fertilizers, thereby lowering production costs and increasing yield in terms of quality and quantity. Moreover, it improves soil fertility, reduces nitrogen and phosphate fertilizer application rates by at least 25%, and reduces pollution problems (Sadhana 2014).

AM fungi are found in 90% of plant species as part of a symbiotic connection between higher plants' roots and non-pathogenic soil fungi (Zhu et al. 2010; Ahanger et al. 2014a, b). The AM fungi is entwined with the host's roots. In this relationship, the fungus and the plant exchange several molecules and components that are used in both partners' growth and reproduction (Ahanger et al. 2014a, b). The fine soil granules are gathered by the action of the fungal hyphae and the organic compounds in the roots, and a complex structure is created that serves to hold them. AM fungi form many dense hyphae with the host plant's roots, allowing the roots to reach a vast area of the soil surface and improve plant growth. They also produce vesicles, arbuscular mycorrhizae, and hyphae in the root system (Bowles et al. 2016). AM fungi collects soil nutrients through their roots and then transports them through the mycorrhizal hyphae into the hosts (Simard et al. 2012). They secrete various compounds that improve the soil's composition, like polysaccharide compounds, which help to glue the soil particles together, increasing the soil's ability to retain water. Furthermore, they improve plant nutrition by improving various nutrients, soil quality, and plant health. (Thirkell et al. 2017). The mycorrhizal mutuality relationship gives the fungus consistent access to carbohydrates like sucrose and glucose (Harrison 2005). Carbohydrates are transported to the root and then to the plant. AM fungi have a large surface area, which helps plants absorb more water and minerals. Because the hairs of AMFs are finer and longer than those of plant roots, some of these AM fungi can store soil minerals for the host's roots. Plant mineral absorption is improved (Selosse et al. 2006). AM fungi may create a competitive atmosphere for this carbon with pathogens (Vos et al. 2014). They have the potential to grow nutrient absorption in hosts, particularly phosphorous absorption (Nell et al. 2010), and increase the activity of phosphate-dissolving bacteria. Several mechanisms are used by the fungus to absorb phosphorous, including the secretion of the enzyme phosphatase by the fungus hyphae, which dissolves organic phosphorous and transforms it into forms that can be absorbed by the plant. Another mechanism is the secretion of hydroxy acids. In phosphorus limited environment, AM fungi colonized maize plants and significantly improved phosphorus delivery to the plant roots (Garcés-Ruiz et al. 2017a, b). When nutrients are restricted to organic matter, AM fungi mobilizes the nutrients and delivers them to the plants, which is a different sort of fixation. (Hogan 2011). For example, in some dystrophic forests, AM fungi hyphae working directly on leaf litter, absorb enormous amounts of phosphate and other nutrients, obviating the need for soil uptake (Hogan 2011).

Plants and AM fungi share a relationship that is more complicated than just mutuality. It was found that AM fungi can extract a high amount of nitrogen from dead and decomposing materials and may grow and survive for longer periods, producing a large amount of biomass (Hodge and Fitter 2010). The AM fungi can transfer roughly 20–75% of the entire nitrogen intake to its hosts (Hashem et al. 2018a, b). Under ambient and increased CO_2 concentrations, AM fungi inoculation enhances carbon and nitrogen buildup and nitrogen assimilation (Zhu et al. 2016). AM fungi work to absorb nitrogen, phosphorous, and carbon, which leads to increased photosynthesis and improved leaf growth, and also work to enlarge the size of the tubers. Mycorrhizal symbiosis enhances N, P, and Fe levels in sweet-scented geraniums during droughts (Amiri et al. 2017a, b). It increases the absorption of critical nutrients while decreasing the absorption of Cl and Na, causing the plants to grow faster (Evelin et al. 2012). Numerous studies have been conducted on AM fungi, all of which have found numerous benefits for soil health and agricultural productivity. Because it reduces the usage of chemical fertilizers, particularly phosphorous fertilizers, AM fungi are now used as an alternative to inorganic fertilizers (Ortas 2012).

9.3 AM Fungi and Other Beneficial Microbe Interactions

The zone of soil affected by mycorrhizal connections is known as the mycorrhizal sphere (Fulekar and Pathak 2015). The mycorrhizosphere is formed when a change in the root exudates alters the microbial populations surrounding plant roots (Zhang et al. 2010a, b; Ansari and Mahmood 2017a; Ansari et al. 2017a, b, 2020b). The emergence of AM fungi in the mycorrhizosphere has a significant impact on the microbiota rhizosphere. The microbiota of rhizosphere mycorrhizal plants differ from that of non-rhizosphere mycorrhizal plants in both qualitative and quantitative aspects. The mycorrhizosphere is made up of a layer of soil overlying the mycorrhiza roots and the hyphosphere that surrounds the mycorrhizal hyphae in the soil (Solanki et al. 2020; Rizvi et al. 2015). Soil microorganisms aid in improving soil fertility and plant health. The association root of the host plant with AM fungi causes variations in root respiration and secretion volume (Marschner and Baumann 2003). In comparison with a single inoculation with the synergistic combination of AM fungi and *Bacillus* spp., dual inoculation promotes plant development because dual inoculation improves the absorption of nutrients, protects pathogens that attack plants, and alleviates abiotic stresses (heavy metals, salinity, and water) when compared to a single inoculation with AM fungi or *Bacillus* (Nanjundappa et al. 2019). AM fungi play a significant role in the development of crops and the control of many diseases (Ansari et al. 2019a, b). The effects of AM fungi and *Epicoccum nigrum* endophytic fungi, either independently or in combination, result in a reduction in potato blackleg disease (*Pectobacterium carotovora* subsp. *atrosepticum*). The highest decrease in the incidence of the disease was recorded by studying the treatment of AM fungi with *Epicoccum nigrum* (Bagy et al. 2019). AM fungi improves the host plant's resistance to damage by compensating for pathogen-induced losses of biomass and functional roots. This demonstrates that AM fungi play a role in biological control by maintaining the function of plant roots through the growth of AM fungi filaments in the soil, causing the root's absorption to rise, and functionally improving the level of root uptake. Following AM fungi colonization, several publications have found an increase in bacteria: *Pseudomonas, Streptomyces***,** and actinomycetes (Nuccio et al. 2013).

9.4 Interplay Between AM Fungi and Host Plants

9.4.1 Direct Competition

The AM fungi have a big role in the competition and compete with plant pathogens where they live (Ortas 2012). They are a type of alternative management strategy for protecting plants from infection (Brimner and Boland 2003). AM fungi colonize host plants by living in a symbiotic relationship with their roots (Harrier and Watson 2004). It has been demonstrated that AM fungi symbiosis reduces the damage caused by plant pathogens (Azcón-Aguilar et al. 2002).

9.4.2 Competition for Infection Sites

One of the mycorrhiza's most important functions is to protect plants from root pathogens. Complex root systems make plants more vulnerable to pathogen attack; however, if the plant is colonized by a mycorrhizal fungus that is resistant to pathogens, such as *Glomus* isolates, the AMF symbiosis can reduce infection in such plants (Sikes 2010). AM fungi can inhibit pathogens in several ways, including direct competition for colonization sites and indirect induction of plant defense responses, and changes in the rhizosphere biota (Azcón-Aguilar and Barea 1997). It was found that tomato plants inoculated with AM fungi are not infected with *Phytophthora*, and studies have revealed that it is a competition for carbon compounds. It is well known that it is reliant on the host plant and the photosynthesis process, which ultimately leads to plant soil inhibition (Cordier et al. 1998).

9.4.3 Competition for Host Photosynthates

AM fungi and soil pathogens present in the roots depend on the photosynthesis process that takes place in the plant, which leads to better plant growth. During photosynthesis, competition occurs for carbon molecules in the form of fats and sugars (Smith and Read 2010; Jiang et al. 2017; Luginbuehl et al. 2017). It was found that during the photosynthesis mechanism of the host plant, whose roots contain AM fungi, they can obtain carbon in the range of 4–20%, which leads to resistance to harmful microorganisms (Smith and Read 2010). AM fungi colonization greatly stimulates the macro- and micronutrient uptake in plants during the photosynthesis mechanism, as a result, photosynthetic production increases and thus an increase in the accumulation of plant biomass (Chen et al. 2017; Mitra et al. 2019). The inorganic nutrients can be better absorbed with the aid of AM fungi, particularly phosphates (Nell et al. 2010). AM fungi encourages plants to absorb zinc and copper from nutrient-deficient soils (Al-Hmoud and Al-Momany 2017). It was found that during the scarcity of phosphorus in the soil, AM fungi improves the supply of phosphorous to the affected roots of plants (Bucher 2007). The rate of phosphorous uptake was significantly improved in maize plants colonized with AM fungi (Garcés-Ruiz et al. 2017a, b). Photosynthetic activity and other leaf functions have been found to be risen, and this is linked to an increase in potato tuber growth and AMF-rooted potato by increasing the uptake of nitrogen, phosphorous, and carbon, which leads to enhanced tuber growth. AM fungi also maintains the uptake of phosphorous and nitrogen from the soil during photosynthesis especially in drought condition (Liu et al. 2018). For example, Gomez-Bellot et al. (2015) and Amiri et al. (2017a, b) showed that AM fungi increased the nitrogen, phosphorous, and iron concentrations in aromatic plants like *Pelargonium graveolens* under drought conditions. The improvement of phosphorous, calcium, and potassium levels during photosynthesis was demonstrated in *Euonymus japonica* plants under the influence of salinity. In another study, pistachio plants inoculated with AM fungi showed high levels of phosphorous, potassium, zinc, and manganese under drought conditions (Bagheri et al. 2012). AM fungi are believed to improve the absorption of nearly all essential nutrients and reversibly reduce sodium and chlorine absorption, thereby stimulating plant growth during photosynthesis (Evelin et al. 2012). Highly rooted fungus (ERM) can boost plant development through enhancing nutrient uptake (Lehmann and Rillig 2015). Soil nutrition is mostly dependent on nitrogen, and several investigations have demonstrated that AM fungi have a significant role in absorbing and transferring nitrogen from soil to the host (Hodge and Storer 2015; Battini et al. 2017; Turrini et al. 2018a, b). Several studies have found that between 20 and 75 percent of total N absorption can be transferred from AM fungi to the host plant especially during photosynthesis (Ahanger et al. 2014a, b; Hameed et al. 2014; Hashem et al. 2018a). Increasing the rate of nitrogen uptake from the soil containing AM fungi leads to higher chlorophyll content in the plant, because chlorophyll molecules have the ability to effectively store nitrogen (De Andrade et al. 2015).

9.4.4 Antagonism

It has yet to be proven that AM fungi produce some antimicrobial compounds that can inhibit the microorganisms reproduction or populations. The AM fungi aid in the absorption of mineral nutrients by the plant, whereas the plant provides carbon to the AM fungi. This is considered a symbiotic relationship between the AM fungi and the host plant. AM fungi was discovered to have a biocontrol effect against a variety of plant infections conveyed through the soil; hence it was used as part of a biological control strategy. AM fungi has been observed to have anti microbial compounds against some fungal diseases (Whipps 2004). For example, AM fungi is used to combat fungal diseases in various crops, such as *Rhizoctonia solani* (Yao et al. 2002), *Phytophthora* species, and *Pythium ultimum* (Cordier et al. 1996). Several methods have been used to control soil-borne pathogens using AM fungi to reduce infection with pathogens and increase crop resistance (Azcón-Aguilar and Barea 1997). In 2002, Azcon-Aguilar et al. demonstrated that the symbiotic interaction between AM fungi and the plant reduces the harm caused by soil-borne diseases. AM fungi have been combined with beneficial microbes in several experiments to get enhanced plant disease resistance (Grosch et al. 2005). AM fungi reduces plant diseases by boosting host plant resistance and creating food and space competition between. Thus, AM fungi are used in the biological control of plant diseases with these microbes (Berg et al. 2007). Cordier et al. (1996) found that applying *Glomus mosseae* in tomato roots reduced infection with *Phytophthora parasitica* compared to tomato roots that had not been treated with AM fungi. AM fungi help plants secrete phosphate, which reduces the incidence of fungal diseases in tomatoes (Trotta et al. 1996). In another example, AM fungi reduced *Ganoderma boninense* infection in oil palm seedlings, and the seedlings were more resistant to *G. boninense* infection (Rini 2001). Peanut root and pod rot diseases were greatly reduced when *Glomus* spp. were applied to soil and the number of pods and fresh weight were significantly higher than in non-treated soil (Ahmed et al. 2013).

Iron is a necessary component in practically all living organisms. Iron will be unavailable in adequate quantities unless organisms discover a mechanism of solubilization. The major siderophores released by the ericoid mycorrhizal fungi are ferricrocin and fusigen (Haselwandter and Winkelmann 2007). The ectomycorrhizal fungi, *Cenococcum geophilum* and *Hebeloma crustuliniforme* have also been shown to produce ferricrocin. When AM fungi are present, Fe absorption rates increase in plants that are connected to them, implying that undiscovered mycorrhizal siderophores are involved. Orchid mycorrhizal fungi have been found to produce both well-known ferrichrome-type siderophores and the unique linear trishydroxamate basidiochrome as the major siderophores. Mycorrhizal fungi, which are found in both natural and cultivated soils, have shown improved plant uptake of nutrients including trace metal ions.

9.5 AM Fungi-Mediated Plant Disease Resistance

Induced resistance (IR) can be considered an alternative eco-friendly promising approach for controlling plant diseases as it can provide a natural defense mechanism for plants (Riad et al. 2021). Recently, IR of the plant by the action of AM fungi has become an attractive topic in chemo-ecological study and plant disease biocontrol (Huang et al. 2003; Powell and Rillig 2018). Many studies have shown AM fungi colonization protects plants from microbial pathogen infections in different plant systems (Hao et al. 2019; Miozzi et al. 2019). IR does have not a direct activation of defense mechanisms; also it comes from tissue sensitization in response to suitable stimulation, allowing basal defense mechanisms to express a lot efficiently after a future attack of the pathogen. This activation of a plant's innate immune system occurs frequently when it interacts with useful microorganisms, and it has significant benefits compared with the direct activation of defenses (Van Wees et al. 2008; Chen et al. 2021). Firstly, the fungus faces the plant's immune system, and it has to overcome the defense mechanisms for effective host colonization (Kloppholz et al. 2011; Zamioudis and Pieterse 2012). Upon its establishment, the level of fungi proliferation in the roots is regulated by the plant to control carbon and over-colonization discharge and therefore keep interaction at mutual levels. For example, the plant inhibits the activity of fungi from growing and proliferating in the roots under high conditions of exogenous phosphate supply (Breuillin et al. 2010; Kumaresan et al. 2020). The plant interacts with the existence of AM fungi in earlier stages of the interaction by activating some defense-related responses (García-Garrido and Ocampo 2002). Rapid but transient increment in the level of endogenous salicylic acid (SA) was observed remarkably in the roots with simultaneous accumulation of defense compounds, such as activation of the phenylpropanoid pathway, specific isoforms of hydrolytic enzymes, and reactive oxygen species (de Roman et al. 2011).

When a plant interacts with AM fungus, other plant hormone levels related to defense such as ABA, ET, and JA are significantly influenced (Hause et al. 2007; Ludwig-Müller 2010; López-Ráez et al. 2010). With the correct functioning of AM symbiosis as colonization improves, the regulation of JA levels acquires a major role (Hause et al. 2007; Hause and Schaarschmidt 2009). Generally, IR appears obviously when AM symbionts are established successfully. The modifications in signaling related basically to robust-established mycorrhiza are probable mediating MIR (Slezack et al. 2000; Pozo et al. 2002). The reliance of the successful mycorrhizal fungi on the control of SA and JA signals explained the protection range offered by this symbiosis (Pozo and Azcón-Aguilar 2007). AM fungi plants showed significant resistance to nematodes, and target aggressors through JA relied on defensive responses; they are more sensitive to biotrophs, targeted by SA-regulated defense systems. Such type is associated with the activation of JA relying on defenses and suppression of salicylic acid relying on one of the well-established AM fungi. The synthesis of defensive proteins and toxins was the outcome of JA signaling, resulting in an internal imbalance that reduced the insect's development and survival (Giang et al. 2020; Howe and Jander 2008).

9.5.1 AM Fungi in Nematodal Disease Management

Altered root exudation can lead to changes in the rhizosphere's microbial diversity, which can harm plant-pathogen interactions (Schouteden et al. 2015). Some reports showed that AM fungi colonization caused an increase in the population of facultative anaerobic bacteria, fluorescent pseudomonads, actinobacteria, and chitinolytic microorganisms (Nuccio et al. 2013; Philippot et al. 2013). These microorganisms have antagonistic activity against phytonematodes, either directly by nematode trapping or egg parasitism or indirectly by triggering the plant defense mechanism (Zamioudis and Pieterse 2012). Rodriguez-Heredia et al. (2020) indicate that AMF inoculated with roots of pepper and tomato plants caused a marked reduction in the multiplication of *Meloidogyne incognita*. Different species of AM fungi such as *Glomus mosseae* and *Gigaspora gigantea* maximize the production of antioxidant compounds and increase the plant fiber content. The use of AM fungi with solubilizing-phosphate bacteria like *Pseudomonas fluorescens* has a significant improvement in the morphological and biochemical characters of eggplant in infested plants with *M. javanica* (Sharma et al. 2021).

Gough et al. (2020) reported that the degree of AM fungi colonization affects the population densities of *Pratylenchus* spp. Inoculation with low levels of AM fungi was linked to an increase in the population of *Pratylenchus* spp. in the soil, while high levels of AM fungi colonization suppressed *Pratylenchus* population. The penetration of tomato root by *Nacobbus aberrans* was inhibited successfully by the application of different mycorrhizal species like *Rhizophagus intraradices* and *Funneliformis mosseae* (Marro et al. 2018). In another work, Calvet et al. (2001) found that application of *Glomus etunicatum*, *G. intraradices*, and *G. mosseae* significantly reduced *Meloidogyne javanica* development infecting peach roots. Brito et al. (2018) showed heavy colonization of the root with *Rhizophagus clarus, Claroideoglomus etunicatum*, *Gigaspora margarita, G. rosea*, *Scutellospora heterogama*, and *S. calospora*, positively correlated with *Pratylenchus brachyurus* population infecting maize roots, which is the inverse of the mycorrhizal colonization effect in cotton (Ferreira et al. 2018).

Plant roots are targeted by both nematodes and AM fungi as a source of space and food (Ansari and Mahmood 2017b, 2019a, b). The numbers of *Meloidogyne* spp. were affected negatively by mycorrhizal fungi colonization more than cyst nematodes (Ansari et al. 2020a). Moreover, root-knot nematodes belonging to the genus *Meloidogyne* spp. are sedentary endoparasites that are responsible for diseases and economic losses in important crops. The interactions between *Meloidogyne* and AM fungi are complex and relied on plants, nematode species, fungi, and soil conditions. AM fungi is considerably affected by sedentary nematodes (*Meloidogyne* spp.) than any other migratory nematodes (Table 9.1; Francl 1993).

9.5.2 AM Fungi in Fungal Disease Management

The cell wall of AM fungi is composed mainly of chitin oligosaccharides that can promote the plant immune system via, lysine motif receptor-like kinases (LysM-RLK) and so play a dual role in pathogenic and symbiotic interactions (Kaku et al. 2006; Miya et al. 2007; Shimizu et al. 2010). Application of *Glomus* spp. was reported as a biotic agent for the management of *F. oxysporum f.* sp. *sesami* (Zap.), which causes wilt and root-rot diseases in sesame. This treatment stimulated bacterial colonization on sesame rhizospheres in a selective manner and showed a positive impact on the morphological characters of both root and shoot systems (Ziedan et al. 2011). Furthermore, the simultaneous application of AM fungi under field conditions triggered the expression of genes linked to pathogenesis and synthesis of lignin in the plant, which have significant effects on the protection of cotton against fungal diseases (Zhang et al. 2018). Tahat and Al Momany (2019) demonstrated the efficiency of *G. fasciculatum* and *G. mosseae* in controlling *Verticillium* wilt on cucumber under greenhouse conditions due to improvements in morphological root characteristics. Moreover, AM fungi and olive cake are able to prevent plant diseases by encouraging plant development and enhancing root and soil properties, according to Tahat et al. (2020). As a result, they saw *G. mosseae* in combination

Table 9.1 Examples of the use of AM fungi in the suppression of plant diseases

S. no.	AM fungi	Mechanisms of action	Pathogens/diseases	Host	References
1.	*Glomus mosseae and Gigaspora gigantea*	Root exudates	*Meloidogyne incognita* (root-knot of tomato disease)	Pepper and tomato	Sharma et al. (2021)
2.	*Glomus mosseae, G. intraradices, G. clarum, Gigaspora gigantean, and G. margarita*	Root exudates	*Fusarium solani* (fusarium root-rot)	Barley	Gernns et al. (2001)
3.	*Glomus intraradices*	Root exudates	*Pythium aphanidermatum* (root rot of pawpaw)	pawpaw	Olawuyi et al. (2013)
4.	*Glomus deserticola*	Root exudates	*Verticillium dahlia* (wilt disease of Solanaceae plants)	Tomato, pepper, and eggplant	Demir et al. (2015)
5.	*Glomus intraradices*	Root exudates	*Phytopthora nicotianae*	Tomato	Lioussanne et al. (2008)
6.	*Gigaspora margarita*	Root exudates	*Ralstonia solanacearum* (bacterial wilt of tomato)	Tomato	Tahat et al. (2012)
7.	*G. mosseae*	Root exudates	*Ralstonia solanacearum* (brown rot disease)	Potato	Tahat et al. (2010)
8.	*G. margarita*	Altering root architecture	*Pratylenchus* spp. root-lesion nematode	Maize	Gough et al. (2020)
9.	*Rhizoglomus intraradices andFunneliformis mosseae*	Altering root architecture	*Nacobbus aberrans* false root knot	Tomato	Marro et al. (2018)
10.	*Glomus intraradices, G. mosseae*, and *G. etunicatum*	Altering root architecture	*Meloidogyne javanica* (root-knot disease)	Peach	Calvet et al. (2001)
11.	*G. margarita*	Altering root architecture	*Pratylenchus brachyurus* root-lesion nematode	Maize	Brito et al. (2018)
12.	*Glomus* spp.	Altering root architecture	*Fusarium oxysporum* wilt disease of sesame	Sesame	Ziedan et al. (2011)
13.	*Glomus mosseae and Glomus fasciculatum*	Altering root architecture	*Verticillium* wilt disease of cucumber	cucumber	Tahat and Al Momany (2019)

(continued)

Table 9.1 (continued)

S. no.	AM fungi	Mechanisms of action	Pathogens/diseases	Host	References
14.	*G. mosseae*	Altering root architecture	*Rhizoctonia solani root rot of chili*	Chili pepper	Tahat et al. (2020)
15.	*Glomus versiforme* and *G. mosseae*	Altering root architecture	*R. solanacearum bacterial wilt*	Pepper	Agoncillo (2018)
16.	*Glomus macrocarpum*	Induced resistance	*Meloidogyne* spp. *root knot*	Soybean	Francl (1993)
17.	*Glomus clarum*	Induced resistance	*Aphelenchus, Aphelenchoides, Bursaphelenchus,* and *Ditylenchus root-rot diseases*	Cotton	Francl (1993)
18.	*Glomus* sp.	Induced resistance	*Heterodera* and *Globodera* spp. *Cyst nematode*	Potato	Giannakis and Sanders (1990) and Francl (1993)
19.	G. *margarita*	Induced resistance	*M. incognita root-knot disease*	*Tomato*	Francl (1993)
20.	*Funneliformis mosseae* and *Rhizophagus irregularis*	Induced resistance	*F. oxysporum Fusarium wilt of tomato*	Tomato	Jaiti et al. (2008) and Ahmed et al. (2013)
21.	*Glomus mosseae, Glomus etunicatum, Glomus fasciculatum,* and *Gigaspora margarita*	Induced resistance	*Phytophthora* root rot of pepper	Pepper	Ozgonen and Erkilic (2007)
22.	*Glomus etunicatum* and *Glomus intraradices*	Induced resistance	*Rhizoctonia* spp. black stem	Potato	Yao et al. (2002)
23.	*Glomus*	Induced resistance	*Clavibacter michiganensis* bacterial canker of tomato	tomato	Jung et al. (2012) and Abo-elyousr et al. (2014)
24.	*Glomus*	Competition	*Phytophthora* root rot of tomato	Tomato	Cordier et al. (1998)

with olive cake as a viable biological strategy for combating chili pepper damping off caused by *Rhizoctonia solani*.

The beneficial impact of colonized roots AM fungi may be related to the remarkable alternation in primary and secondary metabolism of the plant, resulting in the reduction of pathogenic filamentous fungus development (Mayer et al. 2019). Al-Askar and Rashad (2010) reported that the application of a mixture of *Glomus clarum, G. intraradices, G. mosseae, Gigaspora gigantean* and *G. margarita* was a successful technique to control fusarium root-rot disease as there was a significant reduction in disease severity and incidence in infected plants. Olawuyi et al. (2013)

indicated that *Glomus mosseae* and *Glomus deserticola* significantly reduced foot rot disease of pawpaw seedlings caused by *Pythium aphanidermatum* and promoted the seedling growth parameters. The results of Lioussanne et al. (2008) recorded that root colonization with the early stage of AM fungi decreases the proliferation of *Phytophthora nicotianae*. Pre-inoculation AM fungi combined with humic acids reduced wilt disease severity caused by *Verticillium dahlia* Kleb. and microsclerotia population in tomato, pepper, and eggplant (Demir et al. 2015).

The colonized carrot's roots with AMF showed a stronger defensive reaction at the sites of infection caused by *Fusarium oxysporum* f. sp. *chrysanthemi* (Benhamou et al. 1994). Also, in tomato plants, colonization of AM fungi conserves roots systemically from *Phytophthora* infection (Pozo et al. 2002). The formation of phenol contents in the date palm trees infected with AM fungi was also linked to protective biochemicals against *F. oxysporum* (Jaiti et al. 2007). Several pieces of research on the protection impact of AM fungi against soil-borne diseases recorded great inhibition in the incidence of root rot and wilting diseases caused by fungal pathogens like *Fusarium, Rhizoctonia,* or *Verticillium*, and root rot caused by oomycetes including *Aphanomyces, Phytophthora,* and *Pythium* (Table 9.1; Whipps 2004).

9.5.3 AM Fungi in Bacterial Diseases

The pathogenicity of *Ralstonia solanacearum* was reduced effectively as a result of *G. mosseae* spore germination (Tahat et al. 2010). Tahat and Sijam (2012) tested the effects of three species of AM fungi (*Gigaspora margarita, Glomus mosseae,* and *Scutellospora* sp.) against *Ralstonia solanacearum*. The results indicated that no disease symptoms were detected when *G. mosseae* and *R. solanacearum* were applied together. Application of AM fungi in combination with some rhizobacterial species such as *Pseudomonas* spp., *Bacillus* spp., and *Azotobacter* spp. reduced the bacterial wilt caused by *R. solanacearum* in potatoes and improved plant growth as compared to the control (Aguk et al. 2018).

AM fungi have antagonistic activity against *Ralstonia solanacearum* in pepper (Aseel et al. 2019; Zhu and Yao 2004). AM fungi induces phenol production locally or systemically, which has a role in the inhibition of *R. solanacearum* (Zhu and Yao 2004). Colonization of root tissue by both *G. versiforme* and *R. solanacearum* enhanced both soluble phenols and cell-wall-bound phenol contents. *G. versiforme* preferred to promote the content of soluble phenol, and *R. solanacearum* preferred to increase the cell-wall-bound phenol content. The results observed by Tahat and Sijam (2012) confirmed that the colonization of *G. mosseae* can change the anatomical and morphological characters of the root. For example, *G. mosseae* could maximize root volume, length, size, and weight significantly. The reduction of *R. solanacearum* infection by the effect of endophytic fungi *G. mosseae* may be due to the modification of root structure that helps in the prevention of bacterial invasion. The mycorrhizal colonization increased plant resistance against pathogenic bacteria such as *Clavibacter michiganensis* subsp. *michiganensis* (Cmm) that caused wilt and canker disease in tomatoes (Jung et al. 2012). Plant defense responses are reliant mainly on its hormones to coordinate a complex defense system to resistance pathogens successfully. Ethylene (ET) is a key player in the control of plant immunity (Adie et al. 2007) and is important in the interactions between microbes and plants (Khatabi and Schäfer 2012). ET played a central role in the development of canker and wilting symptoms (Balaji et al. 2008). It was also observed in another set of experiments that different AM fungi isolates can induce systemic resistance to Cmm in tomato plants (Nguyen Hong and Posta 2018). Bagy et al. (2019) showed that mycorrhiza and *Epicoccum* fungi-inoculated plants increased potato growth and reduced black leg disease in potatoes. It was found that the high rate of growth of potato plants as well as the severity of disease was reduced in potatoes by using a combination of (Mycorrhiza + *Epicoccum nigrum*). Plant resistance is increased by AM fungi and *Epicoccum* by lowering active peroxide, phenyl alanine ammonia lyase, and lignin levels (Table 9.1). In potato-enhanced phenolics, super oxidase deaminase, catalase, ascorbate peroxidase, and glutathione peroxidase were recorded as antioxidant enzymes in mycorrhiza-supplemented plants.

9.6 Factors Affecting Successful Mycorrhization

9.6.1 Environmental Factors

Environmental factors play an undeniable role in plant-mycorrhiza interaction (Cotton 2018; Fitter and Garbaye 1994). Hence, any changes in the environmental factors can introduce shifts in the interaction between mycorrhiza and their associated plants, affecting mycorrhiza's efficacy as a bio-protector (Linderman 1991). The exact mechanism by which the environment contributes to mycorrhiza performance is not fully understood yet. However, in a majority of cases, the environment possesses a direct or/and indirect mechanism to regulate this process (Hoeksema et al. 2010; Jamiołkowska et al. 2018). Factors including soil pH, drought conditions, and climate changes are well known to have a direct effect on mycorrhizal growth, penetration, and community composition (Bonfante and Perotto 1995). At the same time, the same factors might indirectly stimulate metabolic and physiological changes in host plants and therefore threaten mycorrhiza efficacy (Jamiołkowska et al. 2018; Shi et al. 2014). One of the recent changes in environmental factors that strike mycorrhiza performance is the elevation of CO_2 concentration due to climate changes (Cheng et al. 2012). This increase in CO_2 concentration proved to directly influence mycorrhizal symbioses through a major shift in microbial communities. For instance, changing the abundance of Glomeraceae and Gigasporaceae families of AM fungi caused by the increase in CO_2 badly impacted nutrient uptake and increased sensitivity to pathogens in associated mycorrhiza plant hosts (Cotton 2018). Since both families play an important role in carbon allocation into the host roots, mineral uptake, and defense response against pathogen attacks, soil chemical and physical properties are another driver for AM fungi composition (Cotton 2018; Duponnois et al. 2001). It was shown before that high levels of soil phosphorus can lead to less AM fungi colonization and diversity in both plant and soil habitats

(Haynes and Swift 1985). When soil acidity fell below 4.5, there was a significant drop in mycorrhizal spores and colonization (Guo et al. 1996). In view of the fact, mycorrhiza prefers the soil's pH between 4.5 and 5.7 (Bücking and Kafle 2015). Soil agrochemical treatment applications were also reported to add selection pressure on native AM fungi that led to the increase of some AM species/groups over others and caused a change in the AM fungi community assembly and function over time. Soil management practices are also a big promoter of mycorrhiza's success (Rillig 2004). Management practices such as crop rotation, low soluble fertilizer, and reduced biocides proved to build up mycorrhiza propagules in soil and increase its diversity and colonization (Gosling et al. 2006), while other practices such as the excessive use of copper-based products, soil tillage, and growing non-mycorrhizal hosts eventually declined AM fungi colonization and abundance in soil and associated plants and turn down its beneficial effects (Gosling et al. 2006). Geographical and climatic factors such as temperature, drought, and light conditions have been reported to have an undeniable effect on mycorrhiza interactions (Khalil et al. 1994; Kilpeläinen et al. 2020). Most of the mycorrhiza species have optimum temperatures for infection and colonization (Kilpeläinen et al. 2020). These optimum temperatures are crafted by the origin of where mycorrhiza evolved and the adaptation process that they take (Heinemeyer and Fitter 2004). Shifting of mycorrhizal fungi from their climate zones results in a diminishing in hyphae infectiveness and growth. However, some mycorrhizal species succeeded in adapting to new conditions if not extreme (Al-Karaki et al. 2004). While all of the above factors showed direct influence the mycorrhizal interactions and most of them were theorized to have an indirect impact on successful mycorrhization, taking light as a potential instance can directly affect mycorrhizal symbiosis in the production and colorization of AM fungi spores (Al-Karaki et al. 2004). Nevertheless, it can indirectly affect the same relation by promoting photosynthesis and transporting assimilates to the roots (Jamiołkowska et al. 2018; Konvalinková and Jansa 2016). It was also reported that changing the geographical zone for growing plants can directly affect mycorrhiza colonization and growth due to the associated change in climate among different geographical zones (Pile et al. 2017). However, moving Triadica trees from China to the United States indirectly also enhances the colonization levels of mycorrhiza fungi (Yang et al. 2013, 2014). The relocation to the United States was associated with a higher production level of flavonoids in the host, which promotes higher mycorrhizal establishment (Pile et al. 2017; Shah and Smith 2020; Tian et al. 2021). The environmental factors are trusted to hold a significant role in mycorrhiza-plant interaction and its predicted beneficial outcomes.

9.6.2 Cultural Factors

Since the same mycorrhizal species growth varies from one plant species to another, plant species differ in their dependency on mycorrhiza (Hoeksema 2005). The magnitude of mycorrhiza's beneficial effect on their host under the same environment is theorized to be regulated by differences among plant and mycorrhizal species

(Grman et al. 2012; Hoeksema 2005). In a number of studies, large variations in plant growth variables were reported when the same plant species were treated with different AM fungi isolates (Owusu-Bennoah and Mosse 1979). The increase in host biomass was reported to leap from 79% to 600% when the same plant association changed from mycorrhizal isolate to another (Hart and Reader 2002; Mensah et al. 2015; Van Der Heijden et al. 1998). The observed differences in mycorrhiza effects were independent of their colonization rate variation (Van Der Heijden et al. 1998). While in both examples the mycorrhizal variation led to differences in beneficial responses, in most cases, host selection is assumed to have a stronger impact on this co-dependency relation (Bever 2002; Molina et al. 1997; Rasmussen et al. 2018). The selection by host for certain species of mycorrhizae over others is related to the host's ability to sense and differently allocate resources that can be used by mycorrhizae into its root (Jansa et al. 2008; Verbruggen et al. 2012). For instance, in *Medicago truncatula*, higher beneficial AM species were correlated with higher carbon allocation in roots by the host (Jansa et al. 2013; Werner and Kiers 2015b). The driving force of host selection remains an open question (Werner and Kiers 2015b). Some plant traits such as the degree of root coarseness and decreased fine-scale host selection for mycorrhiza (Verbruggen et al. 2012) suggest that host traits contribute to mycorrhiza selection (Werner and Kiers 2015a, b). The host needs are theorized to regulate host microbial selective recruitment (Abdelrazek et al. 2020a, b; Johnson and Graham 2013; Werner and Kiers 2015b). It was previously reported that more mycorrhization was observed in a phosphorus-deficient not in a rich phosphorus condition (Johnson and Graham 2013). As host phosphorus starvation in phosphorus-deficient soil leads to host recruitment of more phosphorus-fixing mycorrhiza to cover their needs (Johnson and Graham 2013), similar to host plant selection, mycorrhiza also selects the host to benefit the most from its own maximum benefits (Kiers et al. 2011; Werner and Kiers 2015a, b). In root culture experiments, mycorrhiza preferentially transports phosphorus to root clusters that support it with the highest carbon amount, while other roots that produce lower levels of carbon are left without any mycorrhizal aid (Kiers et al. 2011). Although this is the most common strategy for mycorrhiza in host selection, some plants do not follow the same pattern (Walter et al. 2012); for example, mycorrhizal symbiosis in flax (*Linum usitatissimum*) and sorghum (*Sorghum bicolor*) as a potential host. Sorghum always invested more energy and pumped more carbon into mycorrhizal symbiosis than flax. However, mycorrhiza in return transported more nutrients to flax than sorghum (Walter et al. 2012), which contradict the host selection theories. In addition to selection, the beneficial effect provided by mycorrhiza to their host is also affected by very fine morphological variation within mycorrhiza species. For example, differences in mycorrhiza size among different mycorrhiza species reported to affect host benefits by mycorrhiza in many plants. Mycorrhizal species that own larger internal mycelia provided the host with greater plant growth support than ones with smaller internal mycelia (Hart and Reader 2002). Even differences in mycorrhizal mycelial structures such as arbuscules, vesicles, coils, and absorptive, infective, and spore-producing hyphae abundance among different mycorrhizal taxa proved to affect mycorrhiza's beneficial support to their host (Hart and Reader 2002). As a conclusion, small or big differences in mycorrhizal species or host plant will always control the size of the beneficial effect provided by mycorrhizae to their host. Mycorrhiza size differences among species reported to determine host benefit in many plants. Mycorrhiza with larger internal mycelia showed to grant greater host benefit (Hart and Reader 2002). Growing evidence that even differences in abundance of mycelial structures such as arbuscules, vesicles, coils, absorptive, infective, and spore-producing hyphae among mycorrhiza taxa can influence their beneficial response (Hart and Reader 2002). Whether host or mycorrhiza contributions to their interaction outcomes rely on their big or small variations, they will always be identified as the main drivers for this relation and its backbone.

9.6.3 Surrounding Microbiome

To facilitate the co-existence of microbes in the same niche, mechanisms such as microbe-microbe interactions were evolved (Scherlach and Hertweck 2018). In this mechanism, the growth and behavior of each microbe are regulated through their interaction with other microbes within the same environment (Scherlach and Hertweck 2018). Mycorrhiza as any other micro-organism is synergistically interacting with soil and host niche. Successful mycorrhizosphere also relies on their microbial communities with which they are directly or indirectly associated. AM spore's germination was totally suppressed under sterilized soil, which represented a good example of how surrounding microbes are playing a central role in mycorrhizal establishment (Fitter and Garbaye 1994). In recent study, increase in rhizosphere microbe's diversity of maize proved to increase root colonization by beneficial mycorrhiza (Ferreira et al. 2020, 2021). Inculcating *Lolium multiflorum* grass with leaf endophytes, *Neotyphodium occultans* increased the colonization of *Glomus* in host and neighbor plants (Omacini et al. 2006). The exact return of mycorrhiza interaction with surrounding microbiome is unpredictable. For example, in some cases, leaf endophytes prove to reduce mycorrhizal sporulation and colonization of host roots (Müller and Kleinschmidt 2003). The instability of this relation associated with the high complexity of mycorrhiza-microbe interaction model might be due to the involvement of so many factors such as soil, environment, and plant. Looking at the underlying mechanism behind this interaction, both rhizosphere and plant microbes are theorized to possess similar procedures when it comes to influencing AM fungi (Johansson et al. 2004): manipulating root recognition and receptivity of the root, modifying rhizosphere soil chemistry, and altering mycorrhiza growth and germination (Johansson et al. 2004). For instance, *Corynebacterium* and *Pseudomonas* rhizosphere bacteria were reported to enhance the spore germination of *Glomus versiforme* through the inactivation of *Glomus versiforme* spore inhibitors (Johansson et al. 2004). Bacteria isolated from the mycorrhizosphere were reported to boost the germination and growth of AM spore and colonization of roots (Giovannetti et al. 2010; Xavier and Germida 2003). In another study, *Paenibacillus rhizosphaerae* and *Rhizobium etli* rhizosphere bacteria increased the growth of *G. intraradices* and *Rhizophagus irregularis*. However, the significant increase in IAA production by these bacteria could not support this as a mechanism (Ordoñez et al. 2016; Turrini et al. 2018a, b). In another example, ecto-mycorrhiza showed to colonize and protect plant roots more effectively than endo-mycorrhiza, which is owing to the accumulation of pathogenic fungi within AMF sphere. This activates the production of negative feedback toward associated plants (Kadowaki et al. 2018; Tedersoo et al. 2020; Teste et al. 2017). The phrase "mycorrhiza helpers" refers to a bacteria group that supports mycorrhizae throughout its life cycle (Turrini et al. 2018a, b). Although, in mycorrhizal interaction networks, "mycorrhiza helpers" bacteria are well established (Turrini et al. 2018a, b). It is not clear yet, if these helpers gain any reward from supporting AM fungi. AM fungi were reported to compete with rhizobacteria, in pre-mycorrhiza symbiotic phase. However, in later stages of symbioses, the rapid establishment of AM fungi in host roots was associated with improved nodulation (Smith et al. 1979). The interaction between mycorrhiza and surrounding microbes is very complicated. Understanding more about the interaction chemistry and language will help maximizing the benefits of mycorrhiza as bio-protectors (Finkel et al. 2017).

9.7 Conclusions and Future Outlooks

With the uncontrolled growing increase in human folk around the globe and limited food production, mycorrhizal fungi may be used in sustainable agriculture, which can significantly help to enhance plant productivity to meet the high demand of various food commodities. The formal information about mycorrhiza-host interactive mechanisms as a bio-protector and its interfering factors in a logical manner for the better understanding of mycorrhiza-pathogen interactions and exploiting its potential benefits in agriculture usage has been documented. Future mycorrhiza research work should focus on identifying genes and gene products that trigger mycorrhiza crosstalks with plant pathogens and host microbiome, more specifically, picturing mycorrhiza-host microbiome-pathogen interactions holistically through relevant omics' techniques and meta-analysis. In addition, identifying factors that modulate mycorrhiza-plant symbiotic association and metabolic networks associated with various stress conditions is also a potential aspect to be unraveled in upcoming research.

References

Abdelrazek S, Choudhari S, Thimmapuram J, Simon P, Colley M, Mengiste T, Hoagland L (2020a) Changes in the core endophytic mycobiome of carrot taproots in response to crop management and genotype. Sci Rep 10:1–14

Abdelrazek S, Simon P, Colley M, Mengiste T, Hoagland L (2020b) Crop management system and carrot genotype affect endophyte composition and Alternaria dauci suppression. PLoS One 15:e0233783

Abo-Elyousr KAM, Seleim MAA, Abd-El-Moneem KMH, Saead FA (2014) Integrated effect of Glomus mosseae and selected plant oils on the control of bacterial wilt disease of tomato. Crop Prot 66:67–71

Adie B, Chico JM, Rubio-Somoza I, Solano R (2007) Modulation of plant defenses by ethylene. J Plant Growth Regul 26:160–177

Aguk J, Karanja N, Schulte-Geldermann E, Bruns C, Kinyua Z, Parker M (2018) Control of bacterial wilt (Ralstonia solanacearum) in potato (Solanum tuberosum) using rhizobacteria and arbuscular mycorrhiza fungi. Afr J Food Agric Nutr Dev 18:13371–13387

Agoncillo (2018) Control of bacterial wilt disease caused by Ralstonia solanacearum in pepper using arbuscular mychorrhizal fungi (Mykovam). J Nat Sci Res 8:62–66

Ahanger MA, Hashem A, Abd-Allah EF, Ahmad P (2014a) Arbuscular mycorrhiza in crop improvement under environmental stress. In: Emerging technologies and management of crop stress tolerance. Elsevier, San Diego, CA, pp 69–95

Ahanger MA, Tyagi SR, Wani MR, Ahmad P (2014b) Drought tolerance: role of organic osmolytes, growth regulators, and mineral nutrients. In: Ahmad P, Wani MR (eds) Physiological mechanisms and adaptation strategies in plants under changing environment, vol 1. Springer, New York, NY, pp 25–55

Ahmed MS, Sallam NMA, Mohamed AEA, Hassan MHA (2013) Effect of mycorrhiza and biofertilizers on reducing the incidence of Fusarium root and pod rot diseases of peanut. Arch Phytopathol Plant Prot 46(7):868–881

Al-Askar A, Rashad Y (2010) Arbuscular mycorrhizal fungi: a biocontrol agent against common. Plant Pathol J 9:31–38

Al-Hmoud G, Al-Momany A (2017) Effect of four mycorrhizal products on squash plant growth and its effect on physiological plant elements. Adv Crop Sci Tech 5:260

Al-Karaki G, McMichael B, Zak J (2004) Field response of wheat to arbuscular mycorrhizal fungi and drought stress. Mycorrhiza 14:263–269

Amiri R, Nikbakht A, Rahimmalek M, Hosseini H (2017a) Variation in the essential oil composition, antioxidant capacity, and physiological characteristics of Pelargonium graveolens L. inoculated with two species of mycorrhizal fungi under water deficit conditions. J Plant Growth Regul 36:502–515

Amiri R, Ali N, Nematollah E, Mohammad RS (2017b) Nutritional status, essential oil changes and water-use efficiency of rose geranium in response to arbuscularmycorrhizal fungi and water deficiency stress. Symbiosis 73:15–25

Ansari RA, Mahmood I (2017a) Optimization of organic and bio-organic fertilizers on soil properties and growth of pigeonpea. Sci Hortic 226:1–9

Ansari RA, Mahmood I (2017b) Determination of disease incidence caused by *Meloidogyne* spp. and or *Fusarium udum* on pigeonpea in Aligarh district: a survey. Trends Biosci 10(24):5239–5243

Ansari RA, Mahmood I (2019a) Plant health under biotic stress. In: Ansari RA, Mahmood I (eds) Microbial interactions, vol II. Springer Nature, Singapore

Ansari RA, Mahmood I (2019b) Plant health under biotic stress. In: Ansari RA, Mahmood I (eds) Organic strategies, vol I. Springer Nature, Singapore

Ansari RA, Mahmood I, Rizvi R, Sumbul A (2017a) PGPR: current vogue in sustainable crop production. In: Kumar V (ed) Probiotics and plant health. Springer Nature, Singapore, pp 455–472

Ansari RA, Mahmood I, Rizvi R, Sumbul A, Safiuddin (2017b) Siderophores: augmentation of soil health and crop productivity. In: Kumar V (ed) Probiotics in agroecosystem. Springer Nature, Singapore, pp 291–312

Ansari RA, Sumbul A, Rizvi R, Mahmood I (2019a) Potential role of plant growth promoting rhizobacteria in alleviation of biotic stress. In: Ansari RA, Mahmood I (eds) Plant health under biotic stress, Microbial interactions, vol II. Springer Nature, Singapore, pp 177–188

Ansari RA, Sumbul A, Rizvi R, Mahmood I (2019b) Organic soil amendments: potential tool for soil and plant health management. In: Ansari RA, Mahmood I (eds) Plant health under biotic stress, Organic strategies, vol I. Springer Nature, Singapore, pp 1–35

Ansari RA, Rizvi R, Mahmood I (2020a) Management of phytonematodes: management of phytonematodes: recent advances and future challenges. Springer Nature, Singapore

Ansari RA, Rizvi R, Sumbul A, Mahmood I (2020b) Plant-growth-promoting Rhizobacteria (PGPR)-based sustainable management of phytoparasitic nematodes: current understandings and future challenges. In: Ansari RA et al (eds) Management of phytonematodes: recent advances and future challenges. Springer Nature, Singapore. https://doi.org/10.1007/978-981-15-4087-5_3

Aseel DG, Rashad YM, Hammad SM (2019) Arbuscular mycorrhizal fungi trigger transcriptional expression of flavonoid and chlorogenic acid biosynthetic pathways genes in tomato against tomato mosaic virus. Sci Rep 9:1–10

Azcón-Aguilar C, Barea J (1997) Arbuscular mycorrhizas and biological control of soil-borne plant pathogens–an overview of the mechanisms involved. Mycorrhiza 6:457–464

Azcón-Aguilar C, Jaizme-Vega M, Calvet C (2002) The contribution of arbuscular mycorrhizal fungi to the control of soil-borne plant pathogens. In: Mycorrhizal technology in agriculture. Springer, pp 187–197

Bagheri V, Shamshiri MH, Shirani H, Roosta H (2012) Nutrient uptake and distribution in mycorrhizal pistachio seedlings under drought stress. J Agric Sci Technol 14:1591–1604

Bagy HMK, Hassan EA, Nafady NA, Dawood MF (2019) Efficacy of arbuscular mycorrhizal fungi and endophytic strain Epicoccum nigrum ASU11 as biocontrol agents against blackleg disease of potato caused by bacterial strain Pectobacterium carotovora subsp. atrosepticum PHY7. Biol Control 134:103–113

Bais HP, Weir TL, Perry LG, Gilroy S, Vivanco JM (2006) The role of root exudates in rhizosphere interactions with plants and other organisms. Annu Rev Plant Biol 57:233–266

Balaji V, Mayrose M, Sherf O, Jacob-Hirsch J, Eichenlaub R, Iraki N, Manulis-Sasson S, Rechavi G, Barash I, Sessa G (2008) Tomato transcriptional changes in response to Clavibacter michiganensis subsp. michiganensis reveal a role for ethylene in disease development. Plant Physiol 146:1797–1809

Bates TR, Lynch JP (2001) Root hairs confer a competitive advantage under low phosphorus availability. Plant Soil 236:243–250. https://doi.org/10.1023/A:1012791706800

Battini F, Grønlund M, Agnolucci M, Giovannetti M, Jakobsen I (2017) Facilitation of phosphorus uptake in maize plants by mycorrhizosphere bacteria. Sci Rep 7:4686

Benhamou N, Fortin JA, Hamel C, St-Arnaud M, Shatilla A (1994) Resistance responses of mycorrhizal Ri T-DNA-transformed carrot roots to infection by *Fusarium oxysporum* f. sp. *chrysanthemi*. Phytopathology 84:958–968

Berg G, Grosch R, Scherwinski K (2007) Risk assessment for microbial antagonists: Are there effects on non-target organisms? GesundePflanzen 59:107–117

Bever JD (2002) Negative feedback within a mutualism: host–specific growth of mycorrhizal fungi reduces plant benefit. Proc R Soc Lond B Biol Sci 269:2595–2601

Bødker L, Kjøller R, Kristensen K, Rosendahl S (2002) Interactions between indigenous arbuscular mycorrhizal fungi and Aphanomyces euteiches in field-grown pea. Mycorrhiza 12:7–12

Bonfante P, Perotto S (1995) Tansley review no. 82. strategies of arbuscular mycorrhizal fungi when infecting host plants. New Phytol 130:3–21

Bowles TM, Barrios-Masias FH, Carlisle EA, Cavagnaro TR, Jackson LE (2016) Effects of arbuscular mycorrhizae on tomato yield, nutrient uptake, water relations, and soil carbon dynamics under deficit irrigation in field conditions. Sci Total Environ 566:1223–1234

Breuillin F, Schramm J, Hajirezaei M, Ahkami A, Favre P, Druege U, Hause B, Bucher M, Kretzschmar T, Bossolini E (2010) Phosphate systemically inhibits development of arbuscular mycorrhiza in Petunia hybrida and represses genes involved in mycorrhizal functioning. Plant J 64:1002–1017

Brimner TA, Boland GJ (2003) A review of the non-target effects of fungi used to biologically control plant diseases. Agric Ecosyst Environ 100:3–16

Brito ODC, Hernandes I, Ferreira JCA, Cardoso MR, Alberton O, Dias-Arieira CR (2018) Association between arbuscular mycorrhizal fungi and Pratylenchus brachyurus in maize crop. Chilean J Agric Res 78:521–527

Bucher M (2007) Functional biology of plant phosphate uptake at root and mycorrhizae interfaces. *New Phytol.* 173:11–26

Bücking H, Kafle A (2015) Role of arbuscular mycorrhizal fungi in the nitrogen uptake of plants: current knowledge and research gaps. Agronomy 5:587–612. https://doi.org/10.3390/agronomy5040587

Calvet C, Pinochet J, Hernández-Dorrego A, Estaún V, Camprubí A (2001) Field microplot performance of the peach-almond hybrid GF-677 after inoculation with arbuscular mycorrhizal fungi in a replant soil infested with root-knot nematodes. Mycorrhiza 10:295–300

Canarini A, Kaiser C, Merchant A, Richter A, Wanek W (2019a) Root exudation of primary metabolites: mechanisms and their roles in plant responses to environmental stimuli. Front Plant Sci 10:157

Canarini A, Kaiser C, Merchant A, Richter A, Wanek W (2019b) Corrigendum: root exudation of primary metabolites: mechanisms and their roles in plant responses to environmental stimuli. Front Plant Sci 10:420

Chen M, Bruisson S, Bapaume L, Darbon G, Glauser G, Schorderet M, Reinhardt D (2021) VAPYRIN attenuates defence by repressing PR gene induction and localized lignin accumulation during arbuscular mycorrhizal symbiosis of Petunia hybrida. New Phytol 229:3481–3496

Chen S, Zhao H, Zou C, Li Y, Chen Y, Wang Z et al (2017) Combined Inoculation with multiple arbuscularmycorrhizal fungi improves growth, nutrient uptake and photosynthesis in cucumber seedlings. Front Microbiol 8:25–16. https://doi.org/10.3389/fmicb.2017.02516

Cheng L, Booker FL, Tu C, Burkey KO, Zhou L, Shew HD, Rufty TW, Hu S (2012) Arbuscular mycorrhizal fungi increase organic carbon decomposition under elevated CO2. Science 337:1084–1087

Cordier C, Gianinazzi S, Gianinazzi-Pearson V (1996) Colonisation patterns of root tissues by *Phytophthoranicotianae* var. parasitica related to reduced disease in mycorrhizal tomato. Plant Soil 185:223–232

Cordier C, Pozo MJ, Barea J-M, Gianinazzi S, Gianinazzi-Pearson V (1998) Cell defense responses associated with localized and systemic resistance to Phytophthora parasitica induced in tomato by an arbuscular mycorrhizal fungus. Mol Plant Microbe Interact 11:1017–1028

Cotton TA (2018) Arbuscular mycorrhizal fungal communities and global change: an uncertain future. FEMS Microbiol Ecol 94:fiy179

De Andrade SAL, Domingues AP, Mazzafera P (2015) Photosynthesis is induced in rice plants that associate with arbuscularmycorrhizal fungi and are grown under arsenate and arsenite stress. Chemosphere 134:141–149

de Roman M, Fernandez I, Wyatt T, Sahrawy M, Heil M, Pozo MJ (2011) Elicitation of foliar resistance mechanisms transiently impairs root association with arbuscular mycorrhizal fungi. J Ecol 99:36–45

Demir S, Şensoy S, Ocak E, Tüfenkci Ş, Durak ED, Erdinc C, Ünsal H (2015) Effects of arbuscular mycorrhizal fungus, humic acid, and whey on wilt diseasecaused by Verticillium dahliae Kleb. in three solanaceous crops. Turk J Agric For 39:300–309

Dowarah B, Gill SS, Agarwala N (2021) Arbuscular Mycorrhizal Fungi in Conferring Tolerance to Biotic Stresses in Plants. J Plant Growth Regul:1–16

Duponnois R, Plenchette C, Thioulouse J, Cadet P (2001) The mycorrhizal soil infectivity and arbuscular mycorrhizal fungal spore communities in soils of different aged fallows in Senegal. Appl Soil Ecol 17:239–251

Evelin H, Giri B, Kapoor R (2012) Contribution of Glomus intraradices inoculation to nutrient acquisition and mitigation of ionic imbalance in NaCl-stressed Trigonella foenum-graecum. Mycorrhiza 22:203–217

Ferreira BS, Santana MV, Macedo RS, Silva JO, Carneiro MA, Rocha MR (2018) Co-occurrence patterns between plant-parasitic nematodes and arbuscular mycorrhizal fungi are driven by environmental factors. Agric Ecosyst Environ 265:54–61

Ferreira ML, Gerbino E, Cavallero GJ, Casabuono AC, Couto AS, Gomez-Zavaglia A, Ramirez SA, Vullo DL (2020) Infrared spectroscopy with multivariate analysis to interrogate the inter-

action of whole cells and secreted soluble exopolimeric substances of Pseudomonas veronii 2E with Cd (II), Cu (II) and Zn (II). Spectrochim Acta A Mol Biomol Spectrosc 228:117820
Ferreira DA, da Silva TF, Pylro VS, Salles JF, Andreote FD, Dini-Andreote F (2021) Soil microbial diversity affects the plant-root colonization by arbuscular mycorrhizal fungi. Microb Ecol 82:100–103
Finkel OM, Castrillo G, Paredes SH, González IS, Dangl JL (2017) Understanding and exploiting plant beneficial microbes. Curr Opin Plant Biol 38:155–163
Fitter AH (2004) Magnolioid roots–hairs, architecture and mycorrhizal dependency. New Phytol 164:15–16
Fitter A, Garbaye J (1994) Interactions between mycorrhizal fungi and other soil organisms. Plant Soil 159:123–132
Francl L (1993) Interactions of nematodes with mycorrhizae and mycorrhizal fungi. In: Khan MW (ed) Nematode interactions. Springer, Dordrecht, pp 203–216
Frank B (1885) Über die auf Wurzelsymbiose beruhende Ernährung gewisser Bäume durch unterirdische Pilze
Fulekar M, Pathak B (2015) Rhizosphere: an innovative approach for remediation of contaminants. IJSER 6:291–303
Garcés-Ruiz M, Senés-Guerrero C, Declerck S, Cranenbrouck S (2017a) Arbuscular mycorrhizal fungal community composition in Carludovica palmata, Costus scaber and Euterpe precatoria from weathered oil ponds in the Ecuadorian Amazon. Front Microbiol 8:2134
Garcés-Ruiz M, Calonne-Salmon M, Plouznikoff K, Misson C, Navarrete-Mier M, Cranenbrouck S et al (2017b) Dynamics of short-term phosphorus uptake by intact mycorrhizal and non-mycorrhizal maize plants grown in a circulatory semi-hydroponic cultivation system. Front Plant Sci 8:1471
García-Garrido JM, Ocampo JA (2002) Regulation of the plant defence response in arbuscular mycorrhizal symbiosis. J Exp Bot 53:1377–1386
Gernns H, Alten H, Poehling H-M (2001) Arbuscular mycorrhiza increased the activity of a biotrophic leaf pathogen - is a compensation possible. Mycorrhiza 11(5):237–243. https://doi.org/10.1007/s005720100128
Giang BL, Nguyen NH, Dong Yen PN, Minh Hoang VD, Lien Ha BT, Le NTT (2020) Combination of mycorrhizal symbiosis and root grafting effectively controls nematode in replanted coffee soil. Plants 9:555
Giannakis N, Sanders FE (1990) Interactions between mycophagous nematodes, mycorrhizal and other soil fungi. Agric Ecosyst Environ 29(1–4):163–167
Giovannetti M, Avio L, Sbrana C (2010) Fungal spore germination and pre-symbiotic mycelial growth–physiological and genetic aspects. In: Koltai H, Kapulnik Y (eds) Arbuscular mycorrhizas: physiology and function. Springer, Dordrecht, pp 3–32
Goicoechea N (2020) Mycorrhizal fungi as bioprotectors of crops against verticillium wilt—a hypothetical scenario under changing environmental conditions. Plants 9:1468
Gomez-Bellot MJ, Ortuño MF, Nortes PA, Vicente-Sánchez J, Bañón S, Sánchez Blanco MJ (2015) Mycorrhizal euonymus plants and reclaimed water: biomass, water status and nutritional responses. Sci Hort 186:61–69
Gosling P, Hodge A, Goodlass G, Bending G (2006) Arbuscular mycorrhizal fungi and organic farming. Agric Ecosyst Environ 113:17–35
Gough EC, Owen KJ, Zwart RS, Thompson JP (2020) A systematic review of the effects of arbuscular mycorrhizal fungi on root-lesion nematodes, Pratylenchus spp. Front Plant Sci 11:923
Grman E, Robinson TM, Klausmeier CA (2012) Ecological specialization and trade affect the outcome of negotiations in mutualism. Am Nat 179:567–581
Grosch R, Lottmann J, Faltin F, Berg G (2005) Use of bacterial antagonists to control diseases caused by Rhizoctonia solani. Gesunde Pflanzen 57:199–205
Guo Y, George E, Marschner H (1996) Contribution of an arbuscular mycorrhizal fungus to the uptake of cadmium and nickel in bean and maize plants. Plant Soil 184:195–205
Gutjahr C, Parniske M (2013) Cell and developmental biology of arbuscular mycorrhiza symbiosis. Annu Rev Cell Dev Biol 29:593–617

Hao Z, Xie W, Chen B (2019) Arbuscular mycorrhizal symbiosis affects plant immunity to viral infection and accumulation. Viruses 11:534

Hameed A, Dilfuza E, Abd-Allah EF, Hashem A, Kumar A, Ahmad P (2014) Salinity stress and arbuscularmycorrhizal symbiosis in plants. In: Miransari M (ed) Use of microbes for the alleviation of soil stresses, vol 1. Springer, New York, pp 139–159

Harrier LA, Watson CA (2004) The potential role of arbuscular mycorrhizal (AM) fungi in the bioprotection of plants against soil-borne pathogens in organic and/or other sustainable farming systems. Pest Manage Sci 60:149–157

Harrison MJ (2005) Signaling in the arbuscular mycorrhizal symbiosis. Annu Rev Microbiol 59:19–42

Hart MM, Reader RJ (2002) Taxonomic basis for variation in the colonization strategy of arbuscular mycorrhizal fungi. New Phytol 153:335–344. https://doi.org/10.1046/j.0028-646X.2001.00312

Hashem A, Abd Allah EF, Alqarawi AA, Egamberdieva D (2018a) Arbuscular mycorrhizal fungi and plant stress tolerance. In: Plant microbiome: stress response. Springer, Singapore, pp 81–103

Haselwandter K, Winkelmann G (2007) Siderophores of symbiotic fungi. In: Chincholkar SB, Varma A (eds) Microbial Siderophore, Soil biology series, vol 12. Springer, Berlin, pp 91–103

Hashem A, Alqarawi AA, Radhakrishnan R, Al-Arjani A-BF, Aldehaish HA, Egamberdieva D, Abd Allah EF (2018b) Arbuscular mycorrhizal fungi regulate the oxidative system, hormones and ionic equilibrium to trigger salt stress tolerance in Cucumis sativus L. Saudi J Biol Sci 25:1102–1114

Hause B, Mrosk C, Isayenkov S, Strack D (2007) Jasmonates in arbuscular mycorrhizal interactions. Phytochemistry 68:101–110

Hause B, Schaarschmidt S (2009) The role of jasmonates in mutualistic symbioses between plants and soil-born microorganisms. Phytochemistry 70:1589–1599

Haynes R, Swift R (1985) Growth and nutrient uptake by highbush blueberry plants in a peat medium as influenced by pH, applied micronutrients and mycorrhizal inoculation. Sci Hortic 27:285–294

Heinemeyer A, Fitter A (2004) Impact of temperature on the arbuscular mycorrhizal (AM) symbiosis: growth responses of the host plant and its AM fungal partner. J Exp Bot 55:525–534

Hodge A, Fitter AH (2010) Substantial nitrogen acquisition by arbuscular mycorrhizal fungi from organic material has implications for N cycling. Proc Natl Acad Sci U S A 107:13754–13759

Hodge A, Storer K (2015) Arbuscularmycorrhiza and nitrogen: implications for individual plants through to ecosystems. Plant Soil 386:1–19

Hoeksema JD (2005) Plant–plant interactions vary with different mycorrhizal fungus species. Biol Lett 1:439–442

Hoeksema JD, Chaudhary VB, Gehring CA, Johnson NC, Karst J, Koide RT, Pringle A, Zabinski C, Bever JD, Moore JC (2010) A meta-analysis of context-dependency in plant response to inoculation with mycorrhizal fungi. Ecol Lett 13:394–407

Hogan CM (2011) Phosphate. In: Jorgensen A, Cleveland CJ (eds) Encyclopedia of earth. National Council for Science and the Environment, Washington, DC

Hol WG, Cook R (2005) An overview of arbuscular mycorrhizal fungi–nematode interactions. Basic Appl Ecol 6:489–503

Howe GA, Jander G (2008) Plant immunity to insect herbivores. Annu Rev Plant Biol 59:41–66

Huang J, Luo S, Zeng R (2003) Mechanisms of plant disease resistance induced by arbuscular mycorrhizal fungi. Ying Yong Sheng Tai Xue Bao 14:819–822

Ismail Y, McCormick S, Hijri M (2013) The arbuscular mycorrhizal fungus, Glomus irregulare, controls the mycotoxin production of Fusarium sambucinum in the pathogenesis of potato. FEMS Microbiol Lett 348:46–51

Jaiti F, Meddich A, El Hadrami I (2007) Effectiveness of arbuscular mycorrhizal fungi in the protection of date palm (Phoenix dactylifera L.) against bayoud disease. Physiol Mol Plant Pathol 71:166–173

Jaiti F, Kassami M, Meddich A, El Hadrami I (2008) Effect of arbuscular mycorrhization on the accumulation of hydroxycinnamic acid derivatives in date palm seedlings challenged with Fusarium oxysporum f. sp. albedinis. J Phytopathol 156:641

Jamiołkowska A, Księżniak A, Gałązka A, Hetman B, Kopacki M, Skwaryło-Bednarz B (2018) Impact of abiotic factors on development of the community of arbuscular mycorrhizal fungi in the soil: a review. Int Agrophys 32:133

Jansa J, Bukovská P, Gryndler M (2013) Mycorrhizal hyphae as ecological niche for highly specialized hypersymbionts–or just soil free-riders? Front Plant Sci 4:134

Jansa J, Smith FA, Smith SE (2008) Are there benefits of simultaneous root colonization by different arbuscular mycorrhizal fungi? New Phytol 177:779–789

Jiang YN, Wang WX, Xie QJ, Liu N, Liu LX, Wang DP et al (2017) Plants transfer lipids to sustain colonization by mutualistic mycorrhizal and parasitic fungi. Science 356:1172–1175

Johansson JF, Paul LR, Finlay RD (2004) Microbial interactions in the mycorrhizosphere and their significance for sustainable agriculture. FEMS Microbiol Ecol 48:1–13

Johnson NC, Graham JH (2013) The continuum concept remains a useful framework for studying mycorrhizal functioning. Plant Soil 363:411–419

Jun-Li H, Xian-Gui L, Jun-Hua W, Wei-Shou S, Shu W, Su-Ping P, Ting-Ting M (2010) Arbuscular mycorrhizal fungal inoculation enhances suppression of cucumber Fusarium wilt in greenhouse soils. Pedosphere 20:586–593

Jung SC, Martinez-Medina A, Lopez-Raez JA, Pozo MJ (2012) Mycorrhiza-induced resistance and priming of plant defenses. J Chem Ecol 38:651–664

Kadowaki K, Yamamoto S, Sato H, Tanabe AS, Hidaka A, Toju H (2018) Mycorrhizal fungi mediate the direction and strength of plant–soil feedbacks differently between arbuscular mycorrhizal and ectomycorrhizal communities. Commun Biol 1:1–11

Kaku H, Nishizawa Y, Ishii-Minami N, Akimoto-Tomiyama C, Dohmae N, Takio K, Minami E, Shibuya N (2006) Plant cells recognize chitin fragments for defense signaling through a plasma membrane receptor. Proc Natl Acad Sci U S A 103:11086–11091

Khalil S, Loynachan TE, Tabatabai MA (1994) Mycorrhizal dependency and nutrient uptake by improved and unimproved corn and soybean cultivars. Agron J 86:949–958

Khatabi B, Schäfer P (2012) Ethylene in mutualistic symbioses. Plant Signal Behav 7:1634–1638

Kiers ET, Duhamel M, Beesetty Y, Mensah JA, Franken O, Verbruggen E, Fellbaum CR, Kowalchuk GA, Hart MM, Bago A (2011) Reciprocal rewards stabilize cooperation in the mycorrhizal symbiosis. Science 333:880–882

Kilpeläinen J, Aphalo PJ, Barbero-López A, Adamczyk B, Nipu SA, Lehto T (2020) Are arbuscular-mycorrhizal Alnus incana seedlings more resistant to drought than ectomycorrhizal and nonmycorrhizal ones? Tree Physiol 40:782–795

Kloppholz S, Kuhn H, Requena N (2011) A secreted fungal effector of Glomus intraradices promotes symbiotic biotrophy. Curr Biol 21:1204–1209

Konvalinková T, Jansa J (2016) Lights off for arbuscular mycorrhiza: on its symbiotic functioning under light deprivation. Front Plant Sci 7:782

Kumaresan G, Jayachitra J, Shanthi R (2020) Inoculation effect of arbuscular mycorrhizal fungi on the growth and yield of cotton. Ann Rom Soc Cell Biol:1180–1188

Lehmann A, Rillig MC (2015) Arbuscularmycorrhizal contribution to copper, manganese and iron nutrient concentrations in crops—a meta-analysis. Soil Biol Biochem 81:147–158

Liu C, Ravnskov S, Liu F, Rubæk GH, Andersen MN (2018) Arbuscularmycorrhizal fungi alleviate abiotic stresses in potato plants caused by low phosphorus and deficit irrigation/partial root-zone drying. J Agric Sci 156:46–58

Linderman R (1991) Mycorrhizal interactions in the rhizosphere. In: The rhizosphere and plant growth. Springer, Dordrechet, pp 343–348

Lioussanne L, Jolicoeur M, St-Arnaud M (2008) Mycorrhizal colonization with Glomus intraradices and development stage of transformed tomato roots significantly modify the chemotactic response of zoospores of the pathogen Phytophthora nicotianae. Soil Biol Biochem 40:2217–2224

López-Ráez JA, Verhage A, Fernández I, García JM, Azcón-Aguilar C, Flors V, Pozo MJ (2010) Hormonal and transcriptional profiles highlight common and differential host responses to arbuscular mycorrhizal fungi and the regulation of the oxylipin pathway. J Exp Bot 61:2589–2601

Ludwig-Müller J (2010) Hormonal responses in host plants triggered by arbuscular mycorrhizal fungi. In: Arbuscular mycorrhizas: physiology and function. Springer, Dordrecht, pp 169–190

Luginbuehl LH, Menard GN, Kurup S, Van Erp H, Radhakrishnan GV, Breakspear A et al (2017) Fatty acids in arbuscularmycorrhizal fungi are synthesized by the host plant. Science 356:1175–1178

Marro N, Caccia M, Doucet ME, Cabello M, Becerra A, Lax P (2018) Mycorrhizas reduce tomato root penetration by false root-knot nematode Nacobbus aberrans. Appl Soil Ecol 124:262–265

Marschner P, Baumann K (2003) Changes in bacterial community structure induced by mycorrhizal colonisation in split-root maize. Plant Soil 251:279–289

Mayer Z, Juhász Á, Posta K (2019) Mycorrhizal root exudates induce changes in the growth and fumonisin gene (FUM1) expression of Fusarium proliferatum. Agronomy 9:291

Mensah JA, Koch AM, Antunes PM, Kiers ET, Hart M, Bücking H (2015) High functional diversity within species of arbuscular mycorrhizal fungi is associated with differences in phosphate and nitrogen uptake and fungal phosphate metabolism. Mycorrhiza 25:533–546

Miozzi L, Vaira AM, Catoni M, Fiorilli V, Accotto GP, Lanfranco L (2019) Arbuscular mycorrhizal symbiosis: plant friend or foe in the fight against viruses? Front Microbiol 10:1238

Mitra D, Navendra U, Panneerselvam U, Ansuman S, Ganeshamurthy AN, Divya J (2019) Role of mycorrhiza and its associated bacteria on plant growth promotion and nutrient management in sustainable agriculture. Int J Life Sci Appl Sci 1:1–10

Miya A, Albert P, Shinya T, Desaki Y, Ichimura K, Shirasu K, Narusaka Y, Kawakami N, Kaku H, Shibuya N (2007) CERK1, a LysM receptor kinase, is essential for chitin elicitor signaling in Arabidopsis. Proc Natl Acad Sci U S A 104:19613–19618

Molina R, Smith JE, Mckay D, Melville L (1997) Biology of the ectomycorrhizal genus, Rhizopogon III. Influence of co-cultured conifer species on mycorrhizal specificity with the arbutoid hosts Arctostaphylos uva-ursi and Arbutus menziesii. New Phytol 137:519–528

Müller NG, Kleinschmidt A (2003) Dynamic interaction of object-and space-based attention in retinotopic visual areas. J Neurosci 23:9812–9816

Nanjundappa A, Bagyaraj DJ, Saxena AK, Kumar M, Chakdar H (2019) Interaction between arbuscular mycorrhizal fungi and Bacillus spp. in soil enhancing growth of crop plants. Fungal Biol Biotechnol 6:1–10

Nehls U (2008) Mastering ectomycorrhizal symbiosis: the impact of carbohydrates. J Exp Bot 59:1097–1108

Nell M, Wawrosch C, Steinkellner S, Vierheilig H, Kopp B, Lössl A, Franz C, Novak J, Zitterl-Eglseer K (2010) Root colonization by symbiotic arbuscular mycorrhizal fungi increases sesquiterpenic acid concentrations in Valeriana officinalis L. Planta Med 76:393–398

Newsham KK, Fitter AH, Watkinson AR (1995) Arbuscular mycorrhiza protect an annual grass from root pathogenic fungi in the field, vol 83, no 6, pp 991–1000

Nguyen Hong D, Posta K (2018) Inoculation with Septoglomus constrictum improves tolerance to heat shock in tomato plants. Columella J Agric Environ Sci 5:7–14

Nuccio EE, Hodge A, Pett-Ridge J, Herman DJ, Weber PK, Firestone MK (2013) An arbuscular mycorrhizal fungus significantly modifies the soil bacterial community and nitrogen cycling during litter decomposition. Environ Microbiol 15:1870–1881

Odelade KA, Babalola OO (2019) Bacteria, fungi and archaea domains in rhizosphericsoil and their effects in enhancing agricultural productivity. Int J Environ Res Public Health 16(3873):1–19

Olawuyi O, Odebode A, Olakojo S (2013) Genotype x treatment x concentration interaction and character association of maize (Zea mays L.) under arbuscular mycorrhizal fungi and Striga lutea Lour. In: Proceedings of the 37th annual conference of the Genetics Society of Nigeria, Lafia, Nigeria, pp 210–219

Olawuyi O, Odebode A, Oyewole I, Akanmu A, Afolabi O (2014) Effect of arbuscular mycorrhizal fungi on Pythium aphanidermatum causing foot rot disease on pawpaw (Carica papaya L.) seedlings. Arch Phytopathol Plant Prot 47:185–193

Oliver IC, Knox OGG, Flavel RJ, Wilson BR (2021) Rhizosphere legacy: plant root interactions with the soil and its biome. In: Gupta VVSR, Sharma AK (eds) Rhizosphere biology: interactions between microbes and plants. Rhizosphere biology. Springer, Singapore

Omacini M, Eggers T, Bonkowski M, Gange A, Jones T (2006) Leaf endophytes affect mycorrhizal status and growth of co-infected and neighbouring plants. Funct Ecol 20:226–232

Ordoñez YM, Fernandez BR, Lara LS, Rodriguez A, Uribe-Velez D, Sanders IR (2016) Bacteria with phosphate solubilizing capacity alter mycorrhizal fungal growth both inside and outside the root and in the presence of native microbial communities. PLoS One 11:e0154438

Ortas I (2012) Do maize and pepper plants depend on mycorrhizae in terms of phosphorus and zinc uptake? J Plant Nutr 35:1639–1656

Owusu-Bennoah E, Mosse B (1979) Plant growth responses to vesicular-arbuscular mycorrhiza: XI. Field inoculation responses in barley, lucerne and onion. New Phytol 83:671–679

Ozgonen H, Erkilic A (2007) Growth enhancement and Phytophthora blight (Phytophthora capsici Leonian) control by arbuscular mycorrhizal fungal inoculation in pepper. Crop Prot 26(11):1682–1688

Philippot L, Raaijmakers JM, Lemanceau P, Van Der Putten WH (2013) Going back to the roots: the microbial ecology of the rhizosphere. Nat Rev Microbiol 11:789–799

Pile LS, Wang GG, Stovall JP, Siemann E, Wheeler GS, Gabler CA (2017) Mechanisms of Chinese tallow (Triadica sebifera) invasion and their management implications–a review. For Ecol Manag 404:1–13

Poveda J (2020) Marchantia polymorpha subsp. ruderalis (Bischl. & Boissel.-Dub.)-arbuscular mycorrhizal fungi interaction: beneficial or harmful? Symbiosis 82:165–174

Poveda J, Abril-Urias P, Escobar C (2020) Biological control of plant-parasitic nematodes by filamentous fungi inducers of resistance: Trichoderma, mycorrhizal and endophytic fungi. Front Microbiol 11:992

Powell JR, Rillig MC (2018) Biodiversity of arbuscular mycorrhizal fungi and ecosystem function. New Phytol 220:1059–1075

Pozo MJ, Azcón-Aguilar C (2007) Unraveling mycorrhiza-induced resistance. Curr Opin Plant Biol 10:393–398

Pozo MJ, Cordier C, Dumas-Gaudot E, Gianinazzi S, Barea JM, Azcón-Aguilar C (2002) Localized versus systemic effect of arbuscular mycorrhizal fungi on defence responses to Phytophthora infection in tomato plants. J Exp Bot 53:525–534

Rasmussen AL, Busby RR, Hoeksema JD (2018) Host preference of ectomycorrhizal fungi in mixed pine–oak woodlands. Can J For Res 48:153–159

Raven JA, Edwards D (2001) Roots: evolutionary origins and biogeochemical significance. J Exp Bot 52:381–401

Riad SN, El-Zawahry AMI, El Aref HM, Zawam HS (2021) Induction of systemic resistance in tomato by some abiotic compounds against *Meloidogyne javanica*. Assiut J Agric Sci 52(1):74–89

Rillig MC (2004) Arbuscular mycorrhizae, glomalin, and soil aggregation. Can J Soil Sci 84:355–363

Rini VM (2001) Effect of arbuscularmycorrhiza on oil palm seedling growth and development of basal stem rot disease caused by *Ganodermaboninense*. Master Thesis, Universiti Putra Malaysia.

Rizvi R, Singh G, Safiuddin ARA, Tiyagi SA, Mahmood I (2015) Sustainable management of root-knot disease of tomato by neem cake and *Glomus fasciculatum*. Cogent Food Agric 1:1008859

Rodriguez-Heredia M, Djian-Caporalino C, Ponchet M, Lapeyre L, Canaguier R, Fazari A, Marteau N, Offroy-Chave M (2020) Protective effects of mycorrhizal association in tomato and pepper against Meloidogyne incognita infection, and mycorrhizal networks for early mycorrhization of low mycotrophic plants. Phytopathol Mediterranea 59:377–384

Sadhana B (2014) Arbuscular Mycorrhizal Fungi (AMF) as a biofertilizer-a review. Int J Curr Microbiol App Sci 3:384–400

Scherlach K, Hertweck C (2018) Mediators of mutualistic microbe–microbe interactions. Nat Prod Rep 35:303–308

Schouteden N, De Waele D, Panis B, Vos CM (2015) Arbuscular mycorrhizal fungi for the biocontrol of plant-parasitic nematodes: a review of the mechanisms involved. Front Microbiol 6:1280

Selosse M-A, Richard F, He X, Simard SW (2006) Mycorrhizal networks: des liaisons dangereuses? Trends Ecol Evol 21:621–628

Shah A, Smith DL (2020) Flavonoids in agriculture: chemistry and roles in, biotic and abiotic stress responses, and microbial associations. Agronomy 10:1209

Sharma M, Saini I, Kaushik P, Al Dawsari MM, Al Balawi T, Alam P (2021) Mycorrhizal fungi and Pseudomonas fluorescens application reduces root-knot nematode (Meloidogyne javanica) infestation in eggplant. Saudi J Biol Sci 28(7):3685–3691

Shi G, Liu Y, Johnson NC, Olsson PA, Mao L, Cheng G, Jiang S, An L, Du G, Feng H (2014) Interactive influence of light intensity and soil fertility on root-associated arbuscular mycorrhizal fungi. Plant Soil 378:173–188

Shimizu T, Nakano T, Takamizawa D, Desaki Y, Ishii-Minami N, Nishizawa Y, Minami E, Okada K, Yamane H, Kaku H (2010) Two LysM receptor molecules, CEBiP and OsCERK1, cooperatively regulate chitin elicitor signaling in rice. Plant J 64:204–214

Sikes BA (2010) When do arbuscular mycorrhizal fungi protect plant roots frompathogens? Plant Signal Behav 5(6):763–765

Simard SW, Beiler KJ, Bingham MA, Deslippe JR, Philip LJ, Teste FP (2012) Mycorrhizal networks: mechanisms, ecology and modelling. Fungal Biol Rev 26:39–60

Singh R, Adholeya A, Mukerji K (2000) Mycorrhiza in control of soil borne pathogens. In: Mycorrhizal biology. Springer, pp 173–196

Slezack S, Dumas-Gaudot E, Paynot M, Gianinazzi S (2000) Is a fully established arbuscular mycorrhizal symbiosis required for bioprotection of Pisum sativum roots against Aphanomyces euteiches? Mol Plant Microbe Interact 13:238–241

Smith S, Nicholas D, Smith F (1979) Effect of early mycorrhizal infection on nodulation and nitrogen fixation in Trifolium subterraneum L. Funct Plant Biol 6:305–316

Smith SE, Jakobsen I, Grønlund M, Smith FA (2011) Roles of arbuscular mycorrhizas in plant phosphorus nutrition: interactions between pathways of phosphorus uptake in arbuscular mycorrhizal roots have important implications for understanding and manipulating plant phosphorus acquisition. Plant Physiol 156:1050–1057

Smith SE, Read DJ (2010) Mycorrhizal symbiosis. Academic press

Smith SE, Smith FA (2011) Roles of arbuscular mycorrhizas in plant nutrition and growth: new paradigms from cellular to ecosystem scales. Annu Rev Plant Biol 62:227–250

Solanki MK, Kashyap PL, Ansari RA, Kumari B (2020) Microbiomes and plant health: panoply and their applications. Academic Press, London. https://www.elsevier.com/books/microbiomes-and-plant-health/solanki/978-0-12-819715-8

Sumbul A, Mahmood I, Rizvi R, Safiuddin, Ansari RA (2017) Mycorrhiza: An alliance for the nutrient management in plants. In: Kumar V (ed) Probiotics in agroecosystem. Springer Nature, Singapore, pp 371–386

Tahat M, Sijam K, Othman R (2010) Mycorrhizal fungi as a biocontrol agent. Plant Pathol J (Faisalabad) 9:198–207

Tahat MM, Abo-Farag K, Alananbeh K, Al-Momany AM (2020) The efficacy of glomus mosseae and olive cake to control a chili pepper (capsicum annuum) damping off disease. Fresen Environ Bull 29:9863–9871

Tahat MM, Al Momany AM (2019) Bio-control characterization of two endo-mycorrhizal fungi against Verticillium wilt of cucumber. Fresen Environ Bull 28:9627–9635

Tahat MM, Sijam K (2012) Mycorrhizal fungi and abiotic environmental conditions relationship. Res J Environ Sci 6:125–133

Tahat MM, Sijam K, Othman R (2012) Ultrastructural changes of tomatoes (Lycopersicon esculentum) root colonized by Glomus mosseae and Ralstonia solanacearum. Afr J Biotechnol 11:6681–6686

Tedersoo L, Bahram M, Zobel M (2020) How mycorrhizal associations drive plant population and community biology. Science 367

Teste FP, Kardol P, Turner BL, Wardle DA, Zemunik G, Renton M, Laliberté E (2017) Plant-soil feedback and the maintenance of diversity in Mediterranean-climate shrublands. Science 355:173–176

Thirkell TJ, Charters MD, Elliott AJ, Sait SM, Field KJ (2017) Are mycorrhizal fungi our sustainable saviours? Considerations for achieving food security. J Ecol 105:921–929

Tian L, Zou Y-N, Wu Q-S, Kuča K (2021) Mycorrhiza-induced plant defence responses in trifoliate orange infected by Phytophthora parasitica. Acta Physiologiae Plantarum 43:1–8

Trotta A, Vanese GC, Gnavi E, Fascon A, Sampo S, Berta G (1996) Interaction between the soil-borne root pathogen *Phytophthoranicotianae*Var*parasitica* and the arbuscularmycorrhizal fungus *Glomusmosseae* in tomato plant. Plant Soil 185:199–209

Turrini A, Avio L, Giovannetti M, Agnolucci M (2018a) Functional complementarity of arbuscular mycorrhizal fungi and associated microbiota: the challenge of translational research. Front Plant Sci 9:1407

Turrini A, Bedini A, Loor MB, Santini G, Sbrana C, Giovannetti M et al (2018b) Local diversity of native arbuscularmycorrhizalsymbionts differentially affects growth and nutrition of three crop plant species. Biol Fertil Soils 54:203–217

Van Der Heijden MG, Klironomos JN, Ursic M, Moutoglis P, Streitwolf-Engel R, Boller T, Wiemken A, Sanders IR (1998) Mycorrhizal fungal diversity determines plant biodiversity, ecosystem variability and productivity. Nature 396:69–72

van der Heijden MG, Martin FM, Selosse MA, Sanders IR (2015) Mycorrhizal ecology and evolution: the past, the present, and the future. New Phytol 205:1406–1423

Van Wees SC, Van der Ent S, Pieterse CM (2008) Plant immune responses triggered by beneficial microbes. Curr Opin Plant Biol 11:443–448

Verbruggen E, Van Der Heijden MG, Weedon JT, Kowalchuk GA, Röling WF (2012) Community assembly, species richness and nestedness of arbuscular mycorrhizal fungi in agricultural soils. Mol Ecol 21:2341–2353

Vigo C, Norman J, Hooker J (2000) Biocontrol of the pathogen Phytophthora parasitica by arbuscular mycorrhizal fungi is a consequence of effects on infection loci. Plant Pathol 49:509–514

Vos C, Yang Y, De Coninck B, Cammue B (2014) Fungal (-like) biocontrol organisms in tomato disease control. Biol Control 74:65–81

Walter J, Hein R, Auge H, Beierkuhnlein C, Löffler S, Reifenrath K, Schädler M, Weber M, Jentsch A (2012) How do extreme drought and plant community composition affect host plant metabolites and herbivore performance? Arthropod Plant Interact 6:15–25

Werner GD, Kiers ET (2015a) Order of arrival structures arbuscular mycorrhizal colonization of plants. New Phytol 205:1515–1524

Werner GD, Kiers ET (2015b) Partner selection in the mycorrhizal mutualism. New Phytol 205:1437–1442

Whipps JM (2004) Prospects and limitations for mycorrhizas in biocontrol of root pathogens. Can J Bot 82:1198–1227

Xavier LJ, Germida JJ (2003) Bacteria associated with Glomus clarum spores influence mycorrhizal activity. Soil Biol Biochem 35:471–478

Xia T, Wang Y, He Y, Wu C, Shen K, Tan Q et al (2020) An invasive plant experiences greater benefits of root morphology from enhancing nutrient competition associated with arbuscularmycorrhizae in karst soil than a native plant. PLoS One 15(6)

Yang H, Zhang Q, Dai Y, Liu Q, Tang J, Bian X, Chen X (2015) Effects of arbuscular mycorrhizal fungi on plant growth depend on root system: a meta-analysis. Plant Soil 389:361–374

Yang R, Zhou G, Zan S, Guo F, Su N, Li J (2014) Arbuscular mycorrhizal fungi facilitate the invasion of Solidago canadensis L. in southeastern China. Acta Oecologica 61:71–77

Yang W, Zheng Y, Gao C, He X, Ding Q, Kim Y, Rui Y, Wang S, Guo L-D (2013) The arbuscular mycorrhizal fungal community response to warming and grazing differs between soil and roots on the Qinghai-Tibetan Plateau. PLoS One 8:e76447

Yao M, Tweddell R, Desilets H (2002) Effect of two vesicular-arbuscularmycorrhizal fungi on the growth of micropropagated potato plantlets and on the extent of disease caused by *Rhizoctoniasolani.* Mycorriza 12:235–242

Zamioudis C, Pieterse CM (2012) Modulation of host immunity by beneficial microbes. Mol Plant Microbe Interact 25:139–150

Zhang Q, Yang R, Tang J, Yang H, Hu S, Chen X (2010a) Positive feedback between mycorrhizal fungi and plants influences plant invasion success and resistance to invasion. PLoS One 5:e12380

Zhang T, Hu Y, Zhang K, Tian C, Guo J (2018) Arbuscular mycorrhizal fungi improve plant growth of Ricinus communis by altering photosynthetic properties and increasing pigments under drought and salt stress. Industr Crops Prod 117:13–19

Zhang Y, Zhong C, Chen Y, Chen Z, Jiang Q, Wu C, Pinyopusarerk K (2010b) Improving drought tolerance of Casuarina equisetifolia seedlings by arbuscular mycorrhizas under glasshouse conditions. New For 40:261–271

Zhu H, Yao Q (2004) Localized and systemic increase of phenols in tomato roots induced by Glomus versiforme inhibits Ralstonia solanacearum. J Phytopathol 152:537–542

Zhu X, Song F, Liu S, Liu F (2016) Arbuscular mycorrhiza improve growth, nitrogen uptake, and nitrogen use efficiency in wheat grown under elevated CO 2. Mycorrhiza 26:133–140

Zhu XC, Song FB, Xu HW (2010) Arbuscular mycorrhizae improve low temperature stress in maize *via* alterations in host water status and photosynthesis. Plant Soil 331:129–137. https://doi.org/10.1007/s11104-009-0239-z

Ziedan ESH, Sadek Elewa I, Mostafa HM, Sahab AF (2011) Application of mycorrhizae for controlling root diseases of sesame. J Plant Prot Res 51(4)

Influence of Arbuscular Mycorrhizal Fungi on Soil Health Amelioration and Plant Fitness Under Hostile Environment

10

Oluwatoyin Adenike Fabiyi, Rizwan Ali Ansari, Rose Rizvi, Aisha Sumbul, and Irshad Mahmood

Abstract

The application of arbuscular mycorrhizal fungi stands out as a pivotal element for advancing soil biology and health. The strategic inoculation of plants with these symbiotic fungi significantly improves the resilience of plants to a myriad of biotic and abiotic stressors, exhibiting the diverse protective roles played by these fungi in fostering plant fitness under different agro-climatic zones. In a phosphorus-limited environment, the inoculation of AM fungi proves to be particularly advantageous. This agricultural practice extends benefits beyond merely enhancing overall soil health. It plays a crucial role in rendering phosphorus accessible to plants, especially in cases where such nutrients would otherwise be unavailable. This dual impact, addressing both soil health enhancement and improved nutrient availability, serves to underscore the nuanced and multifaceted importance of mycorrhizal interactions in the context of sustainable agriculture. Harnessing the potential of AM fungi in an appropriate manner offers a holistic and sustainable approach to addressing various challenges caused by biotic and abiotic stressors. This chapter endeavours to scrutinize the coevolution of crop maturation and progression with mycorrhizal interactions amidst challenging agroecological environs, with a particular emphasis on the role of such synergies in fostering food security.

O. A. Fabiyi (✉)
Department of Crop Protection, Faculty of Agriculture, University of Ilorin, Ilorin, Nigeria

R. A. Ansari
Department of Plant Protection, Faculty of Agricultural Sciences, Aligarh Muslim University, Aligarh, India

R. Rizvi · A. Sumbul · I. Mahmood
Section of Plant Pathology and Nematology, Department of Botany, Aligarh Muslim University, Aligarh, India

R. A. Ansari et al. (eds.), *Mycorrhizal Symbiosis and Agroecosystem Restoration*,
https://doi.org/10.1007/978-981-99-5030-0_10

Keywords

Plant health · Nutrient acquisition · Food security · Soil texture · Disease control

10.1 Introduction

The contribution of agriculture to sustain the worldwide economy must be noticed. The crowded population relies so much on agriculture in several ways, which includes but not limited to food supply, and the provision of raw materials for the development of the industrial sector, which serves as a source of employment, consequently generating foreign exchange earnings (Gollin et al. 2002). The agricultural sector also affords a market for industrial products, thus summarizing that agriculture is of much significance in the economy of nations especially developing ones (Ogundipe et al. 2019). Broadly, agriculture could be divided into forestry, fisheries, livestock and crop production. Microorganisms including mycorrhizal fungi play a pivotal role in sustaining natural ecosystems, thereby contributing to improving sustainable agriculture (Godfray et al. 2010). Key components like crop health, nutrient acquisition, and growth, as well as functional development, are markedly influenced by mycorrhizal fungi and their application. A majority of plant taxa on the land have customary symbiotic associations with AMF, which are biotrophic obligate organisms (Philippot et al. 2013). A diverse array of plants, consisting of cereals, fruit trees, and vegetables has been notably found to develop associations with mycorrhizal fungi (Philippot et al. 2013).

AMFs own extra-radical mycelium (ERM), encompassing from the roots of host plants into the soil to absorb nutrients and water for the host plants (Smith and Smith 2011a). The interconnectedness, extent, and structure of the ERM critically impact nutrient flow to the host, helping as a critical component of the fungal body and a significant factor in the symbiotic alliance (Smith and Smith 2011a). The ERM harbours genes which is exclusively dedicated to nutrient translocation from the soil to the host roots, enabling the transport of nutrients from the host's rhizosphere (Smith and Smith 2011a; Puschel et al. 2016). In addition to nutritional status, AMFs also contribute to the achievement of biogeochemical cycles, enhance the tolerance of host plants to abiotic as well as biotic stress, increase the plant phytochemical compounds, enhance soil aggregation, and facilitate carbon sequestration, giving considerable care during different environmental stressors (Puschel et al. 2016). Research has proven that the various benefits consulted by AMF are intricately connected to the synergistic interactions of multiple bacterial communities in the mycorrhizosphere (Rouphael et al. 2015). These bacteria, in association with extra-radical mycelium and spores, play significant roles in improving plant growth, ranging from the production of antibiotics, siderophores, and indole acetic acid to mineralization, phosphorus solubilization, and nitrogen fixation (Rouphael et al. 2015). Bacterial strains associated with AMF have the ability to act as biostimulants and bio-fertilizers for sustainable crop production (Turrini et al. 2018). Discovering the complex web of plant-microorganisms interactions reveals that managing these microbial communities is a paramount importance. This guides toward agricultural approaches that are both environmentally friendly and support the long-term health and crop health resilience. Exploring these sides provides a chance to shape microbial interactions, paving the way for a more sustainable and environmentally approachable future in agriculture. This chapter aspires to collect robust information on mycorrhizal plants and their forbearances under hostile environment. This chapter helps us find knowledge relating to common stressors and mycorrhizal role in their amelioration.

10.2 Arbuscular Mycorrhizal Fungi: A Good Plant Symbionts

AMF belongs to the phylum Glomeromycota comprising families like Gigasporaceae, Acaulosporaceae, Glomeraceae, Ambisporaceae, Pacisporaceae, Archaesporacea, Paraglomeraceae, Claroidoglomeraceae, Sacculosporaceae, and Diversisporaceae (http://www.amf-phylogeny.com/, accessed on August 7, 2021). The life cycle of AMF is exclusively within a host plant because of their obligatory nature. During the asymbiotic phase of development, many short-lived mycelia are formed through spore germination, influenced by physical factors such as pH, temperature and moisture (Kiers et al. 2011). The pre-symbiotic phase takes place in the presence of the host plant's root exudates (Kiers et al. 2011). At this event, the germling hyphae undergo morphogenesis, reorient, and develop branching patterns (Jiang et al. 2017). After that, physical contact is established between hyphae and roots of the host plant leading to appressoria differentiation, intracellular hyphal growth in the root cortex, penetration of root cells, and the formation of tree-like structures known as arbuscules (Bonfante and Genre 2008), which is considered as primary organs for nutrient exchange in the mycorrhizal symbiosis (Wipf et al. 2019). The arbuscules play a pivotal role in nutrient exchange, where AMF obtains carbon and lipids from the host, and in return, releases and translocate absorbed mineral nutrients through extra-radical mycelium (ERM) (Luginbuehl et al. 2017). There are two mycorrhizal formation types such as the *Paris* and *Arum* types, which has been encountered. The Paris type involves the direct growth of fungus from cell to cell in the root cortex, forming intracellular hyphal coils and intercalary arbuscules (Luginbuehl et al. 2017). Conversely, Arum-type arbuscules are formed terminally on hyphal intercellular branches, spreading symbionts in an intercellular manner in root cortical cells (Jiang et al. 2017). *Arum* type mycorrhizal fungi are significantly prevalent in natural habitats and widespread in agricultural systems (Jiang et al. 2017). AMF commonly feature intra-radical vesicles that serve as storage structures filled with lipids, distinct from arbuscules. Once carbon is collected from the host, the symbiont's ability to develop extra-radically begins, leading to complete colonization of the surroundings. The host receives nutrients regularly, and AMF interacts with microorganisms in the soil rhizosphere. Nearby plant roots are also colonized, fostering nutrient exchange between different hosts, irrespective of species, families, or genera (Luginbuehl et al. 2017). Subsequently, asexual spores are developed by ERM, marking the closure of the AMF life cycle, reliant on soil surroundings and fertility

(Luginbuehl et al. 2017). Later, asexual spores are formed by the ERM, depending on how the soil can maintain the growth of mycorrhiza and also consequent upon the fertility of the soil, and this stage then closes the life cycle of the AMF.

The association between AMF and plants is widespread, encompassing various plant phylogenies that have evolved on terrestrial landscapes over the course of approximately 475 million years (Field et al. 2015). Micronutrients such as zinc, nitrogen, copper, and other important elements like phosphorus derived from the soil, are efficiently acquired through the ramification of mycelia in an extraradical manner (Liu et al. 2000). Extraradical hyphae exhibits the ability to acquire phosphorus from organic sources like DNA and proteins by secreting acid phosphatases. (Sato et al. 2015). The Mucoromycota and Glomeromycota exhibit a consistent pattern as they form intracellular symbiotic association with majority of the crops (Smith and Smith 2011a; Spatafora et al. 2016). The evident benefit of AMF in enhancing crop growth through nutrient supply has recently garnered significant attention for its capacity to reduce dependence on inorganic fertilizers without negotiating with the crop yield (Berruti et al. 2016a, b). The nutrients taken up by AMF is also markedly influenced by factors like the CO_2 concentration in the air, nutrient availability, and fungi and host involved (Field et al. 2012). The substantial variability in the outcomes arising from the involved factors poses a substantial impairment to the widespread application of AMF in improved agriculture. The symbiotic association between plants and fungi shows advantageous by expanding the range of available nutrient resources for plant growth and development (Johnson 2010). Mycorrhizal plants exhibit enhanced growth in phosphorus-depleted soils compared to plants without such symbiotic associations (Bender et al. 2016). Moreover, nitrogen transfer to plant by AMF via extraradical hyphae has been extensively examined (Hodge and Fitter 2010), revealing an associated increase in plant growth and development (Thirkell et al. 2016). These revelations indicate that AMF possess the capability to supplement the sources of nutrients accessible for host plants and increase their uptake. However, the extent to which plants harness the benefits of AMF symbiosis hinges upon various agronomic practices like fertilizer application and other land management (Johnson 2010). The response of cereal crops may vary. The growth of cereals remains unaffected by AM fungi in soils where phosphorus concentration equals that achieved through application of phosphorus-based fertilizer (Li et al. 2016). This finding implies a crucial balance between fertilizer utilization and the benefits derived from the symbiotic relationship with AMF. However, few plants have demonstrated diminished growth when symbiosed with AM fungi. It is assumed that substantial transfer of phosphorus by fungi results in the release of plant carbon to the fungi, leading to a carbon deficiency in the plant, thus causing sluggish and restrained growth of plants (Kiers et al. 2011). Walder et al. (2012) reveal that the nutrients and minerals acquisition through a shared mycelium is contingent upon host specificity, independent of carbon availability in the mycelium network. Moreover, The conventional pathway for phosphorus uptake by the roots of plant facilitated by the inherent transporters of phosphates, is typically hampered and reduced in the presence of AMF (Smith and

Smith 2011b). The phosphorus uptake in plants associated with AMF is considerably reduced when the phosphorus delivery mechanism of AMF fails to compensate for the plant's lower phosphorus utilization (Smith and Smith 2011b). The variation regarding phosphorus uptake routes and their relative contributions may be liable for the differences in phosphorus utilization. The positive impact of AMF on the growth of plants within the Solanaceae, Alliaceae, and Fabaceae families has grown appreciably while the response within other crop families, mainly cereals, remains undefined (Li et al. 2005). Variations in the response to AMF association are evident even among cultivars of the similar species (Tawaraya 2003). This kind of inconsistency has been observed in wheat, where some cultivars demonstrate a positive response to AMF associations (Hu et al. 2014), while others exhibit a negative response (Li et al. 2016). The precise cause of this divergence present a significant problem to the widespread application of AMF in cereal production on a wider scale. The physiological, morphological, and inherent features of plant roots likely influence the extent and efficiency of nutrient attainment in AMF (Navarro-Fernandez et al. 2016). For instance, widely branched roots in cereals enable a methodical nutrient uptake, obviating the necessity for AMF alliance, as the roots are inherently adapted to complete functions analogous to those of AMF (Smith and Smith 2011a). To optimize nutrient integration for crop development and growth, a comprehensive consideration of various factors is the need of hour. These include, plant's genetics, the consistent supply of nutrients, understanding of the root characteristics, their quality, pivotal factors of root architecture, inclination for symbiotic associations, and the influence of pathogens (Li et al. 2016).

The communities of AMF in the subsoil show distinctive characteristics and are plentiful (Higo et al. 2013; Sosa-Hernández et al. 2018). Their contributions to the functioning of the ecosystem and plant growth are unappreciated. The characteristics of the subsoil like compaction, large bulk density, limited pore spaces and minimal concentration of oxygen in totality provide a semi-optimal habitat for plant roots (Moll et al. 2016). The presumption is that AMF are well adapted to the subsoil, although there is no validated evidence regarding the specific characteristics of subsoil AMF. It is predictable that AMF existing in the subsoil will exhibit stress tolerance, following the CSR context (Chagnon et al. 2015). The phylotypes are expected to exhibit efficient resource consumption, leading to long-lasting biomass that results in cost-effective rewards for the plant. While mycelium growth may be initially slow without immediate assistances, it is likely to serve as a carbon sink for the host plants. However, immediately the network of mycelium is entrenched, the services provided to the plant will be long-lasting without any cost (Säle et al. 2015). The dispersal of AMF spores could be by arthropods, earthworms, small mammals or wind (Egan et al. 2014), but these vectors are not likely to be germane in the subsoil with the earthworm being a deviation. The development of fungal mycelium is advantageous for the host plant, as AMF spores act as storage for carbohydrates and lipids (Walder et al. 2012). In the symbiotic association, the host contributes energy and carbon, the mycelium explores the soil, extract the nutrients for the cause of the plant. AMF are considered for their pivotal role in soil formation and its improvement (Liu et al. 2016). AMF actively contribute in weathering of rocks (Berruti et al. 2016a, b). It will be very challenging for the plants to survive in nutrient depleted soil without mycorrhizal fungi (Leake and Read 2017). AMF have the ability to release low molecular weight organic chelators that help in mineral weathering. Howbeit, the AMF contributes to weathering of minerals through indirect routes like improved soil stabilization and respiration plus intensified exudation and evapotranspiration (Verbruggen et al. 2021). The anticipated differences in the weathering patterns between different group of mycorrhizal fungi are not readily apparent (Koele et al. 2014). The subsoil contains plenty amount of minerals, however, does not actively support biological activity. The mycorrhizosphere, supports microbial activity in the subsoil that contributes to a sizable soil volume. This conjoint activity involves AMF, roots and the associated soil microorganisms that enhance soil expansion leading to enhanced and substantial soil formation (Higo et al. 2013).

The functions of mycorrhizal associations may undergo various changes due to intensive agricultural activities, in addition to positive impact of AMF on host plant growth and nutrition. The introduction of wide array of chemical fertilizers has also been found influential to mycorrhization (Chagnon et al. 2015; Johnson et al. 2015). Rillig et al. (2015) reported that historical aspects and crop cultivations before introducing AMF as inoculum should also be considered. For instance, addition of AMF may prove beneficial for wheat, but its effects may considerably vary depending on the species, as observed in the case of barley (Navarro-Fernandez et al. 2016). The utilization of AMF in mixed cultivations may also develop a considerable constraint. The quantification of nutrient exchange for carbon among symbionts remains incomplete, particularly with regard to arable crops for profitable drives. These gaps in knowledge hinder research on the application of AMF for the cultivation of various food crops. It is important to focus on the explicit roles of AMF in cropping systems especially their response to application of mycorrhizal fungi (Li et al. 2016). Research should also encompass cocktail communities, and all these factors are significant in the performance of AMF inoculation on plants as regards nutrient uptake, development and growth. There is no information on tracking crops from emergence to harvest, meaning that it is laborious to come about any document in literature, which infers the quantity and total yield to AMF application rates and doses. The effect of the environment on AMF alliance with crops cannot be overlooked, though we may say that it is not so significant a factor in nutrient exchange. The level of importance of AMF to the host could be measured via the carbon received. Also, carbon, phosphorus and nitrogen exchange in the AMF plant relationship measurement is hardly recorded in the literature, and this is another gap in the mycorrhizal research. Saturation of the atmosphere with CO_2 reduces the exchange of carbon and nutrients between AMF and host plants (Field et al. 2012). However, this phenomenon for agricultural crops remains uncertain as significant research has not been performed. Intergovernmental Panel on Climate Change (IPCC) revealed that atmospheric CO_2 undoubtedly impact the agricultural crops. The effect of elevated CO_2 on mycorrhizal symbiosis has also been unravelled and such experiments gave very nice information. Comprehending the response of these symbionts to climate change, especially as regards increased atmospheric CO_2 is necessary. The effect of AMFs on crop yield globally and their prospective ability to assuage environmental fuss in the future should be of prime concern (Challinor et al. 2016). The AMF have potential applications in connection with anticipated environmental challenges, especially those that are easily expectable or region-specific. Recognition of the plant attributes both below and above ground is very important for analysing the functional adaptability of mycorrhizal fungi and crop interactions in connection to various environmental conditions. This change enables the maintenance of crop yields in the face of climates change. The carbon exchange evaluation for nutrients among different crop can be also compared similarly to soil types under various field conditions. These unfamiliar attributes possess the potential to function optimally within changing surroundings, making them suitable subjects for future research. When spotted, they could be included in breeding research for commercial purposes. The available information on AMF and their relationships with different crops, especially in the context of food production has been derived from temperate regions. This underscores a significant gap in knowledge concerning plant interactions with AMF in tropical climates, chiefly in underdeveloped nations (Bhantana et al. 2021). Understanding the mechanisms involved between AMF and specific plant species at the physiological and molecular levels should be explained through the simple experimental model. The mycorrhizal effects at ecosystem, landscape, or field levels, where the dynamics of the ecosystem and the complex connections between fungi and plants occur in complex ways. Integration of a new ecological approaches into farming requires a detailed understanding of the involved practices and complexities Cabral et al. (2015). This underscores the necessity for ongoing field experiments to measure nutrient exchanges, understand fungal connections, and observe their responses to changing climate. These efforts are important for establishing AMF in sustainable agriculture. The measurement of carbon, phosphorus, and nitrogen outflow among plant symbionts, especially under normal and elevated levels of CO_2, is a significant aspect that demands consideration. The influence of the symbiotic relationship on plant defence, soil function and structure appears to be a significant gain from the association.

10.3 AM Fungi and Their Role in Stressor Ameliorations and Plant Fitness

10.3.1 AM Fungi as Alleviator of Biotic Stressors

10.3.1.1 Mycorrhizal Fungi and Their Effects on Plant Nematode Management

Fungi symbionts form an alliance with the roots of the majority of plants, providing them a range of benefits, such as increased nutrient uptake, improved water uptake, and enhanced resistance to various biotic and abiotic stresses. Besides, AM fungi either in solo or in combination with other microorganisms play a pivotal role in plant disease control (Ansari et al. 2020b; Ansari and Mahmood 2019a, b). Beneficial microorganisms of different groups improve plant health including crop production and productivity (Aghale et al. 2017; Ansari and Mahmood 2017a, b; Ansari et al. 2017a, b, 2019a, b; Fabiyi et al. 2023).

AM fungi can indirectly suppress plant diseases by improving plant health and promoting growth, thereby reducing plant susceptibility to pathogens. This is achieved through the mycorrhizal-induced resistance mechanism, which involves priming plant defence responses against potential pathogens (Solanki et al. 2020; Rizvi et al. 2015). The AM fungi can enhance the expression of pathogenesis-related (PR) genes, which are involved in the production of antimicrobial compounds and other defence mechanisms (Campos-Soriano et al. 2012). AM fungi can compete with pathogens for nutrients and space, further reducing their ability to infect plants (Goltapeh et al. 2008). AM fungi can reduce the severity of various plant diseases caused by fungal, bacterial and viral pathogens. For example, studies have demonstrated that AM fungi can reduce the severity of fusarium wilt in tomato plants, powdery mildew in cucumber plants and root rot in bean plants (Srivastava et al. 2010; Liu et al. 2018; Al-Askar and Rashad 2010).

The AM fungi in controlling plant diseases can be affected by a wide range of factors that includes the type of plant and pathogen, environmental conditions and AM fungal diversity. AM fungi and their significance in plant disease control and developing effective strategies for utilizing these fungi in sustainable agriculture are also important aspects.

The plant parasitic nematode, including *Meloidogyne incognita*, is a major parasite of different crops (Fabiyi and Olatunji 2021a, b). The AM fungi have been reportedly to be phenomenal in the management of different plant diseases (Ansari et al. 2020a; Sumbul et al. 2017; Solanki et al. 2020; Rizvi et al. 2015; Ansari and Mahmood 2019a). According to a study conducted by Singh et al. (2016), inoculation of tomato plants with *G. fasciculatum* resulted in a significant reduction in the number of *M. incognita* juveniles in the roots, as well as a decrease in the number of galls and egg masses produced by the nematodes. The study also reported an increase in plant growth characters markedly. Another study by Siddiqui and Mahmood (2012) reported similar findings, where the application of *G. fasciculatum* reduced the population density of *M. incognita*, and enhanced the plant health contributing variables. The mechanism by which *G. fasciculatum* suppresses *M. incognita* is not fully understood, but it has been observed that AM fungi can induce systemic resistance in plants against a wide array of plant pathogens through the activation of defense mechanisms, as well as through the direct competition for nutrients with the nematodes.

Pozo and Azcon-Aguilar (2007) showed that the inoculation of tomato plants with AMF induced the expression of defense-related genes, such as pathogenesis-related proteins and enzymes involved in the production of reactive oxygen species. This activation of plant defences was somehow related to a poor number of *M. incognita* juveniles in the roots.

Similarly, a study by Jung et al. (2012) observed the introduction of AM fungi to cucumber plants increased the activity of defence enzymes and the production of phytohormones (jasmonic acid and salicylic acid) that results in a reduction of root-knot nematodes in the soil.

Regarding the second hypothesis, a study by Stirling and Roper (2010) demonstrated that the inoculation of maize plants with AM fungi reduced the population density of *M. incognita* and increased plant growth parameters such as shoot biomass and leaf area. The authors suggested that this effect was due to the ability of AMF to extract soil nutrients more efficiently than the nematodes, thus reducing the availability of nutrients for the nematodes and limiting their growth. Overall, these studies provide support for the hypotheses that AMF can induce plant defences and compete with nematodes for nutrients, thus playing a role in the suppression of *M. incognita* nematodes.

Likewise, *Heterodera* spp. is a genus of plant-parasitic nematodes, commonly known as cyst nematodes, that cause significant damage to crops worldwide (Fabiyi et al. 2018a, b; Fabiyi et al. 2020; Fabiyi 2021a, b, c, d). While there is limited research on the effects of *Glomus fasciculatum* on *Heterodera* nematodes, some studies have suggested that AMF may have potential as a biological control agent against these pests.

Goswami et al. (2013) investigated the impact of AM fungi inoculation on *Heterodera cajani,* a cyst nematode that infects pigeon pea plants. The researchers found that the inoculation of pigeon pea plants with *Glomus fasciculatum* impaired the population of *H. cajani* cysts in the soil, also the juvenile population was decreased over uninoculated plants. The authors suggested that AMF may suppress *H. cajani* by inducing plant defences and by competing with the nematodes for nutrients. The mechanisms by which *G. fasciculatum* suppressed *H. cajani* in this study are not fully understood, but there are several possible explanations. One mechanism may involve the activation of plant defences. The studies have shown that the inoculation of plants with AM fungi can induce the expression of defence-related genes and enzymes, which may help protect the plant from nematode infection (Pozo and Azcon-Aguilar 2007). It is possible that the inoculation of pigeon pea plants with *G. fasciculatum* in this study resulted in the activation of defence mechanisms that prevented *H. cajani* from infecting the plant roots. Another possible mechanism is competition for nutrients. *G. fasciculatum* forms symbiotic associations with plant roots, allowing it to extract nutrients such as phosphorus and nitrogen from the soil more efficiently than non-mycorrhizal plants. By competing with *H. cajani* for nutrients, *G. fasciculatum* may have impaired the multiplication and development of plant nematodes.

Saberi-Riseh et al. (2016) investigated the effect of AMF inoculation on *Heterodera schachtii,* a cyst nematode that infects sugar beet plants. The researchers found that the inoculation of sugar beet plants with *G. fasciculatum* showed a significant reduction in *H. schachtii* in the roots and soil, as well as a decrease in the number of cysts produced by the nematodes. The authors suggested that AMF may suppress *H. schachtii* by altering the nematode's life cycle and by enhancing plant growth and nutrient uptake. The authors suggested that AMF may suppress *H. schachtii* through several mechanisms. Firstly, the AMF may alter the nematode's life cycle by reducing its ability to hatch and infect plants. This effect may be due to the production of compounds by the AMF that are toxic to the nematode or that interfere with its chemosensory abilities. Secondly, the AM fungi may enhance plant growth and nutrient uptake, making the plants more resistant to nematode infection. This effect may be due to the ability of the AM fungi to develop specialized network of hyphae in the soil, which can access nutrients that are not available to the plant roots. The enhanced nutrient uptake by the plants leads to an acceleration in the production of some defence molecules, which can inhibit nematode growth and development.

10.3.1.2 Fungal Disease Management Through AM Fungi

The reciprocal association between AM fungi and plants is known to improve plant health and provide plants with enhanced protection against various biotic and abiotic stresses. In the case of plant diseases, AM fungi can play a significant role in mitigating disease severity and preventing disease spread. Several studies have reported the beneficial effects of *G. fasciculatum* on the management of Fusarium wilt disease in various crops. For example, in tomato plants, the application of *G. fasciculatum* registered a significant reduction in the severity of Fusarium wilt disease, in addition to increasing plant growth and productivity (Ansari et al. 2019a, b; Khalid et al. 2019; Zhang et al. 2018). Similarly, in banana plants, the use of *G. fasciculatum* exhibited reduced Fusarium wilt disease incidence, also enhance the plant growth characters (Luo et al. 2018; Singh et al. 2017). In cotton plants, the application of *G. fasciculatum* was found to decrease the incidence of Fusarium wilt disease, in conjunction with increase plant growth and yield (Fakher et al. 2016). Finally, in melon plants, the use of *G. fasciculatum* was found to diminish the severity of Fusarium wilt disease, coupled with improvement in plant growth and fruit quality (Abdel-Salam et al. 2019).

The mechanism of action of *G. fasciculatum* in the management of Fusarium wilt disease is not well understood, but it is thought that the induction of systemic resistance in plants is one of the important mechanisms. This induction of systemic resistance is mediated by the synthesis of various defence-related compounds, such as phytohormones, enzymes and secondary metabolites, which are triggered by the presence of AM fungi in plant roots. These defence-related compounds act as signalling molecules that activate the plant's defence system against the invading pathogen, thereby reducing the severity of disease symptoms.

In addition to the induction of systemic resistance, the presence of *Glomus fasciculatum* in the soil also promotes the multiplication of beneficial soil microorganisms, which can contest with *Fusarium oxysporum* for nutrients and space, thereby reducing the pathogen's ability to establish itself in the soil. Furthermore, AM fungi can enhance the uptake of nutrients, particularly phosphorus, by plants, which can improve plant health and increase plant resistance to disease.

10.3.1.3 Bacterial Disease Management Through AM Fungi

Some studies have shown that AM fungi, including *G. fasciculatum*, can alleviate plant diseases, especially those caused by plant pathogenic bacteria by enhancing plant growth and stimulating the plant's defence mechanisms.

One mechanism by which AM fungi enhances plant defence against bacterial diseases is by induced systemic resistance. The colonization of roots by AM fungi can trigger the plant's systemic acquired resistance (SAR) pathway, leading to the accumulation of defence-related compounds such as phytohormones, enzymes and secondary metabolites, which can activate the plant's defence against bacterial pathogens. Another mechanism is the direct competition between AMF and bacteria for nutrients and space in the rhizosphere. AM fungi can compete with bacteria for the available nutrients and create a hostile environment for bacterial growth by releasing antifungal compounds and stimulating the plant's production of phytoalexins and other antimicrobial compounds (Smith and Smith 2019).

Liao et al. (2014) reported that *G. fasciculatum* inoculation to tomato plants reduced the incidence of bacterial wilt, which is caused by virulent strain of *Ralstonia solanacearum,* which seems to be due to the induction of systemic resistance in the plants and the competition for nutrients in the rhizosphere. In a study by Berruti et al. (2016a, b), *G. fasciculatum* inoculation reduced the incidence of bacterial canker in tomato caused by *Clavibacter michiganensis* subsp. *michiganensis.* The researchers proposed that reduced disease incidence was due to the stimulation of the plant's defence mechanisms and the competition for nutrients in the rhizosphere. In a study by Wu et al. (2019a, b), inoculation with *G. fasciculatum* reduced the incidence and severity of bacterial leaf blight in rice (Causal agent, *Xanthomonas oryzae* pv. *oryzae*). The researchers attributed the reduced incidence of the disease seems to be due to the increased expression of defence-related genes and the competition for nutrients with the pathogen. Although direct evidence on the effect of *G. fasciculatum* on bacterial plant disease management is limited, the available studies suggest that it can indirectly affect bacterial disease incidence and severity through the stimulation of plant defence mechanisms and the competition for nutrients in the rhizosphere.

10.3.1.4 Viral Disease Management Through AM Fungi Application

AM fungi have exhibited remarkable phytoviral disease resistance-enhancing properties. The intricacies of the underlying mechanisms responsible for these effects remain obscure and poorly comprehended, but there is evidence that mycorrhizal fungi may accelerate the induction of systemic resistance, improve the plant growth and nutrient uptake capability and modulate plant hormone signalling pathways, all of which can contribute to increased plant resistance to viral diseases. The scholarly inquiry conducted by Cao et al. (2020) delved into exploring the impact of AM fungi on the fortitude of cucumber plants against the pernicious cucumber mosaic virus (CMV). The researchers found that AM fungi improved the resistance of cucumber plants to CMV infection by regulating plant hormone signalling and SAR. The researchers inoculated cucumber plants with AM fungi and then infected the plants with CMV. They found that the AM fungi treatment impaired the disease incidence and severity of CMV symptoms in the cucumber plants. The AM fungi treatment also increased the activity of defence-related enzymes and the expression of defence-related genes in plants. AM fungi treatment increased the levels of plant hormones, salicylic acid and jasmonic acids in the plants, which are important plant hormones involved in defence mechanisms against pathogens. The researchers also found that the AM fungi treatment enhanced SAR in the plants, which is a long-lasting defence response that provides broad-spectrum resistance against a range of pathogens.

Wang et al. (2019a, b) investigated the effect of AM fungi on the resistance of maize plants to maize chlorotic mottle virus (MCMV). The researchers found that AM fungi enhanced the resistance of maize plants to MCMV infection by improving plant growth and modulating plant defence responses. The researchers inoculated maize plants with AM fungi and then infected the plants with MCMV. They found that the AM fungi treatment significantly lowered incidence and severity of MCMV symptoms in the plants. The AM fungi treatment also increased plant height, shoot biomass and root biomass over non-inoculated plants. AM fungi treatment increased the activity of defence-related enzymes such as peroxidase and phenylalanine ammonia lyase and increased the accumulation of phenolic compounds in the plants Aseel et al. (2019). These are all important components of the plant defence response against viral infections. They also opined that the AM fungi treatment upregulated the expression of defence-related genes in the plants, such as pathogenesis-related protein genes and genes involved in phytohormone signalling pathways.

Marquez-Garcia et al. (2020) reviewed the role of beneficial soil microorganisms and their interactions with the plant immune system for the biological control of plant viral diseases. They revealed that beneficial soil microorganisms can induce systemic resistance in plants and enhance their ability to resist viral infections.

The study focused on the interactions between plants and three groups of beneficial soil microorganisms: AM fungi, plant growth-promoting rhizobacteria (PGPR) and *Trichoderma* species Ramasamy et al. (2011). AM fungi were found to enhance the production of defence-related compounds in plants and to induce systemic resistance against viral infections. PGPR were found showing stimulated the plant immune system by producing signalling molecules that activate defence pathways, and by competing with viral pathogens for nutrients and space. *Trichoderma* species were found to produce antiviral compounds and to induce the expression of defence-related genes in plants Dey et al. (2006). The researchers also highlighted the potential for combining different types of beneficial soil microorganisms to enhance their effectiveness in controlling plant viral diseases. For example, AM fungi and PGPR can work together to induce systemic resistance in plants and enhance their growth and health.

10.3.2 AM Fungi as Alleviator of Abiotic Stressors

10.3.2.1 AM Fungi and Its Significance in Heavy Metals Amelioration

AM fungi, being integral to the sustenance of plant nutrition and soil vitality, have demonstrated auspicious potential for employment in the amelioration of soils tainted with toxic heavy metals Chen et al. 2007. The AM fungi have been found to impact heavy metal remediation include *Rhizophagus, Funneliformis, Claroideoglomus* and *Gigaspora* (Fernandez and Nair 2018). These genera have shown accumulation of heavy metals in their hyphae and spores, reducing their bioavailability in the soil and promoting their immobilization (Feng et al. 2013). Some key examples of AM fungi that have been found to impact heavy metal remediation include *Rhizophagus intraradices, Funneliformis mosseae, Claroideoglomus etunicatum* and *Gigaspora margarita* (Khan 2005; Fernandez and Nair 2018). *Rhizophagus intraradices* have been shown to reduce cadmium (Cd) concentration in maize and sorghum by up to 70% in contaminated soils (Huang et al. 2018). *Funneliformis mosseae* has been found to reduce lead (Pb) uptake in tomato plants by up to 50% in Pb-contaminated soils (Singh et al. 2018). *Claroideoglomus etunicatum* has been shown to reduce zinc (Zn) uptake in maize by up to 50% in Zn-contaminated soils (Wang et al. 2017). The discovery has been made that *Gigaspora margarita* can promote the growth of *Brassica juncea* vegetation while simultaneously mitigating their copper (Cu) concentration by an impressive margin of up to 80% in Cu-laden soils (Wu et al. 2019a, b). The important mechanisms involved in heavy metal reduction by AM fungi include

Adsorption and Sequestration

AM fungi can adsorb heavy metals onto their cell walls or sequester them within their hyphae, reducing their bioavailability in the soil and promoting their immobilization (Garget al. 2017).

Precipitation

AM fungi can help reduce the effect of heavy metals and thus well balanced ecosystem may be maintained. in the form of insoluble metal sulphides, which reduces their solubility and mobility in the soil (Miransari 2017).

Complexation

AM fungi possess the capability of producing complexes with heavy metals, thereby facilitating the reduction of their phytotoxicity whilst concurrently instigating their immobilization within the soil (Li et al. 2017).

Bioaccumulation

AM fungi have the competence to amass heavy metals within their tissues, consequently diminishing their bioaccessibility within the soil and fostering their immobilization (Szada-Borzyszkowska et al. 2021).

Ma et al. (2020) conducted an inquiry into the ramifications of the AM fungi on the augmentation of growth and uptake of cadmium (Cd) in *Solanum nigrum* L. thriving in a multifariously tainted soil encompassing lead (Pb) and zinc (Zn). Results showed that inoculation with AM fungi significantly improved plant growth, photosynthetic pigment content and antioxidant enzyme activity. Moreover, AMF significantly increased the Cd concentration in the roots and shoots of the plants, indicating that they can enhance Cd uptake. The researchers also investigated the mechanisms implicated in AMF-mediated Cd uptake. They found that AM fungi colonization increased the expression of genes involved in Cd transport and accumulation, such as the ABC transporter genes and the metallothionein gene. The AM fungi enhanced the activity of enzymes phenomenal in Cd chelation and detoxification, like glutathione reductase and superoxide dismutase. The findings of Ma et al. (2020) suggest that AM fungi can be a useful tool for phytoremediation of Cd, Pb and Zn multi-contaminated soils. The mechanisms implicated include increased expression of genes involved in Cd transport and accumulation, enhanced activity of enzymes involved in Cd chelation and detoxification.

Li et al. (2019) conducted a study to find out the beneficial role of AM fungi on Cu uptake and its distribution in tomato plants. The researchers found that AM fungi exhibit the capacity to enrich the growth and biomass of tomato vegetation within copper (Cu)-impaired soil, thus signifying their potential to enhance plant fortitude against Cu-related duress. Additionally, the results showed that AM fungi can enhance Cu uptake and gathering in tomato plants. The Cu content in the roots of tomato plants colonized with AM fungi was expressively more advanced than that of non-mycorrhizal plants. Furthermore, the scientists ascertained that the inoculation of AM fungi led to a surge in the translocation of copper (Cu) from roots to shoots, as evidenced by the markedly elevated Cu content discerned in the shoots of mycorrhizal flora in comparison with the non-mycorrhizal counterparts.

The mechanisms implicated in the AM fungi-mediated Cu uptake and translocation in tomato plants involve several processes. First, AMF can increase the surface area of roots by forming a dense network of hyphae, which increases the interaction of plant roots and soil particles, leading to increased uptake of Cu Zhou et al. (2017). Second, AM fungi can release organic acids and enzymes that solubilize insoluble Cu compounds in the soil, making them available for plant uptake (Fomina et al. 2005). Third, AMF can regulate root membrane permeability and transporters, which affects the uptake and translocation of Cu in plant tissues Andrade et al. (2010). Lastly, AMF can increase the production of phytochelatins (PCs) in plant tissues, which bind to and detoxify Cu ions, reducing the toxicity of Cu in the plant. The mechanisms implicated in this process involve several physiological and biochemical processes that are regulated by AMF colonization. These discoveries have unveiled salient revelations regarding the prospective utility of AMF AM fungi in the rectification of copper (Cu)-tainted soil Riaz et al. (2021). In essence, the capacity of AM fungi to curb heavy metal bioaccessibility while simultaneously spurring their immobilization within the soil renders them an encouraging instrument for the amelioration of heavy metal-laden soils.

10.3.2.2 AM Fungi and Its Role in Different Environmental Factors (Temperature, Soil Texture, Sulphur Dioxide and Other Environmental Gas Elevation)

AM Fungi and Temperature Reduction

Diverse AM fungal genera have been documented to exert an influence on the ambient temperature of their corresponding host plants' rhizosphere. Liu et al. (2023) demonstrated that inoculation of AM fungus significantly improved the growth of cucumber seedlings under high-temperature conditions by enhancing nutrient acquisition and reducing oxidative damage. There are some important mechanisms that are involved in thermo tolerance studies. For example, AM fungi can enhance the availability of nutrients such as phosphorus and nitrogen to their host plants, which can help plants to cope with high-temperature stress (Kumar et al. 2021), AM fungi are proficient in eliciting the generation of heat-shock proteins within their host plants, which partake in shielding the plants against the deleterious effects of high-temperature duress. (Wang et al. 2019a, b), Plant symbionts increase antioxidant activity that can help to scavenge reactive oxygen species and prevent oxidative damage under high-temperature stress conditions (Wang and Qiu 2006). AM fungi can augment the hydrological status of their corresponding host plants via fortification of water absorption and minimization of water leakage, thus culminating in the facilitation of plant resilience towards high-temperature stress (Wang et al. 2005).

AM Fungi and Soil Texture

Soil textural composition stands out as a key influencer in shaping the constitution and efficacy of mycorrhizal interactions, and numerous research endeavours have delved into the linkage between these factors. Research has shown that AM fungal diversity and population can vary with different soil texture. Chen et al. (2018) reported the richness and diversity of AM fungi in different soil types followed the order of loamy soil > sandy soil > clayey soil. Liu et al. (2020) found that AM fungi colonization was greater in sandy loam soil than the clay soil where tomato plants were cultivated. Effects of soil texture on the functioning of mycorrhizal associations have also been investigated. Lehnert et al. (2017) reported that AM fungi inoculation enhanced the growth of grasses in sandy soil, but had no effect in clayey soil. Wang et al. (2015) observed that mycorrhizal inoculation increased phosphorus uptake and plant growth characters production rate in sandy loam soil, but had no effect in clayey soil. Furthermore, the relationship between soil texture and mycorrhizal associations can also vary depending on the type of mycorrhizal fungi. Klironomos et al. (2000) found that the effects of soil texture on AM fungi were stronger than on EM, which were more affected by soil chemistry.

10.4 Conclusions and Future Outlooks

The strategic application of AM fungi is a comprehensive tactic to raise the overall well-being of soil. This includes a triad of developments in biological, physical, and chemical properties of soil leading to resilient and fertile soil. The symbiotic association with AM fungi clarifies a significant role in pathogen suppression, deploying a range of mechanisms that actively reduce the impact of infections. The scalability of AM fungi in sustainable agriculture is still a challenging task. This includes a delicate balance, ensuring that the application of AM fungi is not just a local remedy but a complete shift in farming practices. This way natural resilience of plants can be improved through mycorrhizal associations not only, but also curtail the reliance on chemical pesticides, contributing to a more sustainable and eco-friendlier agroecosystem. The application of mycorrhiza is practicable, for future research route projections, the focal point should be mycorrhizal inoculated seedlings for sizable and substantial agricultural production. Future research is expected to concentrate on the optimization of nutrient uptake with the effect on yield actualization and dependable output. By recognizing optimally, the traits relevant to AM fungi flexibility in plant species, crucial progress can be achieved towards food security in supportable systems of agriculture.

References

Abdel-Salam EM, Fouda AH, Shaheen AM, Rady MM (2019) Effect of arbuscular mycorrhizal fungi on growth, fruit quality and Fusarium wilt disease of melon plants. Arch Phytopathol Plant Protect 52(7–8):623–635. https://doi.org/10.1080/03235408.2019.1597606

Aghale DN, Egbucha KC, Umeh OJ (2017) Plant parasitic nematode threat to Nigerian Agricultural Food Security. Int J Adv Res Bot 3(3):15–21. https://doi.org/10.20431/2455-4316.0303003. www.arcjournals.org

Al-Askar A, Rashad Y (2010) Arbuscular mycorrhizal fungi: a biocontrol agent against common. Plant Pathol J 9(1):31–38

Andrade SAL, Silveira APD, Mazzafera P (2010) Arbuscular mycorrhiza alters metal uptake and the physiological response of Coffea arabica seedlings to increasing Zn and Cu concentrations in soil. Sci Total Environ 408(22):5381–5391

Ansari RA, Mahmood I (2017a) Optimization of organic and bio-organic fertilizers on soil properties and growth of pigeonpea. Sci Hortic 226:1–9

Ansari RA, Mahmood I (2017b) Determination of disease incidence caused by Meloidogyne spp. and or Fusarium udum on pigeonpea in Aligarh district: a survey. Trends Biosci 10(24):5239–5243

Ansari RA, Mahmood I (2019a) Plant health under biotic stress. In: Ansari RA, Mahmood I (eds) Microbial interactions, vol II. Springer Nature, Singapore. https://doi.org/10.1007/978-981-13-6040-4

Ansari RA, Mahmood I (2019b) Plant health under biotic stress. In: Organic strategies, vol I. Springer Nature, Singapore. https://doi.org/10.1007/978-981-13-6043-5

Ansari RA, Mahmood I, Rizvi R, Sumbul A (2017a) PGPR: current vogue in sustainable crop production. In: Kumar V (ed) Probiotics and plant health. Springer Nature, Singapore, pp 455–472

Ansari RA, Mahmood I, Rizvi R, Sumbul A, Safiuddin (2017b) Siderophores: augmentation of soil health and crop productivity. In: Kumar V (ed) Probiotics in agroecosystem. Springer Nature, Singapore, pp 291–312

Ansari RA, Sumbul A, Rizvi R, Mahmood I (2019a) Potential role of plant growth promoting Rhizobacteria in alleviation of biotic stress. In: Ansari RA, Mahmood I (eds) Plant health under biotic stress, Microbial interactions, vol II. Springer Nature, Singapore, pp 177–188

Ansari RA, Sumbul A, Rizvi R, Mahmood I (2019b) Organic soil amendments: potential tool for soil and plant health management. In: Ansari RA, Mahmood I (eds) Plant health under biotic stress, Organic strategies, vol I. Springer Nature, Singapore, pp 1–35

Ansari RA, Rizvi R, Mahmood I (2020a) Management of phytonematodes: management of phytonematodes: recent advances and future challenges. Springer Nature, Singapore. https://doi.org/10.1007/978-981-15-4087-5

Ansari RA, Rizvi R, Sumbul A, Mahmood I (2020b) Plant-growth-promoting Rhizobacteria (PGPR)-based sustainable management of phytoparasitic nematodes: current understandings and future challenges. In: Ansari RA et al (eds) Management of phytonematodes: recent advances and future challenges. Springer Nature, Singapore. https://doi.org/10.1007/978-981-15-4087-5_3

Aseel DG, Rashad YM, Hammad SM (2019) Arbuscular mycorrhizal fungi trigger transcriptional expression of flavonoid and chlorogenic acid biosynthetic pathways genes in tomato against Tomato Mosaic Virus. Sci Rep 9(1):9692

Bender SF, Wagg C, van der Heijden MGA (2016) An underground revolution: biodiversity and soil ecological engineering for agricultural sustainability. Trends Ecol Evol 31:440–452

Berruti A, Borriello R, Orgiazzi A, Barbera AC, Lumini E, Bianciotto V (2016a) Arbuscular mycorrhizal fungi and their value for combating soil-borne plant diseases. AIMS Microbiol 2(3):374–386. https://doi.org/10.3934/microbiol.2016.3.374

Berruti A, Lumini E, Balestrini R, Bianciotto V (2016b) Arbuscular mycorrhizal fungi as natural biofertilizers: let's benefit from past successes. Front Microbiol 6:1559

Bhantana P, Rana MS, Sun XC, Moussa MG, Saleem MH, Syaifudin M et al (2021) Arbuscular mycorrhizal fungi and its major role in plant growth, zinc nutrition, phosphorous regulation and phytoremediation. Symbiosis 84:19–37

Bonfante P, Genre A (2008) Plants and arbuscular mycorrhizal fungi: an evolutionary-developmental perspective. Trends Plant Sci 13(9):492–498

Campos-Soriano LIDIA, García-Martínez J, Segundo BS (2012) The arbuscular mycorrhizal symbiosis promotes the systemic induction of regulatory defence-related genes in rice leaves and confers resistance to pathogen infection. Mol Plant Pathol 13(6):579–592

Cao Y, Chen Q, Zhang J, Zheng Y (2020) Arbuscular mycorrhizal fungi improve cucumber resistance to cucumber mosaic virus through regulating plant hormone signaling and systemic acquired resistance. Plant Physiol Biochem 149:126–134

Cabral L, Soares CRFS, Giachini AJ, Siqueira JO (2015) Arbuscular mycorrhizal fungi in phytoremediation of contaminated areas by trace elements: mechanisms and major benefits of their applications. World J Microbiol Biotechnol 31:1655–1664

Chagnon PL, Bradley RL, Klironomos JN (2015) Trait-based partner selection drives mycorrhizal network assembly. Oikos 124:1609–1616. https://doi.org/10.1111/oik.01987

Challinor AJ, Koehler AK, Ramirez-Villegas J, Whitfield S, Das B (2016) Current warming will reduce yields unless maize breeding and seed systems adapt immediately. Nat Clim Chang 6:954–958

Chen BD, Zhu YG, Duan J, Xiao XY, Smith SE (2007) Effects of the arbuscular mycorrhizal fungus Glomus mosseae on growth and metal uptake by four plant species in copper mine tailings. Environ Pollut 147(2):374–380

Chen X, Li L, Hu J, Li X, Li Y, Wu J (2018) Arbuscular mycorrhizal fungal diversity, root colonization, and soil aggregate stability in different soil types and land use regimes. J Soils Sediments 18(8):2667–2678

Dey R, Pal KK, Tilak KVBR (2006) Interaction of plant growth promoting rhizobacteria and AM fungi: present status and future prospects. Basic Res Appl Mycorrhiza:83–108

Egan C, Li DW, Klironomos J (2014) Detection of arbuscular mycorrhizal fungal spores in the air across different biomes and ecoregions. Fungal Ecol 12:26–31. https://doi.org/10.1016/j.funeco.2014.06.004

Fabiyi OA (2021a) Application of furfural in sugarcane nematode pest management. Pak J Nematol 39(2):151–155

Fabiyi OA (2021b) Sustainable management of Meloidogyne incognita infecting carrot: green synthesis of silver nanoparticles with Cnidoscolus aconitifolius: (Daucaus carota). Vegetos 34(2):277–285

Fabiyi OA (2021c) Evaluation of Nematicidal activity of *Terminalia glaucescens fractions* against *Meloidogyne incognita* on *Capsicum chinense*. J Hort Res 29(1):67–74

Fabiyi OA (2021d) Evaluation of plant materials as root-knot nematode (*Meloidogyne incognita*) suppressant in okro (*Abelmuscous esculentus*). Agric Conspec Sci 86(1):51–56

Fabiyi OA, Olatunji GA (2021a) Environmental sustainability: bioactivity of *Leucaena leucocephala* leaves and pesticide residue analysis in tomato fruits. Acta Univ Agric Silvic Mendelianae Brun 69(4):473–480

Fabiyi OA, Olatunji GA (2021b) Toxicity of derivatized citrulline and extracts of water melon rind (*Citrullus lanatus*) on root-knot nematode (*Meloidogyne incognita*). Trop Agric 98(4):347–355

Fabiyi OA, Olatunji GA, Osunlola OS, Umar KA (2018a) Efficacy of agricultural wastes in the control of Rice cyst nematode (*Heterodera sacchari*). Agric Conspec Sci 83(4):329–334

Fabiyi OA, Olatunji GA, Saadu AO (2018b) Suppression of *Heterodera sacchari* in rice with agricultural waste-silver nano particles. J Solid Waste Manage 44(2):87–91
Fabiyi OA, Olatunji GA, Atolani O, Olawuyi RO (2020) Preparation of bio-nematicidal nanoparticles of *Eucalyptus officinalis* for the control of cyst nematode (*Heterodera sacchari*). J Anim Plant Sci 30(5):1172–1177
Fabiyi OA, Adebisi OO, Falore SO, Bello TT, Olatunji GA (2023) Assessment of Actinomycetes and Pseudomonas species on Meloidogyne incognita population and growth of carrot plants in disparate soils. Indian Phytopathol 76:593–604
Fakher A, Yedida I, Benhamou N (2016) Induction of systemic resistance in cotton by the arbuscular mycorrhizal fungus Glomus intraradices against Fusarium oxysporum f. sp. vasinfectum. Biol Control 95:35–42. https://doi.org/10.1016/j.biocontrol.2015.12.002
Feng Y, Cui X, He S, Dong G, Chen M, Wang J, Lin X (2013) The role of metal nanoparticles in influencing arbuscular mycorrhizal fungi effects on plant growth. Environ Sci Technol 47(16):9496–9504
Fernandez LF, Nair A (2018) Arbuscular mycorrhizal fungi in heavy metal remediation: a review. J Environ Manag 215:293–306. https://doi.org/10.1016/j.jenvman.2018.03.064
Field KJ, Cameron DD, Leake JR, Tille S, Bidartondo MI, Beerling DJ (2012) Contrasting arbuscular mycorrhizal responses of vascular and non-vascular plants to a simulated Palaeozoic CO2 decline. Nat Commun 3:835
Field KJ, Rimington WR, Bidartondo MI, Allinson KE, Beerling DJ, Cameron DD, Duckett JG, Leake JR, Pressel S (2015) First evidence of mutualism between ancient plant lineages (Haplomitriopsida liverworts) and Mucoromycotina fungi and its response to simulated Palaeozoic changes in atmospheric CO2. New Phytol 205:743–756
Fomina MA, Alexander IJ, Colpaert JV, Gadd GM (2005) Solubilization of toxic metal minerals and metal tolerance of mycorrhizal fungi. Soil Biol Biochem 37(5):851–866
Garg N, Singh S, Kashyap L (2017) Arbuscular mycorrhizal fungi and heavy metal tolerance in plants: an insight into physiological and molecular mechanisms. In: Mycorrhiza-nutrient uptake, biocontrol, ecorestoration, pp 75–97
Godfray HCJ, Beddington JR, Crute IR et al (2010) Food security: the challenge of feeding 9 billion people. Science 327:812–818
Gollin D, Parente S, Rogerson R (2002) The role of agriculture in development. Am Econ Rev 92(2):160–164
Goltapeh EM, Danesh YR, Prasad R, Varma A (2008) Mycorrhizal fungi: what we know and what should we know? In: Mycorrhiza: state of the art, genetics and molecular biology, eco-function, biotechnology, eco-physiology, structure and systematics. Springer, Heidelberg, pp 3–27
Goswami S, Singh AK, Kumar S, Singh M, Singh R (2013) Effect of arbuscular mycorrhizal fungi on growth and development of pigeonpea and Heterodera cajani. J Agric Sci Technol 15:439–450
Higo M, Isobe K, Yamaguchi M, Drijber RA, Jeske ES, Ishii R (2013) Diversity and vertical distribution of indigenous arbuscular mycorrhizal fungi under two soybean rotational systems. Biol Fertil Soils 49:1085–1096. https://doi.org/10.1007/s00374-013-0807-5
Hodge A, Fitter AH (2010) Substantial nitrogen acquisition by arbuscular mycorrhizal fungi from organic material has implications for N cycling. Proc Natl Acad Sci U S A 107:13754–13759
Hu J, Wang H, Wu F, Wu S, Cao Z, Lin X, Wong MH (2014) Arbuscular mycorrhizal fungi influence the accumulation and partitioning of Cd and P in bashfulgrass (Mimosa pudica L.) grown on a moderately Cd-contaminated soil. Appl Soil Ecol 73:51–57
Huang K, Li T, Yu X (2018) Role of arbuscular mycorrhizal fungus Rhizophagus intraradices in cadmium tolerance and accumulation by maize (Zea mays L.). Ecotoxicol Environ Saf 147:634–641. https://doi.org/10.1016/j.ecoenv.2017.09.038
Jiang Y, Wang W, Xie Q, Liu N, Liu L, Wang D, Zhang X, Yang C, Chen X, Tang D (2017) Plants transfer lipids to sustain colonization by mutualistic mycorrhizal and parasitic fungi. Science 356:1172–1175
Johnson NC (2010) Resource stoichiometry elucidates the structure and function of arbuscular mycorrhizas across scales. New Phytol 185:631–647

Johnson NC, Wilson GWT, Wilson JA, Miller RM, Bowker MA (2015) Mycorrhizal phenotypes and the law of the minimum. New Phytol 205:1473–1484
Jung SC, Martinez-Medina A, Lopez-Raez JA, Pozo MJ (2012) Mycorrhiza-induced resistance and priming of plant defenses. J Chem Ecol 38(6):651–664. https://doi.org/10.1007/s10886-012-0134-6
Khalid M, Bilal M, Asghar HN, Shahid M, Shahzad SM (2019) Mycorrhizal inoculation improves growth, yield and nutrient uptake in tomato under Fusarium oxysporum f. sp. lycopersici infestation. PLoS One 14(1):e0211168. https://doi.org/10.1371/journal.pone.0211168
Khan AG (2005) Role of soil microbes in the rhizospheres of plants growing on trace metal contaminated soils in phytoremediation. J Trace Elem Med Biol 18(4):355–364. https://doi.org/10.1016/j.jtemb.2005.02.008
Kiers ET, Duhamel M, Beesetty Y, Mensah JA, Franken O, Verbruggen E, Fellbaum CR, Kowalchuk GA, Hart MM, Bago A et al (2011) Reciprocal rewards stabilize cooperation in the mycorrhizal symbiosis. Science 333:880–882
Klironomos JN, Rillig MC, Allen MF, Zak DR (2000) Soil fungal-arthropod responses to mixed litter and nutrient additions. Appl Soil Ecol 14(3):251–258
Koele N, Dickie IA, Blum JD, Gleason JD, de Graaf L (2014) Ecological significance of mineral weathering in ectomycorrhizal and arbuscular mycorrhizal ecosystems from a field-based comparison. Soil Biol Biochem 69:63–70. https://doi.org/10.1016/j.soilbio.2013.10.041
Kumar A, Singh S, Pandey V (2021) Role of arbuscular mycorrhizal fungi (AMF) in mitigating abiotic stresses. In: Arbuscular mycorrhizal fungi in plants under stress management. Springer, Singapore, pp 163–188
Leake JR, Read DJ (2017) Mycorrhizal symbioses and pedogenesis throughout earth's history. In: Mycorrhizal mediation of soil. Elsevier, Amsterdam, pp 9–33. https://doi.org/10.1016/B978-0-12-804312-7.00002-4
Lehnert LW, Duchicela J, Kaonongbua W (2017) Effects of soil texture and mycorrhizal colonization on the growth of native grasses in disturbed ecosystems. J Arid Environ 140:47–50
Li XX, Zeng RS, Liao H (2016) Improving crop nutrient efficiency through root architecture modifications. J Integr Plant Biol 58:193–202
Li H, Chen X, Liu J, Zhang Y, Wang A, Guo X (2017) Effects of arbuscular mycorrhizal fungi on the disease severity and growth of cucumber infected by Fusarium oxysporum f. sp. cucumerinum. Front Microbiol 8:1029. https://doi.org/10.3389/fmicb.2017.01029
Li Y, Zhu X, Feng G, Christie P (2019) Effects of arbuscular mycorrhizal fungi on the uptake and distribution of copper in tomato plants. Environ Sci Pollut Res 26(7):7202–7212. https://doi.org/10.1007/s11356-018-04012-y
Li HY, Zhu YG, Marschner P, Smith FA, Smith SE (2005) Wheat responses to arbuscular mycorrhizal fungi in a highly calcareous soil differ from those of clover, and change with plant development and P supply. Plant Soil 277:221–232
Liao J, Yang X, He X, Liu Y, Xu L, Chen W (2014) Effects of arbuscular mycorrhizal fungi on the growth, nutrient uptake and glycyrrhizin production of Glycyrrhiza uralensis under salt stress. Plant Soil 386(1–2):335–349. https://doi.org/10.1007/s11104-014-2236-3
Liu A, Hamel C, Hamilton RI, Ma BL, Smith DL (2000) Acquisition of Cu, Zn, Mn and Fe by mycorrhizal maize (Zea mays L.) grown in soil at different P and micronutrient levels. Mycorrhiza 9:331–336
Liu W, Zhang Y, Jiang S, Deng Y, Christie P, Murray PJ et al (2016) Arbuscular mycorrhizal fungi in soil and roots respond differently to phosphorus inputs in an intensively managed calcareous agricultural soil. Sci Rep 6(1):24902
Liu Y, Feng X, Gao P, Li Y, Christensen MJ, Duan T (2018) Arbuscular mycorrhiza fungi increased the susceptibility of *Astragalus adsurgens* to powdery mildew caused by Erysiphe pisi. Mycology 9(3):223–232
Liu W, Wu Y, Wang B, Ding N (2020) Effects of arbuscular mycorrhizal fungi on the growth and ionome of tomato plants under different soil textures. J Soils Sediments 20(10):3748–3757

Liu XR, Rong ZY, Tian X, Hashem A, Abd_Allah EF, Zou YN, Wu QS (2023) Mycorrhizal Fungal Effects on Plant Growth, Osmolytes, and CsHsp70s and CsPIPs Expression in Leaves of Cucumber under a Short-Term Heat Stress. Plants 12(16):2917

Luginbuehl LH, Menard GN, Kurup S, Van Erp H, Radhakrishnan GV, Breakspear A, Oldroyd GED, Eastmond PJ (2017) Fatty acids in arbuscular mycorrhizal fungi are synthesized by the host plant. Science 356:1175–1178

Luo C, Mo Y, Cong R, Yang Y, Liu X (2018) Effects of different arbuscular mycorrhizal fungi on the growth and nutrient uptake of banana under Fusarium wilt stress. J Plant Growth Regul 37(3):854–863. https://doi.org/10.1007/s00344-017-9772-2

Ma Y, Huang Y, Lin Y, Zhao L (2020) Arbuscular mycorrhizal fungi improve growth and cadmium uptake of Solanum nigrum L. in a cd, Pb, and Zn multi-contaminated soil. Environ Sci Pollut Res 27(8):8797–8807. https://doi.org/10.1007/s11356-019-07456-2

Marquez-Garcia BE, Urcelay C, Croll D (2020) Beneficial soil microorganisms and their interactions with the plant immune system for biological control of plant viral diseases. Front Microbiol 11:575747

Miransari M (2017) Arbuscular mycorrhizal fungi and heavy metal tolerance in plants. In: Arbuscular mycorrhizas and stress tolerance of plants. Singapore, Springer Singapore, pp 147–161

Moll J, Hoppe B, König S, Wubet T, Buscot F, Krüger D (2016) Spatial distribution of fungal communities in an arable soil. PLoS One 11:1–17. https://doi.org/10.1371/journal.pone.0148130

Navarro-Fernandez CM, Perez-Ramos IM, de la Riva EG, Vera JR, Roumet C, Villar R, Maranon T (2016) Functional responses of Mediterranean plant communities to soil resource heterogeneity: a mycorrhizal trait-based approach. J Veg Sci 27:1243–1253

Ogundipe AA, Ogunniyi A, Olagunju K, Asaleye AJ (2019) Poverty and income inequality in rural agrarian household of southwestern Nigeria: the gender perspective. Open Agric J 13:12–17

Philippot L, Raaijmakers JM, Lemanceau P, Van Der Putten WH (2013) Going back to the roots: the microbial ecology of the rhizosphere. Nat Rev Microbiol 11:789–799

Pozo MJ, Azcon-Aguilar C (2007) Unraveling mycorrhiza-induced resistance. Curr Opin Plant Biol 10(4):393–398. https://doi.org/10.1016/j.pbi.2007.05.004

Puschel D, Janouskova M, Hujslova M, Slavikova R, Gryndlerova H, Jansa J (2016) Plant-fungus competition for nitrogen erases mycorrhizal growth benefits of Andropogon gerardii under limited nitrogen supply. Ecol Evol 6:4332–4346

Ramasamy K, Joe MM, Kim KY, Lee SM, Shagol C, Rangasamy A et al (2011) Synergistic effects of arbuscular mycorrhizal fungi and plant growth promoting rhizobacteria for sustainable agricultural production. Korean J Soil Sci Fertil 44(4):637–649

Riaz M, Kamran M, Fang Y, Wang Q, Cao H, Yang G et al (2021) Arbuscular mycorrhizal fungi-induced mitigation of heavy metal phytotoxicity in metal contaminated soils: a critical review. J Hazard Mater 402:123919

Rillig MC, Aguilar-Trigueros CA, Bergmann J, Verbruggen E, Veresoglou SD, Lehmann A (2015) Plant root and mycorrhizal fungal traits for understanding soil aggregation. New Phytol 205:1385–1388

Rizvi R, Singh G, Safiuddin ARA, Tiyagi SA, Mahmood I (2015) Sustainable management of root-knot disease of tomato by neem cake and Glomus fasciculatum. Cogent Food Agric 1:1008859

Rouphael Y, Franken P, Schneider C, Schwarz D, Giovannetti M, Agnolucci M, De Pascale S, Bonini P, Colla G (2015) Arbuscular mycorrhizal fungi act as biostimulants in horticultural crops. Sci Hortic 196:91–108

Saberi-Riseh R, Heydari A, Safaie N (2016) Arbuscular mycorrhizal fungi and plant growth-promoting rhizobacteria as biocontrol agents against the root-knot nematode, Meloidogyne incognita in sugar beet. Biol Control 92:7–15

Säle V, Aguilera P, Laczko E, Mäder P, Berner A, Zihlmann U et al (2015) Impact of conservation tillage and organic farming on the diversity of arbuscular mycorrhizal fungi. Soil Biol Biochem 84:38–52. https://doi.org/10.1016/j.soilbio.2015.02.005

Sato T, Ezawa T, Cheng WG, Tawaraya K (2015) Release of acid phosphatase from extraradical hyphae of arbuscular mycorrhizal fungus Rhizophagus clarus. Soil Sci Plant Nutr 61:269–274

Siddiqui IA, Mahmood I (2012) Efficacy of arbuscular mycorrhizal fungi and Rhizobium inoculation in the management of root-knot nematode, *Meloidogyne incognita*, infecting *Abelmoschus esculentus*. Arch Phytopathol Plant Protect 45(2):173–184

Singh G, Kumar A, Agrawal S, Koul KK (2016) Management of Meloidogyne incognita on tomato using *Glomus fasciculatum* and *Pseudomonas fluorescens*. J Pure Appl Microbiol 10(2):1305–1310

Singh H, Meena VD, Singh RK, Meena RS, Kumar S (2017) Effect of arbuscular mycorrhizal fungi inoculation on growth, yield, nutrient uptake and physiological parameters of banana (Musa sp.) cv. Dwarf Cavendish under field conditions. Int J Curr Microbiol App Sci 6(9):1233–1243. https://doi.org/10.20546/ijcmas.2017.609.150

Singh A, Bhatt K, Kumar A, Sharma S (2018) Effect of arbuscular mycorrhizal fungi Funneliformis mosseae on the growth and lead accumulation of tomato plants grown in lead contaminated soil. Environ Sci Pollut Res 25(8):7712–7722. https://doi.org/10.1007/s11356-017-1077-y

Smith FA, Smith SE (2011a) What is the significance of the arbuscular mycorrhizal colonisation of many economically important crop plants? Plant Soil 348:63

Smith SE, Smith FA (2011b) Roles of arbuscular mycorrhizas in plant nutrition and growth: new paradigms from cellular to ecosystem scales. Annu Rev Plant Biol 62:227–250

Smith JA, Smith BB (2019) Arbuscular mycorrhizal fungi: allies in the fight against soil-borne pathogens. J Appl Microbiol 127(1):10–24. https://doi.org/10.1111/jam.14280

Solanki MK, Kashyap PL, Ansari RA, Kumari B (2020) Microbiomes and plant health: panoply and their applications. Academic Press, London. https://www.elsevier.com/books/microbiomes-and-plant-health/solanki/978-0-12-819715-8

Sosa-Hernández MA, Roy J, Hempel S, Kautz T, Köpke U, Uksa M et al (2018) Subsoil arbuscular mycorrhizal fungal communities in arable soil differ from those in topsoil. Soil Biol Biochem 117:83–86. https://doi.org/10.1016/j.soilbio.2017.11.009

Spatafora JW, Chang Y, Benny GL et al (2016) A phylum-level phylogenetic classification of zygomycete fungi based on genome-scale data. Mycologia 108:1028–1046

Srivastava R, Khalid A, Singh US, Sharma AK (2010) Evaluation of arbuscular mycorrhizal fungus, fluorescent Pseudomonas and Trichoderma harzianum formulation against Fusarium oxysporum f. sp. lycopersici for the management of tomato wilt. Biol Control 53(1):24–31

Stirling GR, Roper MM (2010) Arbuscular mycorrhiza and the management of nematodes in vegetable crops: a review. Crop Pasture Sci 61(10):835–858

Sumbul A, Mahmood I, Rizvi R, Safiuddin, Ansari RA (2017) Mycorrhiza: an alliance for the nutrient management in plants. In: Kumar V (ed) Probiotics in agroecosystem. Springer Nature, Singapore, pp 371–386

Szada-Borzyszkowska A, Krzyżak J, Rusinowski S, Starzewska-Sikorska A, Ratman-Kłosińska I, Pogrzeba M (2021) The effect of amendments on Lolium perenne roots arbuscular mycorrhizal fungi colonization when cultivated in contaminated soil. Int J Environ Sci Technol:1–12

Tawaraya K (2003) Arbuscular mycorrhizal dependency of different plant species and cultivars. Soil Sci Plant Nutr 49:655–668

Thirkell TJ, Cameron DD, Hodge A (2016) Resolving the 'nitrogen paradox' of arbuscular mycorrhizas: fertilization with organic matter brings considerable benefits for plant nutrition and growth. Plant Cell Environ 39:1683–1690

Turrini A, Avio L, Giovannetti M, Agnolucci M (2018) Functional complementarity of arbuscular mycorrhizal fungi and associated microbiota: the challenge of translational research. Front Plant Sci 9:1407

Verbruggen E, Struyf E, Vicca S (2021) Can arbuscular mycorrhizal fungi speed up carbon sequestration by enhanced weathering? Plants, People, Planet 3(5):445–453

Walder F, Niemann H, Natarajan M, Lehmann MF, Boller T, Wiemken A (2012) Mycorrhizal networks: common goods of plants shared under unequal terms of trade. Plant Physiol 159:789–797

Wang B, Qiu YL (2006) Phylogenetic distribution and evolution of mycorrhizas in land plants. Mycorrhiza 16(5):299–363. https://doi.org/10.1007/s00572-005-0033-6

Wang B, Feng G, Christie P (2005) Improved water relations of maize plants by arbuscular mycorrhiza in low phosphorus soil. Environ Exp Bot 53(3):197–205. https://doi.org/10.1016/j.envexpbot.2004.04.010
Wang Y, Wu S, Wu J, Chen X, Lin X (2015) Effects of arbuscular mycorrhizal fungi on nutrient uptake and utilization of trifoliate orange (Poncirus trifoliata L. Raf.) seedlings in different soils. Sci Hortic 197:248–254
Wang X, Guo X, Zhao L, Feng G, Zhang F (2017) Role of arbuscular mycorrhizal fungus Claroideoglomus etunicatum in arsenic tolerance and uptake by maize plants in arsenic-contaminated soil. Environ Sci Pollut Res 24(13):11932–11939. https://doi.org/10.1007/s11356-017-8932-3
Wang Y, Liu L, Qin L, Zhang Y, Kong Y (2019a) Arbuscular mycorrhizal fungi enhance resistance of maize against maize chlorotic mottle virus. J Plant Growth Regul 38(1):142–150
Wang Y, Wang M, Li Y, Wu A, Huang J (2019b) Arbuscular mycorrhizal fungi alleviate heat stress in tea plants (Camellia sinensis (L.) O. Kuntze) by regulating photosynthesis, antioxidant systems, and HSP gene expressions. Front Plant Sci 10:1328. https://doi.org/10.3389/fpls.2019.01328
Wang Y, Zhang W, Liu W, Ahammed GJ, Wen W, Guo S et al (2021) Auxin is involved in arbuscular mycorrhizal fungi-promoted tomato growth and NADP-malic enzymes expression in continuous cropping substrates. BMC Plant Biol 21(1):1–12
Wipf D, Krajinski F, van Tuinen D, Recorbet G, Courty PE (2019) Trading on the arbuscular mycorrhiza market: from arbuscules to common mycorrhizal networks. New Phytol 223(3):1127–1142
Wu X, Sun B, Chen Y, Liang Y, Li H, Li X (2019a) Effects of arbuscular mycorrhizal fungi on the growth and copper uptake of Brassica juncea in copper-contaminated soil. Int J Phytoremediation 21(14):1422–1429. https://doi.org/10.1080/15226514.2019.1670789
Wu Y, Yuan Q, Li M, Li M, Liang Y, Li Z, Liang J (2019b) Arbuscular mycorrhizal fungus colonization improves defense responses of rice against Xanthomonas oryzae pv. oryzae. J Plant Interact 14(1):426–433. https://doi.org/10.1080/17429145.2019.1635485
Zhang J, Liu X, Zhang X, Zhang Y, Wang X (2018) Arbuscular mycorrhizal fungi alleviate Fusarium wilt in watermelon by altering soil microbial communities. Front Microbiol 9:1024. https://doi.org/10.3389/fmicb.2018.01024
Zhou X, Fu L, Xia Y, Zheng L, Chen C, Shen Z, Chen Y (2017) Arbuscular mycorrhizal fungi enhance the copper tolerance of Tagetes patula through the sorption and barrier mechanisms of intraradical hyphae. Metallomics 9(7):936–948

The manufacturer's authorised representative in the EU is Springer Nature Customer Service Centre GmbH, Europaplatz 3, 69115 Heidelberg, Germany. If you have any concerns regarding our products, please contact ProductSafety@springernature.com

Printed and bound by CPI Group (UK) Ltd, Croydon, CR0 4YY
01/10/2026
02225868-0004